ANCIENT STARCH RESEARCH

We dedicate this book to Tom Loy (1942–2005).

Archaeologists around the world have benefited greatly from Tom Loy's early insights into the significance of starch residues for understanding prehistory. His study of starch grains on 28,000-year-old stone tools from Kilu Cave (Solomon Islands) opened the way for others to apply elsewhere and further develop the extraction and identification methods that he had pioneered.

Despite an ongoing illness, Thomas Harold Loy, BSc (Redlands), MA (UBC), PhD (ANU), taught at the University of Queensland, where he was a Senior Lecturer in Archaeology, and was working on a book about Ötzi, the famous Glacier Mummy found in 1991. Tom was best known for his pioneering studies of ancient blood residues, and was analysing DNA preserved in blood films found on Ötzi's tools and clothing.

Tom's initial discovery of blood and hair on stone tools from ancient frozen sediments startled the archaeological world when it was published in the journal *Science* in 1983. He applied, for the first time, a wide range of forensic techniques for the study of many animal, plant, and mineral residues, which are now routinely found on archaeological artifacts. Challenged by sceptics, he developed and applied novel methods of taxonomic identification and analysis to artifacts from archaeological sites around the world. He recently convened the 7th Ancient DNA and Associated Biomolecules Conference, held at the University of Queensland in 2004.

He was an inspiring teacher, a fine artist, and a keen musician. As a major contributor in earlier stages of this book, we are sad that he has not seen it published.

ANCIENT STARCH RESEARCH

Robin Torrence
Huw Barton

editors

Left Coast
Press Inc.
Walnut Creek, California

Left Coast
Press Inc.

LEFT COAST PRESS, INC.
1630 North Main Street, #400
Walnut Creek, CA 94596
http://www.LCoastPress.com

ISBN 1-59874-018-0 hardcover

Library of Congress Control Number: 2005932077

Printed in the United States of America

∞™ The paper used in this publication meets the minimum requirements of American National Standard for Information Sciences—Permanence of Paper for Printed Library Materials, ANSI/NISO Z39.48–1992.

06 07 08 09 10 5 4 3 2 1

CONTENTS

LIST OF BOXES, TABLES, AND FIGURES

BOXES

TABLES

FIGURES

LIST OF PLATES

following page 128

PREFACE

This book is intended as a gateway to starch analysis in archaeology. Our aim is to provide students and scholars with the necessary background to undertake their own studies of this rather unique plant microfossil. Many readers will be surprised to find that ancient starch analysis has a long history, although the pace of scholarship has been erratic. It is only in the last two decades, with the publication of a significant body of literature, that ancient starch studies could truly be referred to as a serious subdiscipline of a wider range of paleobotanical approaches to archaeology.

The original impetus for this book comes from Australian scholarship and particularly the work of Tom Loy and Richard Fullagar, who established the study of organic residues in this country. This approach to stone tools is central to archaeology in this part of the world because use wear traces are difficult to interpret due to the coarse nature of the raw materials. However, while the analysis of organic residues had opened many useful lines of inquiry into the past, for various reasons much of the research on ancient starch has been conducted by undergraduate students. Many of the findings and insights gained from their highly original Honours theses were in danger of being lost to scholarship through lack of publication. At the same time, it was also emerging that scholars in other parts of the world were turning their attention to the possibilities of using ancient starch to solve longstanding archaeological problems that were hampered by the limitations unavoidably inherent in all paleobotanical techniques. In 1998 the time was ripe for the small body of active researchers to join together to raise the profile of ancient starch research.

This book is the outcome of a truly collaborative working project beginning with the initiation of the Ancient Starch Research Group (ASRG) at the Australian Museum in August 1998, which was crystallized at its second meeting in February 2000. These gatherings were organized as an opportunity for formal presentations, workshops, and informal discussion about the range of starch research currently undertaken in archaeology as well as the jumping-off point for creative thinking about its future possibilities. The group was composed of scholars from the United States, Japan, and the Netherlands as well as Australia. During the meetings it was decided that the most efficient way forward for ancient starch research was to pool our collective knowledge and experience within a manual for its analysis. Over time the project broadened considerably to aim for a more comprehensive treatment.

The approach to this book is collaborative both in terms of its writing and in our recommendations for approaches to and procedures for the analysis of ancient starch. We see the real strength of this source of data as complementary to existing plant microfossils, such as pollen, phytoliths, and diatoms. Each has its own strengths and weaknesses, due to differences in its plant biology, formation and depositional histories, and susceptibility to decay or preservation in different postdepositional contexts. Like the other categories of material, starch can provide unique information about plant use in the past that is as powerful as that provided by existing techniques. The book attempts to review these aspects of ancient starch and provide a balanced view of its potential for archaeological research.

The multi-authored approach is unusual for a basic text but we think it is one of the book's great strengths. We have tried to go beyond the presentation of a straightforward manual of approaches by linking procedures to real-life archaeological problems. We hope that readers who get interested in ancient starch analysis will pursue suggested ways forward and/or repeat experiments since this will markedly increase the reliability of results and help further identify areas of weakness or strength. Within chapters, there are further individual and collaborative contributions set as BOXES that represent self-contained discussions of relevant topics, examples of applications discussed in the text, the results of experiments addressing specific research problems, and recommended protocols adopted by various members of the group that are tried and tested, but like any new form of analysis, might be improved upon. The BOXES are important because they enable a wide range of researchers at varying stages of their careers to voice their views and showcase their findings.

We are grateful to a great many people for help in bringing this project to fruition, especially the authors who volunteered to write contributions on subjects that required much background research. The book would not have been possible without such dedication. Researching and writing important scientific background material took much longer than expected. We thank all the contributors for their patience and support over the long gestation period. We would also like to thank everyone who contributed to the ASRG workshops and helped get this project up and running. Although not all participants were able to produce written contributions, their enthusiasm and sage advice at the workshops together with subsequent suggestions and critical comments on manuscripts have added immeasurably to the book: James Coil; Linda Conroy; Katia Davis; Anelou van Gjin; Linda Perry; Robyn Stocks; and LuAnn Wandsnider. We especially thank Dolores Piperno, Delwen Samuel, and Peter L. Frazier for guidance and specialist advice. Special thanks to Mark Aldenderfer for setting up and running the ASRG website and portal for exchanging documents across the globe and to Donald Ugent who also maintained a presence for the group on his website.

We are extremely grateful to the Australian Research Council for awarding Robin Torrence a Special Research Initiative Grant, which provided the cornerstone funding that launched the entire project, and to the Wellcome Trust, who funded Huw Barton during the final push to complete the manuscript. We are also grateful to the Australian Museum for hosting both meetings of the ASRG and for providing additional research facilities when discussion led participants to microscopes and laboratories. Over the years Sue Lindsay of the SEM unit has provided equipment, expertise, and good cheer to many of the group. The Anthropology Division and staff from Search and Discover at the Australian Museum provided support during the workshops and writing bees. Also, particular thanks to Ceridwen Cherry for help with references and to Peter White for hospitality at the Sydney University Archaeology Laboratory, refreshments during the workshops, and massive amounts of help with producing the manuscript. Finally, we are grateful to Richard Fullagar for support and comments in the final stages.

Huw Barton
Robin Torrence

Inaugural meeting of the Ancient Starch Research Group.

CHAPTER 1

STARCH AND ARCHAEOLOGY

Robin Torrence

Starch is surprisingly ubiquitous in the archaeological record. It has been found in deposits as much as two million years old and in highly varied settings spread across the globe. It has been recovered from the ideal conditions provided by dry caves or tombs as well as in places where organic preservation is exceedingly poor, such as open sites in the humid tropics. Starch granules have been identified within desiccated and/or charred tubers, preserved bread, as part of residues in ceramic vessels, adhering to the edges of stone tools, inside coprolites, within dental calculus, and in soils taken from rock shelters, open settlements, gardens, and forests. The wide temporal, geographic, and contextual range of the archaeological contexts in which starch has been studied is illustrated by the list in Table 1.1. Starch is clearly plentiful and widespread in contexts dating from ancient to modern, but what contribution can it make to archaeological research?

The archaeological potential of ancient starch was recognised a century ago (Wittmack 1905), but scholars were surprisingly slow to capitalise on this valuable source of information about past environments, plant use, diets, and artifact functions. Ancient starch research has really only become established in the past twenty years or so, but it has rapidly developed as a sophisticated new branch of research. A great deal of basic background work has been completed, methodologies have been put into place, and the pace of research output on ancient starch has significantly increased in recent years.

A rapidly growing number of case studies representing many parts of the world have produced exciting new data about vegetation histories, plant domestication, diet, mobility patterns, and artifact function. Selected as groundbreaking research by *Nature* or *Science*, recent studies incorporating data on ancient starch have reported plant cultivation earlier than expected, which in some cases may represent the first hard evidence for domestication both in Panama (Piperno and Holst 1998; Piperno et al. 2000) and in Uruguay (Iriarte et al. 2004) as well as in Papua New Guinea (Denham et al. 2003), all places thought by many to be far from the supposed heartlands of agriculture. Pioneering research using starch residues preserved on stone tools had already shown that some plants were transported far from their native lands thousands of years before they were supposedly cultivated, as in the case of taro starch granules that were found on stone tools dating to at least 28,000 years ago in the Solomon Islands (Loy et al. 1992). An important innovation for studying past diets has been the extraction of starch from coprolites. Sweet potato and bracken fern starch found in human and canine coprolites is the first direct evidence for these important food sources in prehistoric New Zealand (Horrocks et al. 2004a).

Studies based on stone tool residues have produced surprising and significant results concerning the importance of plants in the past. For instance, Williamson (2004) has shown that the majority of stone tools deposited at the site of Rose Cottage Cave in South Africa during the Middle and Late Stone Ages were used to process plants rather than to butcher animals, as was previously assumed. Her work also challenged ideas about gendered roles in the far distant past. In another study in which starch has contributed to our understanding of the earliest modern human behaviour, Van Peer et al. (2003) found starch granules on grinding slabs dated to around 180,000 BP. This represents one of the earliest instances of complex food processing among our human ancestors. Perry's (2004, in press) study of starches has seriously called into doubt the ascription of particular stone tools as parts of manioc graters and therefore challenged the role of this food source in prehistoric Venezuela. Fullagar (1992, 1993a) questioned the importance of recent ethnography as a source for reconstructing prehistoric stone tool use in Papua New Guinea, because he documented the unexpected dominance of plant processing rather than shaving and cutting of human skin in the assemblages he studied.

Starch extracted from sediments has provided evidence for changes in land use (Therin et al. 1999) and helped reconstruct activity areas on sites in Australia (Balme and Beck 2002) and at Copan in Honduras (Haslam 1999). Starch has also assisted in identifying which plants were grown within various types of agricultural features in New Zealand (Horrocks et al. 2004b).

Finally, an exciting contribution of ancient starch research has been the reconstruction of methods for making beer and bread in ancient Egypt (Samuel 1996a, 1996b, 2000). Surprisingly, Egyptian scholars had got the recipes for these important foods wrong and it was detective work based on damage to starch granules combined with other evidence that produced a more accurate picture of the manufacturing methods for these key sources of food.

This sample of the rapidly expanding number of exciting new case studies demonstrates conclusively that ancient starch has an important role to play in archaeological research, especially when tied in with other paleobotanical and archaeological techniques. Starch analysis can extend the range of our knowledge as well as the ability to make inferences about past human behaviour and environmental variation. Given the substantial body of successful research, the time has clearly come for a comprehensive description of ancient starch research and a guide to methods that will foster further work. This book fills these roles by (1) describing the fundamental principles underlying starch research (e.g., biology, human uses of starch, taphonomy); (2) guiding researchers through the methodology; (3) reviewing the results of significant case studies; and (4) pointing the way to future avenues for research. *Ancient Starch Research* aims to bring the important new field of ancient starch analysis to the attention of a wider range of scholars and to provide them with the information needed to embark on their own research.

STARCH AND THE BIG QUESTIONS

What role did plant resources have in the evolution of the human species? Why and how have plants been managed and transported to new environments? Where, how, and why were plants domesticated and why do the patterns vary in different parts of the world? What is the relationship between the intensification of food production and the rise of complex societies? Finding out exactly which starchy plants were consumed in the past and how dietary choices were made, understanding the processes leading to domestication, and explaining changes in subsistence strategies and the adoption of new foods are high research priorities shared by archaeologists working all over the globe. The recent development of ancient starch research has grown out of the need for relevant data to answer 'big questions' in archaeology. For this reason much of the earlier work has been strongly problem-oriented and led to important new discoveries.

Hather (1991) has shown that tubers (a convenient shorthand that will be used throughout the book for 'swollen vegetative storage organ—root, rhizome, corm, tuber, stolon, etc.—consisting largely of starch-bearing parenchyma' [Hather 1994a: 719]) were utilised as far back as the Paleolithic period in Europe and has argued that they probably continued to form a significant component of the diet even after the introduction of domesticated seed crops. To find out when tubers were first incorporated into hominid diets is an important research agenda because some archaeologists have theorised that starchy tubers were a very significant if not crucial part of past human subsistence patterns. O'Connell et al. (1999) and Wrangham et al. (1999) have proposed that a critical factor in the evolution of humans and the expansion of *Homo erectus* out of Africa was the ability to process wild yams. Cooking could have increased the energetic value of this food and the exploitation of yams also might have enabled populations to survive through the much drier conditions of that time (Box 1.1). The recovery of ancient starch could have an important role to play in testing this theory.

Some time ago Sauer (1952) proposed that the earliest agriculture was based on the domestication of tubers and roots in Southeast Asia. His theory was derived from the many different types and importance of root crops in this region. Matthews (Box 1.2) has also argued that the widespread distribution of one of these, taro, suggests

Table 1.1. The very broad temporal and geographic ranges of archaeological contexts in which starch has been preserved (prepared by Peter J. Matthews and Robin Torrence).

Location	Site	Date (yrs BP)	Site Type	Context	Form	References
South Africa	Sterkfontein, Member 5	c. 2,000,000	Cave complex	Stone tool residues	Granules, raphides	ABC 2004
South Africa	Cave of Hearths	c. 300,000	Cave	Stone tool residues	Granules	Williamson 2000
North Africa	Sai Island	180,000	Open site	Residues on grinding slab	Granules	Van Peer et al. 2003
South Africa	Rose Cottage Cave	c. from 100,000	Cave	Stone tool residues	Granules	Williamson 1997, 2000, 2004
Australia	Ngarrabullgan Cave	>37,000	Rockshelter	Stone tool residues	Granules	Fullagar and David 1997
Australia	Cuddie Springs	c. 30,000	Open site, clay pan	Residues on grinding stones	Granules	Furby et al. 1993; Fullagar and Field 1997; Field and Fullagar 1998; Fullagar et al. in press
Australia	Lake Mungo	c. 30,000	Open site	Stone tool residues	Granules	Loy 1990
Solomon Islands	Kilu Cave	c. 28,000	Rockshelter	Stone tool residues	Granules, raphides	Loy et al. 1992; Loy 1994a
Australia	Jinmium	<20,000	Rockshelter	Residues on stone pounders, in soil	Granules	Fullagar et al. 1996; Atchison and Fullagar 1998; Atchison et al. 2005
Chile	Monte Verde	c. 12,500	Open site, wet, anoxic	Stone tool residues	Macroremains, granules	Ugent 1987, 1997; Ugent et al. 1987; Tuross and Dillehay 1995
Papua New Guinea	Balof	c. 12,000–recent	Rockshelter	Stone tool residues	Granules, raphides	Barton and White 1993
Peru	Zaña Valley	9,200–7,600	Open settlement	On hut floor	Macroremains	Rossen et al. 1996
Panama	Aguadulce	8,000–5,000	Rockshelter	Residues on grinding stones	Granules	Piperno and Holst 1998; Piperno et al. 2000
Papua New Guinea	Bitokara Mission	>6,000–recent	Open site	Stone tool residues, in soil	Granules, raphides	Fullagar 1992, 1993a; Fullagar et al.1998; Kealhofer et al. 1999; Therin et al. 1999
Papua New Guinea	Garua Island, site FAO	>6,000–recent	Open site	Stone tools residues, in soil	Granules	Barton et al. 1998; Kealhofer et al. 1999; Therin et al. 1999
Austria	Haslabjoch Pass	c. 5,200	Open site	Residues on copper axe	Granules, macerated cellulose	Loy 1994b, 1998a, 1998b
Panama	Monagrillo	5,000–3,500	Shell mound	Residues on grinding stones	Granules	Piperno and Holst 1998
USA, Texas	Hinds Cave	5,000–2,000	Rockshelter	Stone tool residues	Granules	Shafer and Holloway 1979
Australia	Widgingarri Shelters 1, 2	c. 4,970–<1,700	Rockshelters	Stone tool residues	Granules	Wallis 1994; Wallis and O'Connor 1998
Egypt	Various	c. 4,000–3,200	Tombs	In preserved bread	Granules	Samuel 1994
Peru	Casma Valley	3,800–500	Open middens	In soil	Macroremains	Ugent et al. 1981, 1982, 1984, 1986
Egypt	Deir el-Medina	3,550–3,300	Workmen's village	Residues on pottery	Granules	Samuel 1996a
Egypt	Amarna	c. 3,350	Workmen's village	Residues on pottery	Granules	Samuel 1996a
Spain, Mallorca	S'Aigua Dolca;Urbiola; St Antoni de Portmany;Prat de la Riba	Neolithic/ Chalcolithic;Bronze Age;Iron Age;Late Roman	Megalith;Cave;Open site;Necropolis	Dental calculus	Granules	Juan-Tresserras et al. 1997
Papua New Guinea	Kamgot	c. 3,200	Open site	Residues on pottery	Granules, raphides	Crowther 2001, 2005
Spain	Geno	c. 3,100	Houses	Residues on pottery	Granules	Juan-Tresserras 1998
Peru	Paracas Necropolis	c. 2,390	Tomb	As grave offering	Macroremains	Ugent and Peterson 1988
Panama	La Mula	c. 2,300	Village	Stone tool residues	Granules	Piperno and Holst 1998
Australia	Toorbul	c. 2,000	Open site	Residues on stone pounder	Granules	Robertson 1996
Australia	Queensland, various	c. 2,000–1,000	Various	Residues on stone pounders	Granules	Hall et al. 1989
Panama	Cerro Juan Diaz	c. 1,300	Village	Residue on grinding stone	Granules	Piperno and Holst 1998
Peru	Pachacamac	1,000–500	Open midden	In soil	Macroremains	Ugent and Peterson 1988
South Africa	Little Muck	c. 1,000	Rockshelter	Residues on stone tools	Granules	Williamson 2000
Honduras	Copan	2,050–1,920	Pit	Residues on stone tools	Granules	Haslam 1999
Venezuela	Pozo Azul Norte	1,520–1,080	Open site	Residues on stone tools	Granules	Perry 2000
Papua New Guinea	West New Britain	<500	Gardens, regrowth forest	In soil	Granules	Lentfer et al. 2002
Venezuela	Los Mangos del Parguaza	700–200	Open site	Residues on stone tools	Granules	Perry 2000, 2004
New Zealand	Triangle Flat	<500	Open site	Coprolites	Granules, xylem	Horrocks 2004; Horrocks et al. 2004a
Australia	Victoria	c. 150	Open site	Residues on bottle glass	Granules	Wolski and Loy 1999

Box 1.1. Tubers And Human Evolution

James F. O'Connell

Many important developments in human evolution are thought to have been connected with changes in diet. Those that occurred prior to the late Pleistocene are typically linked with increased consumption of meat, usually from large animal prey (Klein 1999). Although changes in the use of plant foods are sometimes also mentioned as contributing factors, this possibility is rarely taken seriously, mainly because of a perceived absence of archaeological evidence. A common argument is that while large animal bones are an obvious part of the paleolithic record, plant food remains are not; hence plants 'cannot' have been an important component of early human diets, let alone the catalyst for major evolutionary change (e.g., Kaplan et al. 2000).

Arguments about the evolution of early African *Homo erectus* (aka *H. ergaster*) are a case in point. Dated to about 1.8 Ma, this form differs from earlier hominids in its larger body size, reduced sexual dimorphism, simplified digestive anatomy, greatly increased geographical range, delayed maturity, increased longevity, and probable earlier age at weaning (O'Connell et al. 1999; Wood and Collard 1999). Collectively, this suite of traits marks the most important development in human evolutionary history between the adoption of bipedalism (>4.0 Ma) and the appearance of modern humans (~100 ka). Changes in body size, digestive anatomy, and geographical distribution all point to a shift in diet. The life history changes further suggest the emergence of a pattern of long-term offspring provisioning, similar if not identical to that typical of all modern humans, but unknown among other primates. The first challenge is to provide a theoretically well-warranted hypothesis that links these various developments. The second is to devise an archaeological test of the new hypothesis.

Conventional approaches to the problem routinely emphasize increased meat-eating as the key behavioral innovation. More meat, especially from large ungulates, is said to have provided the dietary 'surplus' that fuelled changes in body size, underwrote delayed maturity and other shifts in life history, and facilitated the use of a broader range of habitats. The argument draws apparent support from the archaeological record, specifically the repeated association of stone tools and humanly processed large animal bones in sites of Plio-Pleistocene age, which are roughly contemporary with the first appearance of earliest *H. erectus* (Klein 1999).

Archaeological data notwithstanding, there are good reasons to be skeptical of the argument based on meat eating. In order to support early weaning and a long period of juvenile development, provisions must be available on a reliable, near-daily basis; yet modern hunter-gatherers, equipped with traditional weapons, cannot provide meat on anything like that schedule. Records for the modern Hadza, living in a game-rich habitat in northern Tanzania similar to that occupied by early *H. erectus*, indicate an average success rate of only one large animal every thirty hunter-days (Hawkes et al. 1991). Groups of six to eight men commonly co-resident in a single camp often go for weeks without making a single kill, despite hunting every day. Less well armed, *H. erectus* is unlikely to have done as well, let alone much better. Meat may have been an important part of early human diets, but its consumption was probably not central to the emergence of early humans.

Plant foods, particularly underground storage organs ('tubers' in common parlance), have long been identified as a potentially important food source for early *H. erectus* (e.g., Coursey 1973); but until recently no argument linking their use to changes in morphology and life history has been available. Hawkes and colleagues now provide one (Hawkes et al. 1998; O'Connell et al. 1999; see also Wrangham et al. 1999). Specifically, they suggest that a trend toward increased seasonality, roughly contem-

Box 1.1 (continued)

Figure 1.1. Hadza great-grandmother Wande pauses in the course of a tuber collecting trip in the Eyasi Basin, northern Tanzania during the early dry season, 1986. Older women may have played an important role in human evolution by providing tubers that enabled young children to be weaned at an early age. *Photo by James O'Connell*

porary with the first appearance of *H. erectus*, reduced the availability of foods (especially fruit) that very young children could take on their own, forcing women to provision their offspring through lean seasons. If women toward the end of their own reproductive careers provided some of the necessary provisions, especially to their grandchildren, then their daughters might have been able to wean those grandchildren earlier and move to the next pregnancy sooner (Figure 1.1). Not only would fecundity be increased, at no cost in weanling survivorship, but selection would likely also favor increased longevity, which would in turn promote delayed maturity and larger adult body size, consistent with general mammalian patterns (Hawkes et al. 1998).

Tubers may have been among the resources used to provision young children so that they could be weaned at an earlier age. Tubers occur at high densities (up to 10 tonnes/hectare), are rich in carbohydrates (up to 90 percent dry weight), and are widely exploited by modern temperate and tropical foragers; yet they are often not taken by younger children, mainly because of their occurrence at reasonable depths below ground surface, the complexity of their carbohydrate content, or the chemical defenses they contain. In contrast, adults contend with most of these problems quite easily and often gain returns high enough to support themselves and one or two others, with relatively little day-to-day variance. If a similar situation occurred in the past, then tubers could have been the crucial resource that provisioned young children and thereby enabled *H. erectus* populations to be so successful.

The difficulty faced by archaeologists is to test the suggestion that tubers actually played this role for *H. erectus*. This problem is now under attack in three stages. First, botanical and ethnographic sources are being reviewed to establish the range of tubers potentially available for use and the costs in time and nutritional benefits gained from exploiting them (e.g., Thoms 1989; Wandsnider 1997). As this effort proceeds, gaps in the data will be identified and filled, mainly by additional field and lab work. Once completed, this exercise should provide the basis for a second step: predictions about which tubers are likely to have been exploited under various conditions, and what the archaeological consequences of their use might have been, for example, with respect to site location relative to habitat type, trace element signatures in the bones of hominid consumers, the composition of resulting archaeological assemblages, or the residues likely to be left on collecting and processing gear. This should enable the third and final step: searching the archaeological record for evidence of the predicted patterns. Here is where ancient starch preserved in sediments or as residues on stone tools could provide crucial evidence in support of the theory that plant eating in general and tubers in particular played a major role in human evolution. Work completed so far suggests that the prospects for learning something new and exciting about the human past by focusing on tubers are very promising.

Box 1.2. How And When Did Taro Become The Most Widespread Starchy Food Crop In The World?

Peter J. Matthews

Before the modern era of rapid international transport, taro (*Colocasia esculenta*) was the most widespread starchy food crop in the world (Figure 1.2). Historical and botanical records show that it was grown and used in tropical and temperate regions in both hemispheres, from Northeast Asia to Southern Europe, from West Africa to India, and from Southeast Asia to Eastern Polynesia. The enormous premodern range of taro far exceeded the geographical distributions of better known crops such as rice, wheat, and maize. Taro is still grown and used in these regions and is now also common in the Americas (Figure 1.3). Although eaten by hundreds of millions of people resident on most continents and on innumerable small islands, taro has been little studied because it is a major crop in only a few places, most notably in the tropical Pacific. Many basic questions concerning the history of this plant remain unanswered. How and when did taro spread so widely? Did domestication take place in a single location or in many places? What natural and human processes were involved in the dispersal of this crop?

Figure 1.2. Taro *(Colocasia esculenta)* is one of the most important and widespread starchy food sources in the world. *Photo by Carol Lentfer*

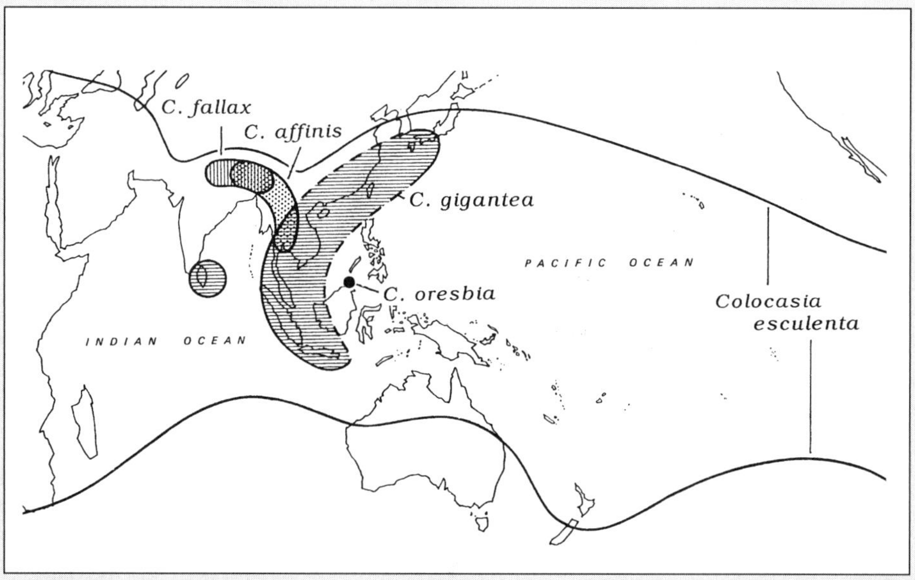

Figure 1.3. The distributions of taro *(C. esculenta)* in cultivation and of other *Colocasia* species in the wild (from Matthews 1991; cf. Matthews 2003). Dotted lines indicate where the boundaries for *C. gigantea* are most uncertain. Note that *C. oresbia*, a wild species in Sabah, Malaysia, was recently discovered and described by Hay (1996).

either that humans began to manipulate and move this crop a very long time ago or that it has spread incredibly rapidly in recent times.

The appearance of taro as a starchy residue on stone tools dating as old as 28,000 years ago (Loy et al. 1992) and at the controversial site of Kuk Swamp where a multidisciplinary project has argued for the presence of cultivation possibly as early as 10,000 years ago (Denham et al. 2003, 2004a, 2004b) highlights the possibility of early domestication for taro (Box 1.3). Many scholars were quick to dismiss Sauer's views, but others have continued to stress the importance of tubers in the history of plant domestication (e.g., Harris

Box 1.2 (continued)

On present evidence, it seems most likely that the natural range of taro (before its use by humans) extended from India to southern China, Southeast Asia, northern Australia, and New Guinea. No one has seriously studied wild taro populations in Africa and so the possibility of independent domestication there cannot be ruled out. The apparently natural wild type of taro is characterized by extreme acridity, long, thin stolons (instead of short, starchy sidecorms), and simple coloration (green or mostly green, and without variegation). The acrid factors and other chemicals that occur in the plant help protect taro against microbial diseases, insect pests, and large herbivores such as birds and pigs. Wild-type taros are common in habitats that appear natural, most notably alongside waterfalls and streams in wet, tropical rainforest (Southeast Asia, Australia, Papua New Guinea), or stream banks, springs, and seepages in the savanna woodlands (in the monsoon zone of northern Australia) (Matthews 1990, 1991, 1997, 2003). Although wild-type taros are toxic, they can be eaten after thorough processing. In future studies, comparisons of starch grain morphologies in wild and cultivated forms of taro combined with archaeological studies could indicate where and when domestication took place (Plates 1–4).

To understand the history of taro, one can also look at where its nearest relatives are found, since it must have evolved from among these species or their common ancestors. The nearest relatives are other species of *Colocasia*, which grow wild in the region from northeast India to southern China, Borneo, and Sumatra. This is its minimum range. It is in this region that we might expect to find the earliest archaeological evidence for using taro. Since modern taro is always peeled before eating to avoid the toxins and the peel is never eaten, we might find discarded peelings or the remnants of starchy tissue on scraping tools.

How and when did taro spread so far beyond its natural range in Asia, Southeast Asia, and the western Pacific? From the long distances involved, we can hypothesize either that taro was domesticated a very long time ago (before cereal crops such as rice and wheat?) and then moved to other areas, or that it was domesticated relatively recently and moved quickly, perhaps because it was very easy to grow and could be transplanted by vegetative means and/or because it was often exchanged or moved with people who were migrating.

The tentative identification of taro starch grains on stone tools from the Solomon Islands, in a context dated to 28,000 BP (Loy et al. 1992), can be interpreted in at least two ways. Firstly, taro might be one of the earliest cultivated plants in the world or its natural range extended to island Melanesia, where it was available as a wild food source for the earliest human arrivals. To choose between these and other interpretations, more research is needed in the realms of archaeology and natural science.

The history of the domestication and cultivation of taro throughout its enormous geographical range is a very important topic that has been neglected by archaeologists and botanists alike, possibly because direct evidence has been so difficult to obtain. The study of ancient starch offers a new way to study the domestication and spread of taro around the world. Results from this new approach could produce some interesting surprises concerning the origins and development of agriculture.

Box 1.3. What Really Happened At Kuk Swamp?

Robin Torrence

From many points of view, Kuk Swamp, located in the Highlands of Papua New Guinea, is among the most important and controversial archaeological sites ever studied. The archaeological evidence at Kuk, largely comprised of ditches and drains, has been recently redated as beginning around 10,000 years ago (Denham et al. 2003, 2004a, 2004b) (Figure 1.4). The integration of research on starch granules found on stone tools at the site is addressing significant questions about the origins of agriculture in the Pacific region and the relationship between changes in the type of agricultural crops and social intensity.

Figure 1.4. A sequence of water control features at Kuk Swamp, Papua New Guinea. Phase 2 grey clay 'islands' underlie later cross-cutting water control channels which define a series of mounds. Bananas may have been grown on the islands. In later phases taro was probably grown in the ditches and, most recently, sweet potato on the mounds between them. *Photo by Peter White*

Basing his inferences on changes in the configurations of the drainage system at Kuk, backed up by palynological, geomorphological, and ethnobotanical studies, the excavator Jack Golson proposed that the building of large drains in the earliest phase signalled the use of the swamp for the cultivation of crops, including taro (*Colocasia esculenta*) (Figure 1.2), sugar cane (*Saccharum officinarum*), and bananas (Australimusa varieties), which he and others proposed had been domesticated in the Papua New Guinea lowlands (cf. Bayliss-Smith 1996: 508–10). From about 4,000 BP (phase 3), agriculture was inferred by Golson to have become increasingly intensive through time (the 'Colocasian revolution') on the basis of changes made to the drainage systems. Another major shift occurred at c. 2,000 BP when the drainage system was organised into

Figure 1.5. Tuberous roots of the sweet potato (*Ipomoea batatas*). This species is grown mostly along the coast and forested lowland regions of Peru, at elevations ranging from 0 to 2,500 m. The same species was introduced into Highland New Guinea during the latest phase at Kuk Swamp and is still the dominant crop in that region. *Photo by Carlos Ochoa*

Box 1.3 (continued)

a grid pattern (cf. summaries in Bayliss-Smith 1996; Bayliss-Smith and Golson 1999; Golson 1977, 1981). The reorganisation of the swamp has been used as evidence for a further intensification of production which was hypothesized to have been caused by new social systems based on competition for status and prestige (cf. Golson and Gardner 1990; Bayliss-Smith and Golson 1999: 209).

Finally, at around 300 years ago, it is assumed that sweet potato (*Ipomoea batatas*) (Figure 1.5) was introduced to the region, both on the basis of changes in the nature of the ditch systems, but also because of the finding of charred tubers (e.g., Golson 1977, 1982). The adoption of sweet potato in the Highlands caused, in Golson's view (1977, 1982; Golson and Gardner 1990), major social changes known as the *Ipomoean Revolution*. He proposed that the previous control of political power by a few groups who were able to produce more taro and pigs because of their access to swampland agriculture would no longer have been feasible. The arrival of the sweet potato, which can be grown very successfully in a much broader range of environmental settings than taro, would have enabled people outside the swamp area to increase their production substantially and compete for prestige more effectively. According to Golson's (1982) proposal, the well-known Big Man system, a relatively egalitarian system in which success in production and exchange led to high status, was a consequence of the adoption of sweet potato. The change to the Big Man political system is an interesting reverse of what is expected to happen with agricultural intensification because it was essentially a case of devolution from power based on control over the swamp to open competition, rather than a change to a more power-based form of political organisation.

The controversies surrounding these reconstructions of agricultural history at Kuk Swamp centre around several issues. The first problem is whether the drains, especially in phases 1 and 2, are the products of human labour and whether their presence necessarily means the swamp was used for agriculture. If one accepted a date of 10,000 BP for cultivation at Kuk, the date for the origins of agriculture in Papua New Guinea would be comparable to other parts of the Old World. This conclusion has been unacceptable to some scholars who believed that domestication is unlikely to happen in more than a very few places (cf. Harris 1996: 6). The subtext to this view is that since Papua New Guinea never developed 'civilisation,' it is also unlikely that agriculture was initiated there. A second and related question is whether the crops growing in the earliest phase were domesticates or wild plants, and if the former, had they been domesticated locally or were they imported from Southeast Asia, which has long been the favoured centre of domestication, especially for taro (Bayliss-Smith 1996: 508; Smith 1998: 143; cf. Box 1.2). If the taro grown in Kuk swamp at this early date was domesticated, then the process would have taken place even earlier since wild taro cannot grow at this altitude. This possibility seriously threatens models of agricultural origins in which the Near East and China are favourites for the 'earliest.' Bayliss-Smith (1996: 517) suggested, for example, that taro was not grown at Kuk until after 4,000 BP, making it easier for it to have been introduced from Southeast Asia.

Third, when did sweet potato arrive? This is connected to a fourth question: where did it come from? Since sweet potato was originally domesticated in South America (see Figure 1.5), it had to find its way to Kuk Swamp before the Europeans arrived in the 1930s to witness it as the dominant crop in the region (Yen 1974). Scholars debate whether sweet potato was introduced from Polynesia—where it was cultivated since at least 1,000 BP (Hather and Kirch 1991)—or whether it found its way to the Highlands through trade with lowlanders who had acquired it from early European visi-

Box 1.3 (continued)

tors to the region (cf. Bayliss-Smith 1996: 517). Those who favour a Polynesian introduction suggest that sweet potato had entered Papua New Guinea significantly earlier than European contact (e.g., Scaglion and Soto 1994). At present consensus appears to favour European introduction primarily on the basis of oral history, although there is still no well-dated evidence at Kuk Swamp or any other archaeological context (Ballard et al. 2005).

Obviously, finding answers to the questions posed by Golson's tantalising interpretations of the evidence at Kuk Swamp is extremely important to world archaeology, particularly for an understanding of the origins of plant domestication and agriculture, but also for testing theories which link intensification of food production with social evolution (e.g., the Colocasian and Ipomoean revolutions). The answers to these questions largely lie in finding out what was actually grown at Kuk Swamp in each of the phases of ditch construction. Especially critical is the date for the arrival of taro and distinguishing whether it was a Southeast Asian or Melanesian domesticate. The sweet potato debate could also be resolved with good dates, but information on whether the crops were more closely related to Polynesian or European-introduced types is also an issue (Ballard et al. 2005). It would also be helpful to know about the contemporary dryland patterns of agriculture. The problem archaeologists faced was that macroremains for the key crops (taro and sweet potato, as well as yams and Pandanas which were also important) are not well preserved in the Kuk Swamp, these plants do not produce significant quantities of pollen, and they have no phytoliths. As Tim Bayliss-Smith (1996: 516) summed up the situation,

> It is clear that we remain in almost total ignorance about the origins of New Guinea agriculture. . . . Equally, much uncertainty remains about the subsequent evolution of agriculture. Neither the ignorance nor the uncertainty will be resolved simply by more information being obtained about a wider range of sites, unless better ways can be devised to generate information about the plant species that were being managed or cultivated.

A multidisciplinary study combining evidence from new archaeological excavations, geomorphology, micromorphology, paleobotany (pollen, charcoal, phytoliths), and entomology have provided evidence for significant modification of the swamp and the possible cultivation of bananas beginning at least by 10,000 BP, although the new data have also generated further debates (e.g., Denham et al. 2004a, 2004b). The recovery of starch granules identified as *Colocasia esculenta* from stone tools (Plates 1, 2) have made a critical contribution since they confirm the presence of this plant from the earliest levels of Kuk (Fullagar et al. 2005; cf. Denham et al. 2003). At this stage the status of the crop as domesticated or wild or as introduced or indigenous is still not certain (Denham et al. 2004a; Fullagar et al. 2005), but the preliminary findings confirm the importance of starch analysis as one of a number of methods for studying the origins of agriculture.

Having shown the potential of ancient starch for addressing key issues in archaeology, additional analyses of ancient starch will hopefully help better resolve the very tantalising and important questions surrounding the issue of an independent centre for the domestication of plants in Oceania and also contribute to understanding the role of agricultural intensification in social change and the origin of the classic Big Man system.

1981). New evidence from the New World also suggests that aspects of Sauer's theory should be resurrected (e.g., Piperno and Pearsall 1998; Piperno et al. 2000).

The white potato is one of the most important sources of food in the modern world and has had a profound effect on the history of modern world population growth and migration. Yet the early history of its domestication in the South American Andes and its role in the development of civilisation in that region are very poorly known. Another South American domesticate, the sweet potato, was eventually transported several thousand kilometers away to the Polynesian Islands where it was adopted by at least 1,000 years ago (Yen 1974; Hather 1992, 1994a; Hather and Kirch 1991) and became the staple food for many people in that region. At European contact the sweet potato formed the basis of very intensive agricultural systems and high population densities in Highland New Guinea. Was sweet potato a European introduction to mainland New Guinea? What effects did it have on the development of social organisation in the Highlands? These and other questions about the arrival of sweet potato are hotly debated issues, but have been difficult to resolve in the absence of preserved macroremains (e.g., Ballard et al. 2005; Bayliss-Smith 1996; Golson 1977, 1982) (Box 1.3). Ancient starch may offer a way to move these questions beyond speculation.

Taking a slightly different tack, we can be confident that starch was a vital source of energy in most past human diets. All plants make starch

Figure 1.6. Garu village market, Papua New Guinea, where underground storage organs are the most important source of daily food. *Photo by Carol Lentfer*

as part of the daily process of photosynthesis, but those that store it in relatively large quantities as seeds or tubers are the most important for humans, since these comprise an excellent source of calories and carbohydrates. Although seeds, such as wheat, maize, and rice, tend to dominate subsistence strategies in temperate and other seasonal environments, roots and tubers have been the major foods for people in the wet tropics. Starchy plants have also been used as raw materials for making tools, clothes, housing, and other items such as papyrus scrolls (Leach and Tait 2000). The starchy parts of plants can also provide useful poisons, pigments, adhesives, fragrances, and drugs.

According to Borgstrom (1973), the starchy granules of wheat, rice, and maize combined with the tubers potato, yam, and cassava provide 80 percent of the world's caloric intake. Large areas of the world are almost completely dependent on starchy plants for their daily food, many of which are roots and tubers (Figure 1.6). Data collected by the Food and Agriculture Organization of the United Nations (1990: 26, 129) show that in the tropical rainforest of Sub-Saharan Africa tubers make up 78 percent of human caloric intake. Overall, they estimate that just one tuber, cassava, comprises 37 percent of calories consumed in this heavily populated region. Tubers account for 15 to 43 percent of caloric intake of the Pacific nations (FAO 1990: 26) with most of the rest accounted for by imported foods. Before easy access to store-bought rice, the percentage would have been markedly higher. Given what we know about the modern world, starch has probably been critical to human subsistence as a major source of carbohydrates since earliest times. With the development of ancient starch analysis, archaeology now has another powerful method for answering 'big questions' about past human diets and subsistence strategies, as well as material culture.

HISTORICAL BACKGROUND

Ancient starch analysis is not a new technique within archaeology, although its potential was not appreciated until its rapid expansion in the past ten to twenty years. Modern systematic archaeological studies of starch were pioneered by Donald Ugent (1994, 1997; Ugent et al. 1981, 1982, 1983, 1984, 1986, 1987; Ugent and Peterson 1988),

Box 1.4. Early Starch Research

Robin Torrence

Donald Ugent was the first modern scientist to use ancient starch within a systematic program of archaeological research on plant use and domestication. He began his pioneering studies of ancient starch by researching remains of dessicated tubers recovered from refuse middens at numerous sites located in Casma Valley within the Peruvian coastal desert (Figure 1.7). Starch granules ranging in date from 2,250 to 295 B.C. were recovered from the interior parts of many dried and some partially charred tuber or rhizome skins. Associated evidence indicated that root crops had been baked next to wood fires. Ugent's team noted that the starches were often well preserved. Although some granules had deep fissures or were cracked right through, the majority had retained their structural and chemical properties. Most granules were birefringent under cross-polarised light and reacted normally to the stain iodine-potassium-iodide (IKI) (Figure 1.8) (cf. Chapter 2; Box 7.2).

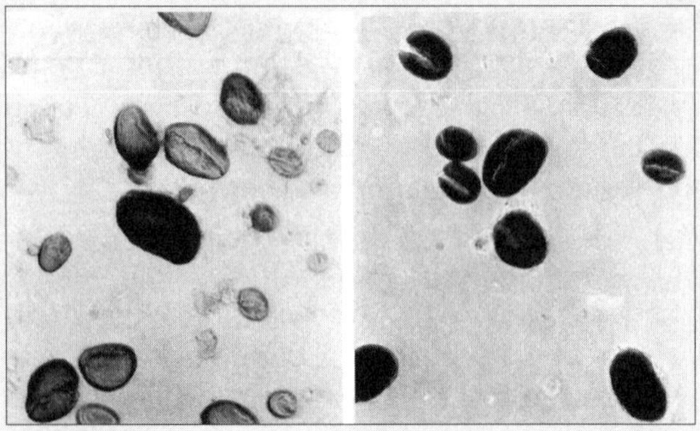

Figure 1.8. Starch granules from preserved tubers of white potato recovered from Huaynuma, Casma Valley, Peru and dating to c. 2,000 BP. Despite their age they still react with iodine-potassium-iodide and methylene blue stains; (left) largest granule is 96 microns; (right) largest granule is 125 microns. *Reproduced with permission from Ugent et al. 1982: figure 3*

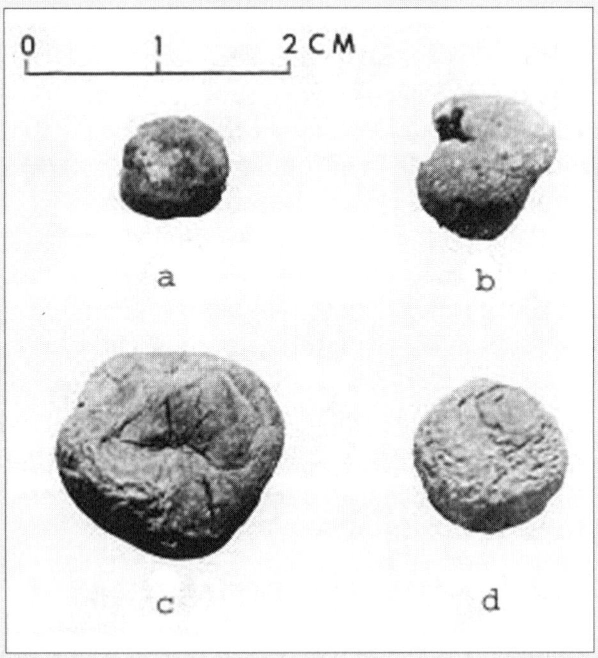

Figure 1.7. Archaeological specimens of sweet potato (*Ipomoea batatas*) from two sites in the Casma Valley (lower left corner, Huaynuma; remaining specimens from Pampa de las Llamas-Moxeke). *Reproduced with permission from Ugent et al. 1982: figure 2*

Over many years Ugent and his colleagues employed a wide range of techniques for studying ancient starch within dessicated macroremains in order to identify remains of potato *(Solanum tuberosum)*, sweet potato *(Ipomoea batatas)*, achira or canna *(Canna* sp.) and manioc *(Manihot esculenta)*. Scanning electron micrographs and optical phase contrast photos of sweet potato starch from the initial period (1,800–1,500 B.C.) were among the first techniques used to study the granules (Ugent et al. 1981) (Figure 1.9). Later publications featured the iodine-stained granules of archaeological white potatoes (Figure 1.8) (Ugent et al. 1982) and sweet potatoes (Ugent et al. 1983) (cf. Box 7.2); phase contrast, polarized light, and SEM micrographs of achira (Ugent et al. 1984) (Plates 5, 6); polarized light, and SEM micrographs of manioc (Ugent et al. 1986) (Plates 7, 22). The differentiation of an-

whose work is reviewed in Box 1.4. Over many years he and his associates have used the diagnostic characteristics of starch to amplify the identification of desiccated remains of tubers recovered from sites in South America. Ugent's seminal studies conclusively showed first, that ancient starch has been well preserved, and second, that through its distinctive properties starch can be used to identify the kinds of plants used in the past. These results, based on the association of starch with macroremains, provided the basis for subsequent studies which have relied solely on the starch granules themselves. In addition, Ugent (1994) tested out a range of methods that still form the core of ancient starch research, including various kinds of microscopy, staining, chromatography, and spectrophotometry.

The next major development in starch research was the realization that the individual granules can be studied in their own right apart from the rest of the plant. The focus on granules is largely derived from research based on microscopic residues found on stone tools (cf. Chapter 9). Studies of chipped stone inspired by Tom Loy (1994a; Loy et al. 1992) and Richard Fullagar (1998) in Australia and parallel work in the New World by Dolores Piperno and Irene Holst (1998, 2004; Piperno et al. 2000) were critical innovations. Meanwhile in the Old World, Delwen Samuel (1996a, 1996b, 2000) continued Ugent's focus on macroremains, but she extended the method to include processed foods such as bread and beer. From its initial focus on food remains, ancient starch research has now diversified into the broad range of research areas represented in this book.

UNIQUE PROPERTIES

Ancient starch analysis has much in common with paleobotanical research based on other plant fossils, particularly pollen and phytoliths. Many of the methodologies described in this book have been largely modelled and built on these established and successful fields of scholarship. Due to their unique properties, starch granules can provide important supplementary information to that gained from these better established disciplines.

Capturing large quantities of high-quality starch is often why people collect and cultivate plants in the first place. In contrast to pollen and

Box 1.4 (continued)

cient starch types was also successfully accomplished using paper and thin-layer chromatography (TLC) and UV Spectrophotometry (Ugent 1994). These studies have convincingly demonstrated that despite the potentially damaging effects of cultural practices (e.g., food processing and cooking), not only has ancient starch been preserved, but also the starch granules can be assigned to plant taxa.

In the most recent study, Ugent (1997; Ugent et al. 1987) has identified ancient starch granules attached to the skin peelings of an endemic Chilean wild potato *(Solanum maglia)* dated to about 12,500 years ago at the site of Monte Verde in Chile (Plate 8). Having shown that wild potato has diagnostic starch granules, this study illustrates the potential contribution of ancient starch analysis for monitoring the history of tuber domestication. All future research on ancient starch builds on the solid groundwork established by Ugent's methods and the exciting potential created by the results of his South American research.

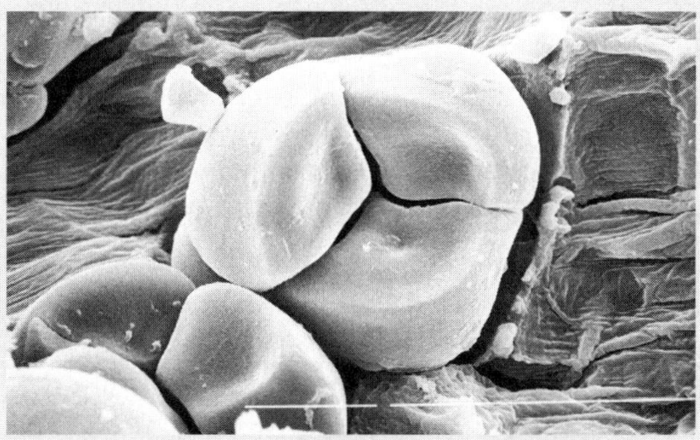

Figure 1.9. SEM micrograph of a starch granule from an archaeological sample of sweet potato (*Ipomoea batatas*) collected at Pampas de las Llamas-Moxeke. Gold-palladium coated, x2000. *Photo by Donald Ugent*

phytoliths which are often derived from plant parts that are not consumed, starch storage organs such as seeds, roots, and tubers comprise the food source. Archaeological traces of tubers and roots have been so difficult to obtain because they are consumed directly and these plants do not generally contain phytoliths. In addition, people often deliberately manipulate the plants so that they yield little or no pollen or harvest them before pollen is produced. In fact, Hather (1994a: 721) has proposed that it is the ability of particular tubers to be 'vegetatively reproduced through stem tissue' that has led to their domestication. Unlike the other plant microfossils, starch is directly linked to food consumption. For most types of tubers and roots which are the major sources of food for many parts of the world (e.g., potatoes, taro, yam, manioc), the preservation of starch granules is likely to be the only macro- or microscopic evidence of their existence.

Piperno (1988: ix) has noted that when phytolith analysis began, it was seriously hindered because the basic background work had not been successfully completed. Fortunately, ancient starch analysis does not have to face these difficulties because modern starch has been thoroughly studied by botanists and within the modern food industry. A great deal of research into starch formation in the plant has already been carried out. Its biochemical properties are well known and there is a large body of work on established techniques that archaeologists can adopt and/or modify to build relevant methods. For example, it is so well accepted that starch granules can be diagnostic to plant taxa that this critical information appears in a popular source, 'The granules of starch present in any plant have size, shape, and markings characteristic of the species of the plant in which the starch is made' (Microsoft Encarta 1993–7). There is even a whole journal devoted to starch research (*Die Stärke/Starch*).

REDRESSING THE BALANCE

Since seed crops have long been assumed to have been a critical element in the rise of civilisations (e.g., Diamond 1997), archaeologists have invested much effort into studying their domestication. In particular, a great deal of research has been carried out into the history of wheat, rice, and maize (e.g., Harris and Hillman 1989; Harris 1996; Piperno and Pearsall 1998). The methods used by consumers to process seeds often involve charring and/or storage which enhance their preservation in the archaeological record. Many of these plants also have distinctive pollen and phytoliths which can be recovered from archaeological and paleoenvironmental contexts. In contrast, roots and tubers are much more difficult to study because the majority do not make pollen or phytoliths and most involve limited processing. They also preserve very poorly in archaeological contexts and so have been found only in unusual circumstances, for example as waterlogged, carbonised, desiccated, or mineralised remains (e.g., Hather 1991, 1992, 1994b; Ugent 1994; Ugent et al. 1981, 1982, 1984, 1987). Fortunately, tubers produce starch in abundance. Recent research has successfully begun to redress the lack of research on the domestication of roots and tubers (e.g., Piperno et al. 2000; Perry 2002b; Denham et al. 2003; Matthews 2003).

The absence of 'smoking gun' type evidence (i.e., macroremains such as seeds or dessicated tubers) has meant that the whole history of plant domestication in large areas of the world, including the tropical Americas, Africa, and Asia, has been thought to be almost beyond knowing (e.g., Smith 1998: 143). For example, due to the lack of evidence, Smith (1998: 12–13) did not consider the heartlands of root crops in Southeast Asia and the rain forest of South America as centers of plant domestication, but relegated them to the role of satellites to other areas that were dominated by seed crops. It is not surprising, then, that many of the contributors to this book work in Oceania, Southeast Asia, or Central and South America, areas where tubers have comprised the bulk of the diet during much of the past.

A COLLABORATIVE EFFORT

At its inaugural meeting in 1998, the Ancient Starch Research Group reviewed the state of the field. Although the seminar concluded that there was an urgent need for a comprehensive guide to starch research because it would greatly benefit archaeology, no single individual felt capable of writing the definitive summary. This is not surprising given that the required knowledge and expertise covers a broad spectrum and the field is expanding rapidly, whereas most researchers are relative newcomers to the discipline. The group therefore decided to pool the knowledge and experiences of the participants together with others not present whose advice, assistance, and case studies were sought. Although the editors have had to take a strong hand to produce a coherent book, many scholars contributed their knowledge, experience, and case studies. To preserve the collaborative nature of the project, many voices are represented in the text and especially in the discussion of particular topics presented in the text boxes. The boxes are especially important because they offer insights into how the new field has been developing.

The long-term collaborations that led to *Ancient Starch Research* have created a broader and deeper review than a single scholar could manage. Chapters written by several scholars supplemented with text boxes written by additional contributors are somewhat unusual for a comprehensive manual. One scholarly press rejected our proposal because it felt the book would not be authoritative enough, and they were correct. Theirs is a different view of how to build research strength than we have advocated. Our multi-authored approach is intended to capture the knowledge and experience of a significant proportion of the scholars actively researching ancient starch as well as to highlight the diversity and dynamic character of this growing field.

The book provides a comprehensive guide to methods, but it also incorporates a large number of new ideas, experiments, and case studies, many of which have been rescued from obscurity in unpublished notes, student theses, or the grey literature. The team decided to include summaries of recent and ongoing research because they can accelerate the development of starch research by showcasing its potential spectrum, by signposting areas that would benefit from ancient starch research, and through sharing information as widely as possible. In addition, this book is not intended to be a definitive 'how to' manual. Although many methods are presented, the text encourages researchers to use the basic knowledge to devise approaches that are geared toward a particular research question. The intention is to stimulate innovation and creative applications. By establishing the basic groundwork for ancient starch analysis, we hope that others will now push onward, tackle the areas that have been identified as needing further development, and create whole new applications that have not yet been imagined.

PRINCIPLES, PATHWAYS, AND APPLICATIONS

Ancient Starch Research is comprised of three sections. Descriptions and examples of basic principles about starch and microscopy and the pathways of starch from the plant to the archaeologist are presented. The basic background is then followed by the discussion of a range of archaeological and paleoenvironmental applications which are illustrated with case studies. We start off in Chapter 2 with a basic introduction to the biology of starch and a description of the morphological and biochemical properties that affect how it was used in the past, and why and how it persisted, and that are necessary for its identification. Since starch is a microscopic substance, Chapter 3 presents a basic introduction to the kinds of microscope techniques necessary for monitoring and describing it. The chapter is especially designed to assist newcomers to paleobotany or to residue analysis.

Having covered the basic principles required to detect and describe starch, the next two chapters consider the life history of starch and the formation processes that lead to its preservation in the archaeological or environmental record. Chapter 4 introduces the concept of starch pathways as a useful way to understand the life history of starch granules from the plant to the archaeological context. It discusses how and why plants are chosen for exploitation, the wide range of ways they are used and modified, and identifies sources of information or techniques archaeologists can use to model these processes. The importance of ethnoarchaeology and experimentation in developing means for reconstructing starch pathways are especially highlighted by the case studies in Chapter 4. The fairly frequent use of these in early starch research has set this book slightly apart from many previous discussions of paleobotany. The second part of the life history of starch, which forms the subject of Chapter 5, begins after it has entered into the archaeological record. This takes us into the realm of postdepositional processes or *taphonomy*. Chapter 5 assesses current knowledge derived from biochemical and archaeological studies and experiments that helps us understand how and why starch is preserved in various contexts and provides the basis for predicting its occurrence in new contexts.

Whereas the first five chapters consider the background theory and knowledge that one should acquire before undertaking research involving the analysis of ancient starch, the second half of the book discusses ways that starch can contribute to archaeological and environmental research. Methods for recovering starch from soils and artifacts, the identification of taxa, and the description and interpretation of results are presented and illustrated. Since the field is still young, the authors make recommendations for how methods could be further improved and expanded.

The most important initial condition for starch analysis is an appropriate reference collection. Chapter 6 leads off the second half of the book by describing the interactive procedure between posing questions and the type of reference collection that is required to answer them. The basic procedures for setting up a reference collection, including obtaining plants and making slide mounts, are detailed and illustrated by several case studies.

Next, Chapter 7 considers how one utilises reference material in association with published sources to describe starch granules and to identify the plant from which they were derived. The chapter echoes the rest of the book because it argues that the nature of the questions asked dictate the level of identification required. A number of potential scenarios of differing levels of specificity are described, ranging from the recognition of material as starch to the specification of domesticated vs. wild forms of the same species.

Having established a set of methods for identifying starch, Chapters 8 through 10 describe techniques for making inferences about the past based on the recovery of ancient starch. Chapter 8 describes approaches that extract starch from sediments. Methods for reconstructing past environments both in the broad sense and also on a smaller scale as, for instance, in the recognition of different activity areas within a single archaeological site, are explained and exemplified through case studies. Starch recovered from sediments can also contribute to an understanding of the history of local vegetation in relation to climatic change and/or human impacts in much the same way that phytoliths have done (e.g., Piperno 1988, 1998: 200–18; Kealhofer and Penny 1998). Variation over time or through space within and between localities or archaeological sites can all be explored by focusing on differences in the starch compositions.

In Chapter 9 the focus of research is centered on starch which has been preserved on artifacts in the form of residues. Within this broad field there are two basic aims. The first is to identify the kinds of plants present at a particular place or site, for example as part of a study of early plant domestication, whereas the second is particularly concerned with the interpretation of how artifacts were used. Most of the case studies are about stone tools, since these have previously received the most attention. The authors make the very important point, echoed from the focus on actualistic research in Chapter 4, that experimentation provides the basis for all analyses of residues. They also emphasise that studies of starch should be integrated into a larger research program that also considers tool design and use-wear.

As detailed in Chapter 10, starch altered through cooking and other forms of food preparation or as part of craft activities can also be detected within ancient food remains, residues on artifacts, or within material culture. This information is relevant for understanding food technology and other industries. For example, Samuel (1996a, 1996b, 2000) has reconstructed the techniques and procedures used in bread making and brewing beer in ancient Egypt. Her work showed that previous interpretations based on texts and illustrations had made fundamental errors.

The concluding chapter of the book assesses the current state of ancient starch research and suggests places where improvements are warranted. It looks to future areas where ancient starch research is likely to contribute answers to archaeology's 'big questions.' For example, it seems clear from what we know about starch pathways that a wider range of artifacts recovered from an expanded number of contexts and including museum collections are stores of ancient starch. Poorly explored sources of starch, such as within teeth or coprolites, also have high potential. One of the strengths of ancient starch is that it can complement and add to existing paleobotanical research. The most promising area of future research is the combination of multiple types of plant macro- and microfossils in conjunction with each other. Since each represents different parts of the plant, they often involve different life histories, and therefore they record different aspects of environments and plant use.

A BRIGHT FUTURE

The future for ancient starch research looks very bright, especially given the high level of interest, collaboration, and cooperation that went into the writing of this book. The increasing interest in starch research among younger scholars and especially those working in the Third World, who appreciate from firsthand experience the importance of starchy foods, is notable. It seems obvious that if an important variable, such as the use of tubers, has been left out of archaeological interpretations of past human behaviour because of the paucity of data, major revisions will be required when its presence is reinstated through the creation of a new methodology. Since starch research has enabled archaeologists to monitor more effectively a wider range of starchy foods than previously, it is not surprising that new patterns of plant use have been revealed and previous reconstructions of human behaviour have been shown to require revision. The three main areas of starch analysis—environmental reconstruction, the study of plant use, and the analysis of artifact function—will benefit from further development and refinement, but already they have made useful and, in some cases, extremely significant contributions to archaeology. More advances are certain to follow.

Since starch analysis can contribute to a large number of archaeological research problems, ranging from diet to environmental reconstruction, and level of specificity—from 'is it plant or animal' to 'is it a domesticated or wild form of a specific plant'—then a very broad suite of methods will be required. Not all of these are completely in place and there is still a great deal more work that needs to be done. On the other hand, there will never be a definitive set of methods, because the kinds of questions raised by archaeologists will respond to the answers that are possible as well as to changes in theoretical direction within the discipline. For this reason, the book has emphasised the process of devising methods that are specifically directed to the types of questions answered, rather than advocating a particular approach or type of technique.

Piperno's (1988: ix) summary of the state of the art concerning phytolith analysis could equally apply to ancient starch analysis in 2005:

> There exists a group of plant microfossils that exhibits all the attributes necessary to achieve legitimacy and prominence in paleoethnobotany and paleoecology: production in large numbers, durability in ancient sediments, and sufficient morphological specificity to allow identification of a wide range of taxa.

Since starch has all the characteristics listed by Piperno, this class of data should also take its place in the tool kits of paleobotanists, archaeologists, and environmental scholars. To this end the book presents the background data necessary for understanding how starch is produced, used, discarded, and preserved in the archaeological record. It also summarises previous studies and presents new cases that have utilized this information in a creative and productive way. We hope our efforts will help pave the way for the growth of ancient starch studies by tantalising scholars to pursue new avenues of research.

CHAPTER 2

BIOLOGY OF STARCH

Beth Gott, Huw Barton, Delwen Samuel, and Robin Torrence

This chapter provides relevant general background on the biology of starch necessary for ancient starch research. We begin by summarising current botanical knowledge about how and where starch is formed and stored in plants. We then examine properties of starch granules. An understanding of particular biological processes is fundamental for archaeological and paleoenvironmental research on ancient starch. For example, the processes of starch formation within the plant affect the nature and occurrence of the morphological traits that are used to classify starch granules; variability in the abundance and location of starch within different plant types is the basis for the occurrence of starch granules found in tool residues; and the biochemical properties of starch are affected by the archaeological contexts in which it is preserved. Additional background information about starch is presented in Chapter 5 (Taphonomy) and Chapter 7 (Identification).

Two forms of ancient starch have been recognised. The first, generally termed native starch, comprises raw, unmodified starch derived directly from plants. Native starch has been recovered from sediments within archaeological sites (cf. Chapter 8) and within residues on many types of artifacts, such as grinding and pounding tools and stone knives (cf. Chapter 9). To date it is the best understood and most widely studied form of ancient starch. Native starch can be altered through various cultural processes, especially those that involve pounding, abrasion, and heating, as described in Chapter 4. The resulting modified starch has been identified within ancient foods such as bread (Samuel 1996a, 1996b, 2000) and beer (Juan-Tresserras 1998). Modified starch has also been detected in manufactured artifacts such as some Egyptian papyri (de Bignicourt and Flieder 1996; Wiedemann and Bayer 1983) and a Romano-British cosmetic cream (Evershed et al. 2004). A review of processed starch preservation and archaeological examples of its occurrence are discussed in Chapter 10.

THE PRODUCTION OF STARCH

Starch is produced in different parts of the plant as a form of energy storage. Ultimately the starch story begins with the process of photosynthesis in which the energy of sunlight is converted into a solid form of potential energy. This process takes place within the chloroplasts (green plastids which colour the plant). The energy of light starts a series of reactions that split water into hydrogen and oxygen, and then recombine the free hydrogen with absorbed carbon dioxide to form glucose. This simple sugar provides the basic building block for all substances that the plant requires, such as protein, fat, and complex carbohydrates including cellulose and starch. Some of the glucose building blocks are transported from the chloroplasts to specialised organs of the plant where, within another specialised unit called an amyloplast (starch plastid), it is converted into reserve or storage starch, an insoluble product designed for long-term storage (Figure 2.1). The formation of a starch granule starts at a point called the hilum with additional layers laid down successively. Under normal growing conditions, one layer is added to the granule each day (Tester 1997: 166).

When the plant needs energy, the storage starch is converted back into sugar and transferred to parts of the plant where it is required. When the rate of photosynthesis is high during the day, small starch granules of indeterminate shape and about 1 micron in diameter may be formed within the chloroplast. These are called temporary, transitory, or transient starch granules (Buléon et al. 1998: 94). During the night the transient starch granules are reconverted to sugar which is moved to other locations in the plant and used as energy or transformed into storage starch in the amyloplasts (Raven et al. 1999: 126). Transient starch granules are unlikely to be identifiable to plant taxa, because 'their shapes are not species specific and are likely to be determined simply by the space available at the site where they are formed' (Sivak and Preiss

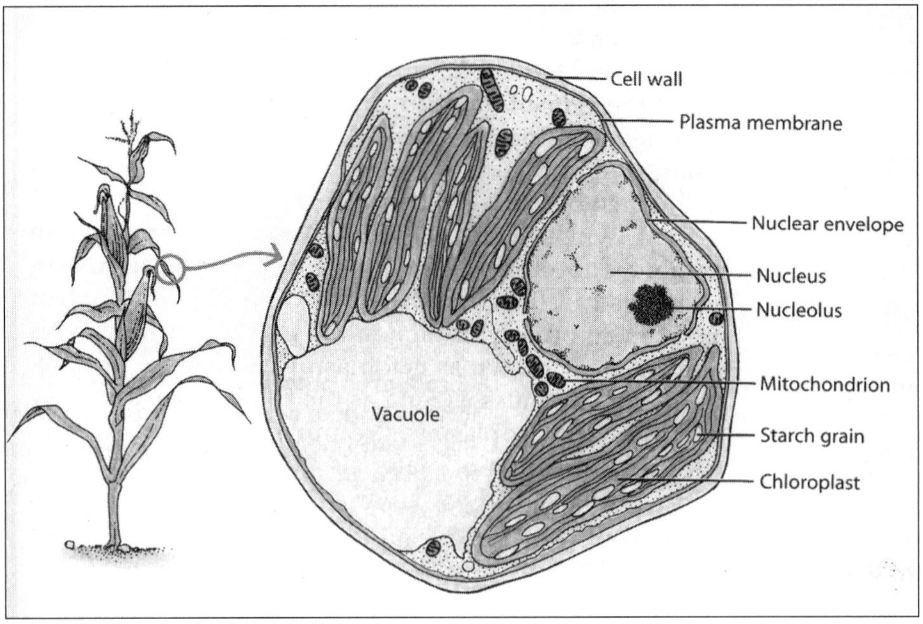

Figure 2.1. Cell from the leaf of a maize plant showing starch granules that are produced during the day within the chloroplast. *Reproduced with permission from Raven et al. 1999: 45*

1998a: 4). Also, their biochemical make-up is somewhat different from that of storage starch (Matheson 1996; Tomlinson et al. 1997). Although no study has been made of the occurrence of transient starch within archaeological contexts, it seems possible that some ancient starch samples, particularly those extracted from sediments, could contain significant quantities of nondiagnostic, transient starch granules derived from leaves and other plant parts containing chloroplasts.

LOCATIONS OF STARCH STORAGE

The location and abundance of reserve starch within plants is important knowledge for ancient starch researchers because it provides the basis for understanding how humans might have used plants in the past. Knowing the locations of starch storage is also a key to predicting the possible routes by which starch could have entered the archaeological record: e.g., through consumption, use of plant material in tool use, housing, etc. These models are considered more fully in Chapters 4 and 5, which discuss cultural and natural formation processes, respectively.

Starch is found in almost every kind of tissue within most of the green plants, e.g., 'leaves, fruit, pollen grains, roots, shoots and stems' (Sivak and Preiss 1998a: 1). The reserve starch, which humans mainly exploit for food, is largely concentrated within the storage organs such as roots, tubers, fruits, and seeds (cf. Guilbot and Mercier 1985: 210). In this book we use the term storage organ in a generic sense, to mean a vegetative organ of the plant that has become swollen due to the storage of reserve food. There exists a wide variety of subterranean plant parts used for starch storage. Botanists use specific terms to distinguish among these, although definitions vary widely: root tuber (e.g., sweet potato, Figure 1.5); stem tuber (e.g., potato); tap-root (e.g., carrot); corm (e.g., taro, Figure 1.2); bulb (e.g., onion); and rhizome (e.g., achira, Figure 2.2). Hather (1993) has proposed a detailed classification scheme. For our purposes the widely used shorthand term for all these variants, underground storage organ (USO), is adequate.

The location and density of starch can vary within the storage organ. For example, starch is more concentrated in the interior of the tuber than in the outer parts (Jenner 1982). In potatoes (*Solanum* sp.) there is more starch in the older, basal end of the tubers than in the younger, distal tissues. Yams and cassava also have different patterns

Figure 2.2. Freshly harvested rhizomes and flower buds of an achira plant (*Canna tuerckheimii*). The starchy rhizomes are covered by triangular scales and adventitious roots that are generally trimmed off before baking. *Photo by Carlos Ochoa*

Figure 2.3. Flowers and tubers of the Leopard orchid (*Diuris pardina*). *Photo by Beth Gott*

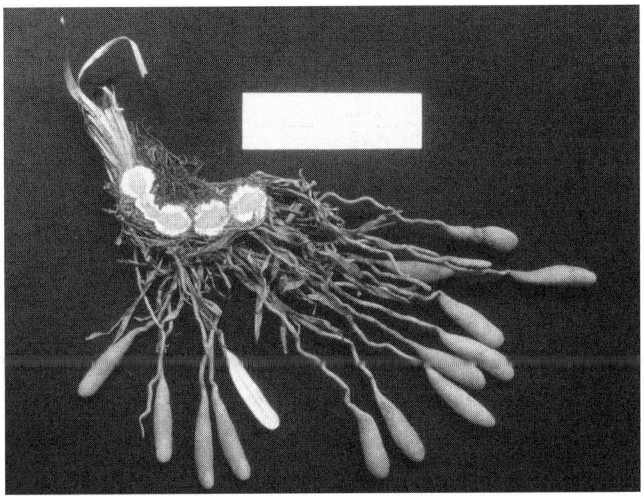

Figure 2.4. Abundant starch is stored in the native Australian tubers of water ribbons (*Triglochin procera*) (cf. Plates 11, 12 for examples of the starch granules). *Photo by Beth Gott*

of starch distribution within their USOs (Sivak and Preiss 1998a: 3).

Although the edible parts of plants have traditionally been of most interest to archaeologists, there are other locations of minor starch storage that people in the past have exploited for nonfood purposes. Starch will appear in residues on tools from some of these uses as, for example, those used to manufacture wooden artifacts or to prepare string from plant fibres. To emphasize this point further, we present some ethnographic examples of plant processing and use by Australian Aboriginal people. These illustrate the diversity of starch storage in plants and point to the many ways that starch from plants might enter the archaeological record (cf. Gott 1997, 2000).

Underground Storage Organs

Plants with starchy underground organs have been very important for the human diet in many parts of the world. The importance of USOs in present-day hunter-gatherer societies indicates these plant resources were used long before the first domestication of plants and the appearance of agriculture. Recently, it has been suggested that exploitation of USOs for food played a key role in hominid evolution (e.g., Box 1.1; Wrangham et al. 1999). The starch content of USOs such as potato, yam, sweet potato, and manioc/cassava is as high as 65 to 90 percent of dry weight (Sivak and Preiss 1998a: 3). The most commonly used tubers are so soft they would not normally be processed except by cooking and perhaps peeling. Good examples are the tubers of Australian orchids (Figure 2.3) and the water ribbon sedges (Figure 2.4).

A number of USOs contain toxins and therefore require substantial amounts of time, effort, and technology to make them suitable for human consumption. It is likely that starch residues are deposited on tools during the preparation of toxic starchy resources. Bitter varieties of manioc (cassava) (*Manihot esculenta*) are perhaps the most widely consumed example in this category (Figure 2.5). In South America, for example, manioc is grated and then squeezed through a basket to

Figure 2.5. The fleshy, starchy roots of the bitter varieties of manioc (*Manihot esculenta*) require special processing to remove their poisonous sap before they can be eaten. However, sweet varieties of manioc that can be made edible by baking are also known from the coast of Peru. *Photo by Carlos Ochoa*

Figure 2.6. Starchy fibres from *Typha* sp. used by Australian Aboriginal people for food and for making string. *Photo by Beth Gott*

leach out the cyanide, and then the flour that remains is dried and sifted before cooking. In Australia the seeds and stem tissue of cycads (*Cycas*, *Zamia*, *Macrozamia* species) require relatively lengthy processing before they can be eaten. Many yams, which are a common food source, also require processing (cf. case studies in Chapter 4).

Other USOs, especially swollen tap roots and rhizomes, are fibrous. Since these are processed to free starch from the fibre, they are also likely to create residues on tools. Young roots of trees such as the Australian kurrajong (*Brachychiton populneus*, Backhouse MS: 67) and fibrous storage organs found in an Australian bulrush or cattail called Cumbungi (*Typha* sp.; see Box 4.3, Figure 2.6) and ferns (*Blechnum* and *Pteridium* species, cf. Box 4.5, Plate 34) were pounded by Australian Aboriginal people to break up the fibres (Hodgkinson 1845: 225) or were sometimes chewed. Some trees store starch between other tissues within the root bark and root sapwood. The root bark of various *Eucalyptus* species, known as 'mallee,' was roasted and pounded for food (Eyre 1845: II, 250) and the woody roots were shaped for implements (Smyth 1878: 301).

A few plant species store carbohydrate in the form of fructan rather than starch (Sivak and Preiss 1998a: 5). These polysaccharides do not form solid granules and thus demand different methods of archaeological analysis (cf. Wandsnider 1997). Plants that store fructans are most often found in temperate regions. Good examples of these types of food sources include the Jerusalem artichoke (*Helianthus tuberosus*), which was consumed by Native Americans (Wandsnider 1997), and Murnong or Daisy Yam (*Microseris lanceolata*) (Gott 1983) and a number of small lilies that were important to Aboriginal people in southeastern Australia (Incoll et al. 1989).

Above-ground Stems

The stems of some species of nonwoody plants, such as the palms and cycads that are important food sources in many areas of the world, contain a soft starchy pith, e.g., sago, *Metroxylon sagu* in New Guinea and *Eugeissona utilis* in Borneo (Plate 23). In some cases the outer layer (cortex) of young or small stems can store starch. The stems of some plant families, such as the mallows (Malvaceae), were used for fibres. In particular, two rice-flower species, *Pimelea axiflora* and *Pimelea pauciflora*, were stripped by Australian Aborigines to provide the fine string for nets used to catch Bogong moths, an important source of seasonal food (Helms 1895: 393, 396). In plants with larger, woody stems, branches, and trunks, starch is sometimes also found in outer tissues—i.e., 'bark' in common parlance—which can be easily stripped from the tree. The bark of several wattle trees (*Acacia* species) was used for buckets, coarse string, and medicine (Dawson 1881: 14, 57) and the bark of white mallee (*Eucalyptus dumosa*) was beaten out into a mat (Woods 1879: 43).

Starch is also stored in the sapwood of trees, which is the soft, light-coloured, wet wood as distinct from the dry, dark, heartwood. When wooden articles such as spears, digging sticks, shields, clubs, and boomerangs are manufactured, starch granules may be incorporated into the residues left on the shaping tools. The stem bases of grasses, lilies, and sedges can also be starchy and were cut out using tools. Exudates such as sap, gum, or resins were widely used as adhesives and also contain starch granules picked up from stem tissues (Figure 2.7, cf. Box 9.4).

Flower Heads, Seeds, and Fruits

Flower heads and pollen from many plants contain storage starch that serves as an energy reserve. In the case of the Arums, starch stored within the spadix is used to generate heat as part of the process of volatising scent. A remarkable example of pollen being consumed (presumably as a source of starch) is that of *Typha orientalis* in New Zealand, where the pollen was collected and made into steamed 'bread' (Prendergast et al. 2000).

Seeds, which are a major food source for humans and animals, need a food store to nourish the developing seedling. Starch often plays this role. Grass seeds are clearly important in agricultural societies, having given us all our modern cereals. Starch is laid down in the short period between fertilisation and ripening of the seed. Since the water content of the seeds is quite low, only around 10 percent (Thomas and Corden 1977: 8, 20, 21), they are easily stored. In most parts of the world, seeds were usually processed on grindstones, sometimes with the addition of water.

Within seeds, oils and proteins are associated with starch to varying degrees. Pulses such as lentils, broad beans, and chickpeas as well as seeds of the *Acacia* genus (all in the Fabaceae family) are good examples of protein and starch-rich seeds (Figure 2.8). In many important food sources, the principal seed storage is lipid (oil or fat) accompanied by significant amounts of protein but very little starch, e.g., flax (*Linum* sp.) and various nuts (Vaughan 1970). In some seeds, such as soybean and rape, transient starch is accumulated early in development but replaced prior to maturity by oils (Wang et al. 1998: 486).

Other reproductive plant parts can also be important sources of starch or starch-derived compounds. Unripe fruits often contain starch that is converted to sugars as the fruit ripens. For example, some fruits, such as bananas and plantains, retain large quantities of starch even at maturity, as much as 90 percent of the dry weight (Sivak and Preiss 1998a: 1). The spore cases of a small Australian fern, Nardoo (*Marsilea* species) are rich in starch and were widely used in arid Australia (Gott 1997, Figure 6.12, Plate 62).

Figure 2.7. Resin from *Acacia dealbata* was widely used by Aboriginal people in Australia as an adhesive. *Photo by Beth Gott*

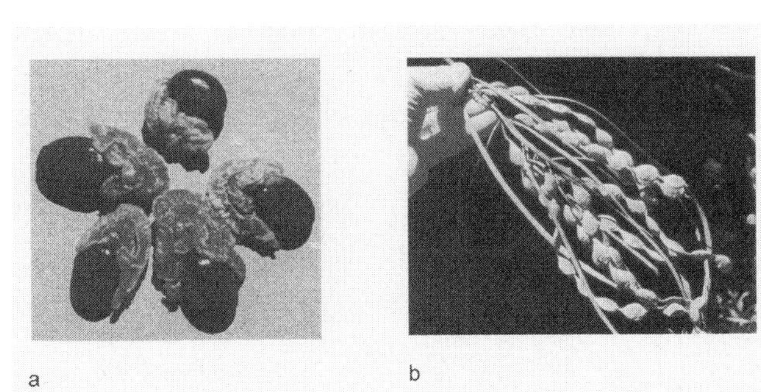

Figure 2.8. *Acacia stenophylla* and *Acacia salicina* seeds were probably used widely in the past by Aboriginal Australians. *Photos by Beth Gott*

Minor Sources

Storage starch is found in the tissues of some leaves. In addition to the transient starch granules in the leaves, they could form a significant component of the archaeological record in cases where leaves are used for roofing or bedding. In the wet zone of tropical Australia, for example, banana leaves were often used in this fashion, and throughout the world palms, grasses, and reeds have been widely used for thatching. A range of algae and microorganisms are also known to produce small quantities of starch in different forms (Shannon and Garwood 1984: 27–28; Raven et al. 1999: 44–45). These are, however, highly unlikely to enter the archaeological record.

STARCH GRANULE MORPHOLOGY

Starch granule morphology is largely dependent on the genetic composition of the plant, but size and shape can be modified by both the internal and external environments of the plant (Nikuni 1978; Oliveira et al. 1994; Haase and Plate 1996). Regardless of environmental factors, the shape and size of the starch granule as well as a number of additional properties discussed below are often characteristic of the plant taxon. Just as with other microfossils (e.g., pollen, phytoliths), however, not all starch granules in a plant are diagnostic. Methods for detailed description of granules and identification procedures are presented in Chapter 7.

Some granules are highly diagnostic, whereas others are less useful for identifying the plant species. Any given population of starch granules varies to some extent so that no two granules are exactly alike (Banks and Greenwood 1975). In some species there may be several distinct forms of starch. Granules from the same plant may differ according to the organ in which they are deposited. The form also changes over time according to the individual maturity of the granule and the age of the tissue or organ in which the starch is located (Shannon and Garwood 1984: 37).

Although starch granules vary in shape and size, the general properties shown in Figure 2.9 are typical of many starch granules. Generally, starch granules can be distinguished with relative ease from other microscopic plant fossils, diatoms, and other organic and inorganic substances (see

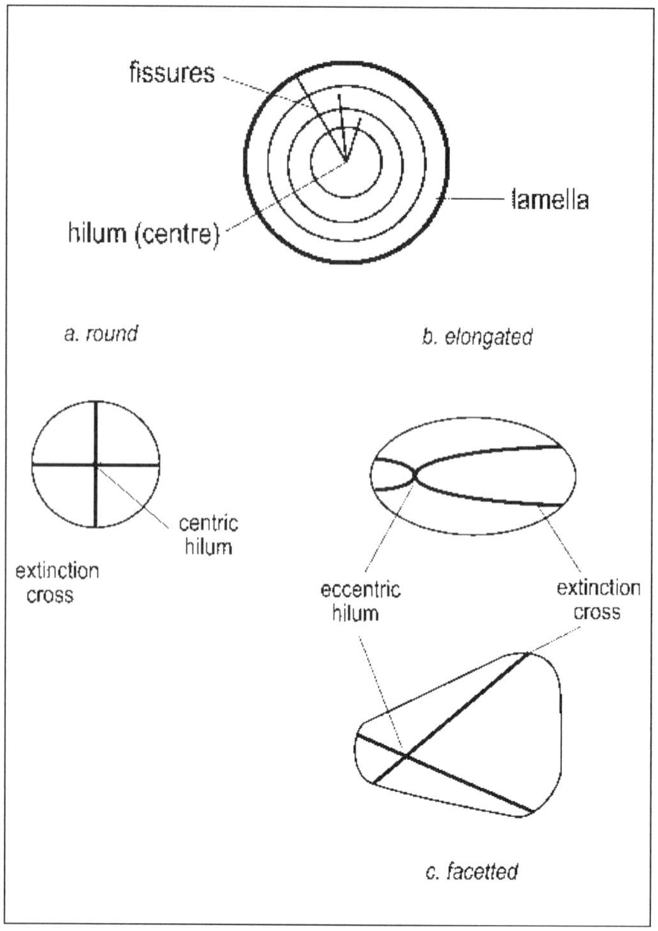

Figure 2.9. Basic characteristics of a starch granule. *Illustration by Fiona Roberts*

Box 7.3). The *hilum*, which is the core of the granule and the starting point from which the granule grows, is usually less organised than the rest of the granule and may contain nonstarch material from the amyloplast (French 1984: 189). Most commonly it is situated near the middle of the granule, but it can be eccentric, i.e., towards one end of the granule, as in the potato or yam (Plates 9, 10). In some species there are *fissures* with various shapes that emanate from the hilum (Plate 16). *Lamellae*, which are growth layers, are visible in many large granules and in granules of some plant species (e.g., Plates 9, 10). Striations, ridges, and central *vacuoles* (also known as open hila) are also found in some species (Plate 23). Fannon et al. (1992) report *pores* all over the surfaces of starch granules from cultivars in the subfamily Panicoideae (corn, sorghum, and millet). The large granules of wheat, rye, and barley have *equatorial grooves*, along which pores are also sometimes found, but no pores have been detected on starches from root and tuber cultivars (Fannon et al. 1992: 287). It is possible that pores are present on the surfaces of starches from wild taxa, particularly those in and related to the grass subfamily Panicoideae, but no relevant investigations have been reported.

Granule Types

Starch granules are classified as simple, compound, or semi-compound depending on how they are formed in the amyloplast. Simple granules have only one component in the amyloplast. In contrast, compound granules are comprised of several separate parts, known as subgranules or granula, which have formed simultaneously within a single amyloplast (Banks and Greenwood 1975: 251) (Plates 13, 34), but each separate granula of a compound granule still exhibits a polarising cross (see Birefringence, below). Examples of taxa with compound granules are cassava, sweet potato, quinoa (Chenopodiaceae), rice, oats, and some other important edible grasses such as *Eragrostis* sp. (Plates 63, 64) (Tateoka 1962). When compound granules are milled, they can break up into the separate subgranules.

Some starch-bearing plants produce semi-compound granules, as in the case of the bulb of *Scilla ovatifolia* and the edible seeds of *Amaranthus retroflexus* (American pigweed) (Lineback 1984: 16; Shannon and Garwood 1984: 35). These begin as compound granules, but the separate subgranules become fused together by a surrounding layer of amorphous starch. The resulting semi-compound granule has one exterior surface but two or more hila (French 1984: 186). French (1984: 184, 186) and Lineback (1984: 16) use the term 'pseudo-compound granules' for single granules that appear compound but have several fissures (defined below), as, for example, in wrinkled peas.

Size and Shape

Starch granules come in a wide range of shapes, which assists identification; discs, spheres, ovals, elongated, rounded, kidney-shaped, polyhedral, and irregular forms are common. Granules high in amylose (see "Molecular Building Blocks" below) tend to be elongated and irregular (Buléon et al. 1998: 95). Cereals in the tribe Triticeae (that is, wheat, rye, and barley) display two distinct granule forms: larger lenticular or disc-shaped and smaller spherical granules (Shannon and Garwood 1984: 35; Jane et al. 1994; Sivak and Preiss 1998b: 13) (e.g., Plate 14). For wheat, in particular, patterns and timing of growth within the starch storage tissue (endosperm) account for the different forms. The larger A-type starch granules form soon after flowering and may continue growing throughout the period of seed grain filling, whereas the B-type granules are produced some days later and remain considerably smaller (Stoddard 1999: 145). Because of their size, the large A-type granules contribute by far the majority of the volume of starch in the starchy endosperm, but generally only represent a small percentage of total granule numbers (Shannon and Garwood 1984: 35–36).

The size of starch granules generally varies from about 1 micron to 100 microns, although those in the rhizome of the canna plant can reach up to 175 microns (Sivak and Preiss 1998b: 13). Granules at the lower end of the range are most common (Moss 1976: 22–27; Jane et al. 1994). Size is positively correlated with the amount of water in the granule (Atkin et al. 1999: 170).

The particular storage site in the plant can influence the size and shape of starch granules. For example, the starch at the periphery of the stem pith of *Dieffenbachia* consists of small, round granules, whereas towards its centre the starch granules become increasingly elongated (Badenhuizen 1965). Noda et al. (1992a) have also found that starch granules from the inner tissue of the sweet potato were generally larger than those from the peel and the outer tissue adjacent to the peel. It is generally thought that granules with flat faces are the result of crowding within cells where the faces have been pressed against other granules. Such crowding may be characteristic for some plant species (e.g., Plates 11, 12).

The age of the granule influences its size such that younger granules will be smaller than older ones. A young, actively accumulating plant organ will contain more young (and therefore smaller) granules than a mature storage organ (Plates 19, 20). A number of starch researchers have shown that starch granule size increases with the age of the plant up to a certain point where it stabilizes (e.g., Moorthy and Ramanujam 1986; Noda et al. 1992b).

The general nutritional status of the plant may affect granule size. Stressed plants produce less carbohydrate and, consequently, fewer and often smaller starch granules. Gott and Field have observed that plants in their reference collections obtained under drought conditions are characterized by fewer, smaller starch granules than those from individuals of the same species growing in more favourable conditions (Plates 17, 18). Oliveira et al. (1994) also report that the relative proportion of the two main granule types, as well as surface area and volume of the granules, varied between barley seeds grown in different environments.

STARCH GRANULE STRUCTURE

The physico-chemical structure of starch granules has been a topic of study for over a century, but many aspects have yet to be satisfactorily resolved because of its immense complexity. We therefore limit this discussion to basic principles. Useful recent references for further reading include Guilbot and Mercier (1985); Zobel (1988); Oates (1997); Donald et al. (1997); Buléon et al. (1998); Sivak and Preiss (1998b); Atkin et al. (1999); and Tester et al. (2004).

Molecular Building Blocks

Starch is a semicrystalline substance composed of crystalline and noncrystalline regions made up of different compounds (Figure 2.10). The building blocks of the granule are units of the simple sugar glucose, that have been formed into two different glucose chains which include the essentially unbranched molecule called *amylose*, and a branched molecule called *amylopectin*. Some starch literature discusses the presence of a third 'intermediate material,' but some experts now think this is an artifact of processing (Tester 1997: 166) or perhaps immature amylopectin clusters from granule surfaces (Oates 1997: 379).

The glucose units of the amylose and amylopectin molecules can be considered as a ring made up of six carbon atoms conventionally numbered 1 to 6. In amylose, the first and fourth carbon atom of each glucose unit are linked

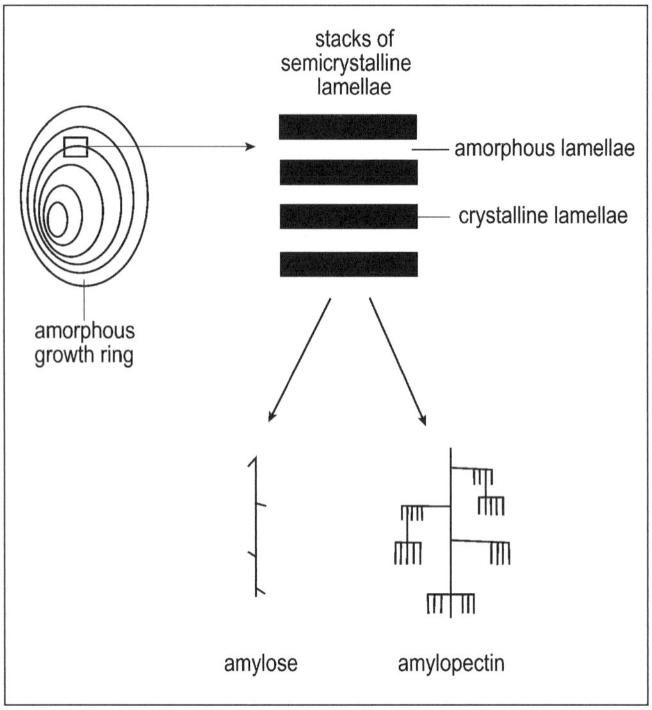

Figure 2.10. Schematic diagram of a starch granule: (a) growth rings emanating from the hilum; (b) close up of a stack of alternating semicrystalline amylopectin and amorphous amylose; (c) branching pattern of the amylopectin chains. *Based on Donald et al. 1997: figure 1; Buléon et al. 1998: figure 5*

(conventionally called an α-1,4 bond). For amylopectin, most glucose units are also joined by their first and fourth carbon atoms, but about 4 to 6 percent of the glucose units have an additional link at the sixth carbon atom which forms the branch point (an α-1,6 bond). Amylopectin is one of the largest biomolecules known, made up of about 600,000 glucose units (Sivak and Preiss 1998b: 20). Amylose chains are made up of about 1,500 units of glucose and are usually present in starch granules in lower proportions than amylopectin.

The amylose-amylopectin ratio in starch granules affects certain physical properties such as gelatinisation and reactions to stains (described below and in Chapter 5). The ratio is controlled by both inherited genetic and environmental factors. Amylose content commonly ranges from 20 to 30 percent in many economic plants (Lineback 1984: 16; Buléon et al. 1998: 95; Hoover 2001: 254). In some species the amount of amylose increases as the granule develops (Tester 1997: 163) or within different granule forms, as, for example, in barley where the larger type granules contain more amylose than the smaller type granules (Kang et al. 1985).

Granule Microstructure

Native starch is a mosaic of hard and soft material that is arranged in alternating crystalline and amorphous shells about 120 to 400 nm thick. These shells are sometimes visible as growth rings when starch granules are observed with optical microscopy as, for example, in potato starch. The hard layers are clearly visible with electron microscopy when starch is attacked with acid or amylase enzymes (Plate 15), which preferentially break down the amorphous layers (French 1984: 195–96; Gallant et al. 1997: 178; Sivak and Preiss 1998b: 29).

The acid- and enzyme-susceptible, amorphous shells of the starch granule have been much less studied than the crystalline areas. It is thought that amylose is largely concentrated in this part of the granule and that an increased interaction of amylose and amylopectin in these layers disrupts the crystalline organisation of the amylopectin side chains (Jenkins and Donald 1995; Gallant et al. 1997: 185). However, the distribution of amylose in relation to the amorphous and dense granule shells is not well understood (Tester 1997: 166). There is no sharp boundary between the amorphous and crystalline shells; it appears that the starch chains, particularly of amylopectin, run continuously between them (Guilbot and Mercier 1985: 256). Gallant et al. (1997: 187) have obtained microscopic evidence to suggest that there are serpentine channels running through the starch granule, perhaps made up of amorphous, enzyme-permeable starch molecules.

Few studies have been made of the surface of starch granules. Some starches are known to have higher concentrations of amylose towards the granule surface (Gallant et al. 1997: 186). Stark and Lynn (1992: 10) suggest that amylopectin clusters and amylose chain ends protrude in 'islands' from the surface of granules as foundations for the next growth layer of the granule. Images of the surface of potato and wheat starch have been obtained with atomic force microscopy, indicating differences in topology. In these cases potato starch granules appear to have distinct protrusions arising from the surface while wheat starch is smoother (Baldwin et al. 1997).

Minor Components

Starch granules from various plant species contain trace amounts of other constituents which may be associated with either the surface or the interior of the granule and which can affect the characteristics of the starch (Guilbot and Mercier 1985; Sivak and Preiss 1998b: 30–32). Lipids, known to occur at low levels in cereal starches (c. 0.5–1 percent) and by implication also in wild seeds, appear to be closely associated with the amylose fraction. Trace amounts of protein have been found associated with starch granules (Stark and Lynn 1992: 7). Many starches also contain phosphorus. In cereal starch phosphorus is mostly associated with lipids, but in potatoes and other starches it is linked to the glucose units close to the branch points of amylopectin.

STARCH GRANULE CHARACTERISTICS

Birefringence

Viewed with polarising microscopy, starch granules show strong *birefringence*, which means that they appear bright white against a black background (e.g., Plates 3, 12). This pattern occurs because the semicrystalline arrangement of starch molecules causes the polarised light to travel at different velocities through the granules. Starch granules are thus *anisotropic* (Sterling 1984: 194). Birefringence of starch granules has been attributed to their semicrystalline nature, but is due to a highly ordered molecular structure (Banks and Greenwood 1975: 247). Sterling (1984: 193–96), Dziezak (1988: 114–15) and Yiu (1993: 124) provide additional information about polarising microscopy and starch birefringence.

Chemical extraction and drying of granules can result in reduced birefringence, less-marked lamellae, and centrally developed cracks at the hilum (Atkin et al. 1999: 171). This has implications for both reference collections and identification of ancient starch granules. For example, the cracks should be carefully distinguished from the fissures deriving from the hilum that are found in native starches of many species, such as manioc (cf. Plates 16, 21, 31, 32).

One of the most important methods for starch identification is the use of polarising microscopy to detect the *extinction cross* within the bright image of the granule. The extinction cross is a dark, 'Maltese cross' pattern with its centre at the hilum of the granule, as seen in Plate 12. An important property of the cross is that it rotates when the polariser is turned. The relative length of the arms of the cross and the angle at which they meet at the hilum vary widely between different plant species.

Iodine Staining

Starch reacts to several stains in particular ways, which is useful for identification (Plate 24, cf. Chapter 5 and Box 7.2). For example, pure amylose gives a deep-blue colour when stained with iodine-potassium-iodide, while amylopectin gives a reddish colour. The relative proportions of these components, which vary from species to species, determine the ultimate colour of stained starch granules. For example, potato starch stains purple, indicating a relatively high amount of amylose, whereas starch with very high amylopectin content, such as waxy maize and the tubers of Australian terrestrial orchids, turn a more red colour when stained with iodine.

Gelatinisation

Starch granules are insoluble but are permeable to water. In contact with water at low temperatures, they swell. This process is reversible. Native starch can hold up to 30 percent of its dry weight in moisture. Hydration increases the birefringence of the granules (Atkin et al. 1999: 169). When heat is applied in the presence of water, swelling is reversible up until the point of *gelatinisation*, defined as a sudden and nonreversible change that usually occurs at temperatures above 50° C. Gelatinisation involves a breakdown in the structure of the starch granule that is initially characterised by the loss of birefringence, alteration of typical granule shape, and ultimately leads to the formation of an amorphous, jelly-like mass (pasted starch) (Banks and Greenwood 1975: 260; Galliard and Bowler 1987; Fredriksson 1998: 25) (Plates 42, 57–59). Technically, gelatinisation results from the disruption of hydrogen bonds that hold the linear molecular chains of amylose and amylopectin together in their ordered arrangement (Galliard and Bowler 1987; Fredriksson 1998). In addition, various salts, acids, and alkali can cause gelatinisation to occur at room temperature (Blanshard 1987).

Gelatinisation temperatures and times differ between different types of starch, although results

Table 2.1. Temperatures from onset to conclusion of gelatinisation for some native starches.

Common name	Starch source	Type	Gelatinisation temperature range °C[a]
Bitter Yam[c]	*Dioscorea dumetorum*	Tuber	65.5–72.5
Coco Yam[c]	*Xanthosoma sagittifolium*	Rhizome	66–81.8
Greater Yam[c]	*Dioscorea alata*	Tuber	65–71.5
Potato[c]	*Solanum tuberosum* (42 genotypes)	Tuber	63.5–71.7
Rice[a]	*Oryza* sp.	Cereal	61–77.5
Rice[b]	*Oryza* sp.	Cereal	65–73
Barley[d]	*Hordeum vulgare*	Cereal	52–69.7
Wheat[e]	*Triticum* sp.	Cereal	51.2–76
Maize[e]	*Zea mays*	Cereal	62.3–84
Sweet potato[c]	*Ipomoea batatas* (44 genotypes)	Tuber	64.6–84.6
Tapioca[c]	*Manihot esculenta* (5 varieties)	Root	57–76.1
Taro[c]	*Colocasia esculenta* (5 varieties)	Rhizome	43–63.3
White Yam[c]	*Dioscorea rotundata*	Tuber	63.5–71

[a] Leach (1965)
[b] Whistler and Smart (1953)
[c] Hoover (2001)
[d] Tang et al. (2001)
[e] Jenkins and Donald (1998)

have varied among researchers (Table 2.1). Increased amylose content in starch tends to raise the gelatinisation temperature of the starch. Since starch granules vary in shape and size and each one has a complex, three-dimensional structure, the gelatinisation of starch dispersed in water generally occurs over a range of temperatures, rather than at a single temperature or as a homogenous event (Leach 1965; Sivak and Preiss 1998b; Banks and Greenwood 1975: 260). Experiments with cooked starchy tubers have shown that within any sample of gelatinised starch, some granules appear to resist swelling and do not become gelatinised. The reasons for this are unclear but may involve variation in the internal structure and surface characteristics of starch granules (Galliard and Bowler 1987). The behaviour of starch during gelatinisation varies between plant species. Sivak and Preiss (1998b) note that potato starch swells greatly and then bursts open (e.g., Plate 30), while cereal starches swell but tend not to burst (e.g., Plate 57), and high-amylose maize starch granules do not swell very much (e.g., Plate 32).

X-ray Diffraction Patterns

Native starches show three main X-ray diffraction patterns, called A-, B-, and C-types, which are caused by differences in the crystalline regions of the amylopectin molecules (Pérez and Imberty 1996: 17) and which relate to botanical differences (Gallant et al. 1992: S6; Buléon et al. 1998: 98–99). In general, cereal starches usually give the A-type diffraction, while tuber starches generally show the B-type pattern (although some tropical tubers have A-type starches) and some root and seed starches give the C patterns (Banks and Greenwood 1975: 242; Gallant et al. 1982: 255; Zobel 1988; Pérez and Imberty 1996: 17). Starch granules with the A-type X-ray diffraction pattern are formed from a very compactly arranged double helix, while the same arrangement in the B-type starch forms a more open, hexagonal channel arrangement that is filled with water molecules (Pérez and Imberty 1996: 19). The C-type diffraction pattern is caused by a mixture of A- and B-type crystallite arrangements (Buléon et al. 1998: 98). Another pattern, the V-type, can appear in gelatinised starches (Gallant et al. 1992: S7).

Although X-ray diffraction patterns have been used to classify starches, these types are unlikely to be useful for ancient starch research. Fairly large sample sizes are needed for analysis and the types are too general for specific identification. Understanding the structural basis of the different diffraction patterns is useful, however, because the varying structures affect how different starches react to the cultural and natural modifications described in Chapters 4 and 5.

SUMMARY

Starch is produced by plants and stored in the form of small granules. It occurs in many plant parts, but is most abundant in storage tissues, such as seeds and the variously termed underground storage organs. The shape and size of storage starch granules are characteristic of the plant species in which they occur, although some granule forms may be less diagnostic than others. The nature and abundance of starch granules may differ within a single plant (e.g., transient versus storage starch) or storage organ and may vary among individuals of the same species due to differences in plant maturity or environmental factors. This range of variation is likely to be narrower than that caused by genetic differences.

The structure of starch granules is highly complex. Native starch granules contain alternating hard (crystalline) and soft (amorphous) material that is together composed of the long-glucose-chain molecules amylose and amylopectin. The ratios of these molecules vary according to botanical source and influence some important starch granule characteristics. All starch granules share some basic features, useful for identification and classification. These include size, shape, surface characteristics, presence of the hilum, reaction with iodine, and the characteristic birefringence and extinction cross when viewed with polarised light.

CHAPTER 3

MICROSCOPY

Huw Barton and Richard Fullagar

Ancient starch analysis requires basic knowledge about microscopy together with practical skills in using some microscope techniques. The choice of particular equipment and microscopic techniques will depend on the nature of the material under study and the questions being asked of it. This chapter provides a basic introduction to the methods in microscopy that are most relevant for starch research. Starch granules are generally smaller than 100 microns; most are less than 30 microns. Their small size means that they are best observed under compound microscopes fitted with various lighting conditions and magnifications from about x200 to x800. The analysis of starch extracted from sediments (Chapter 8) can be executed solely with a compound transmitted-light microscope, whereas the study of macroremains (cf. Box 1.4; Chapter 10) or residues adhering to artifacts (Chapter 9) requires a more complex set of procedures and a wider range of microscopy techniques associated with increasingly magnified views. In these cases the analyst begins with an examination of the artifacts or macroremains under lower magnification stereomicroscopes before moving to higher magnification compound microscopes. Both types of analysis (somewhat unfortunately dubbed 'low-power' and 'high-power' use-wear analysis by Odell 2004: 142) can also proceed from the light microscopes to the enhanced viewing available with scanning electron microscopes and other approaches such as UV spectrophotometry (Ugent 1994).

LOW MAGNIFICATION

Research focused on either residues adhering to artifacts or macroremains should begin at lower magnifications with a stereomicroscope (sometimes called a 'dissecting microscope'). As described in more detail in Chapter 9, data obtained at this stage of observation are mostly qualitative but are essential for structuring the next stage of analysis. It is often easier to relate tool residues to use-wear and artifact modification at low rather than at high magnification. In addition, the nature and spatial distribution of residues are useful for identifying tool function and help to understand how it was formed and deposited on the tool.

A stereomicroscope with external oblique illumination permits reflected light to pass along two paths, one for each eye, and therefore provides a three-dimensional image of the sample. Powerful cold light sources with optical cables are recommended for this purpose because they allow changes in the direction of light, creating varying light and shadow that enhance contrast and relief on artifacts or macroremains. The stereomicroscope allows low magnifications (usually up to about x100, without additional lower lenses) of relatively large samples and specimens. In practice, magnifications of between x10 and x50 are usually adequate for locating potential residues. Note that individual starch granules are usually not the prime focus of data collection at this stage.

Low magnifications (x50 and x100) on compound microscopes with vertical incident darkfield (low-angled) illumination are also useful for observing the distribution of residues and any surface relief. Because the light direction is fixed, some care must be taken to position the artifact on the microscope stage. It is common to orient the surface under study so that it is flat, but the artifact can be positioned differently, although this will affect (and sometimes enhance) the image of features.

At low power it is much easier to detect and describe features, such as the presence of plant tissues and inclusions adhering to or embedded in the body of the sample or residue. At lower magnification it is sometimes easier to see internal structures of macroremains or gather useful information about tool residues such as the direction of smearing, linear patterns, or features related to ongoing contact with the worked object after the residue was deposited. Such observations also help distinguish what was deposited on the artifact surface during use, as opposed to 'contaminants,' as well as other extraneous material unrelated to use, such as adhering sediment and organic tissue acquired from incidental contact. The low power approach also helps clarify the nature of contact. For example, variations in the distribution of residues resulting from different modes of use (such as cutting, drilling, or scraping hard or soft types of material) are useful indicators for interpreting tool function.

HIGH MAGNIFICATION

Although possible with stereomicroscopes, the modern light microscopes commonly used for high-power magnification have an internal light source often with both transmitted and reflected illumination. In the type known as a compound microscope, the internal light source is focused by a complex system of lenses. With these microscopes the main limitation on sample resolution at high magnification is the wavelength of light itself, which can provide maximum useful magnification of up to x1000. This is usually good enough to resolve clear images of starch granules down to a few microns in diameter. Figure 3.1 shows the main attachments and light paths used for observing starch granules under the two major arrangements called reflected and transmitted light.

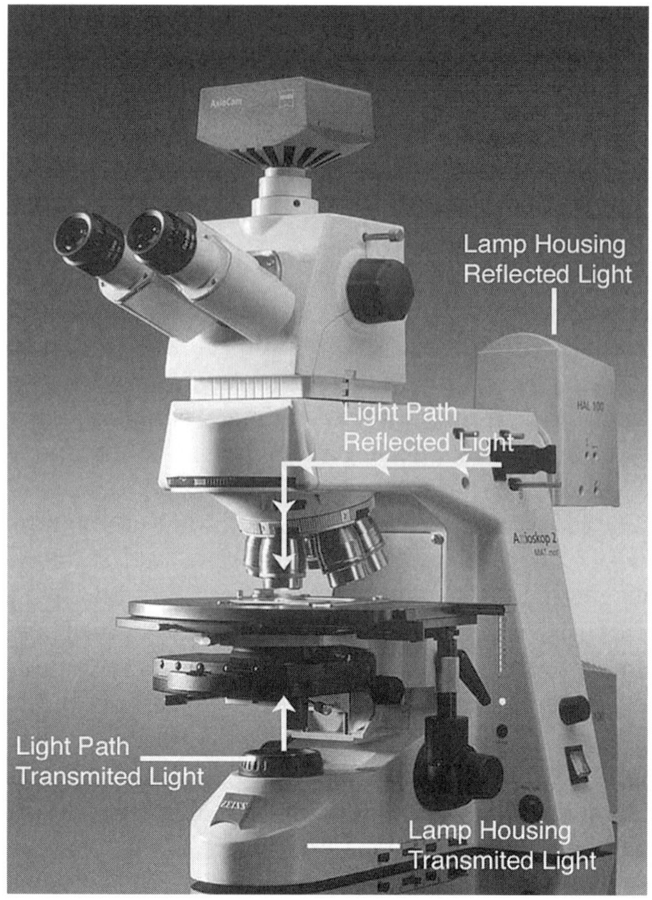

Figure 3.1. The compound light microscope showing the different paths for transmitted and reflected light. *Image courtesy of Zeiss Microscopes*

Transmitted Light

Transmitted brightfield illumination is ideal for observing many biological materials mounted on a glass microscope slide, including thin sections, and, of course, starch extracted from archaeological sediments, artifact residues, or macroremains. Brightfield refers to the use of visible light for illumination so that, in the absence of filters, specimens are seen against a bright background. In this mode, light is focused vertically through a condenser before passing through the specimen, an objective lens, and up to the eyepiece (where it is usually split for binocular, but not stereoscopic, viewing).

On reaching the specimen, the intensity of light may be altered by absorption or diffraction thereby inducing changes in contrast visible to the human eye; such objects are termed amplitude specimens. Specimens that are transparent and do not absorb light but induce a change in the phase of light passing through them are termed phase specimens. These specimens are either invisible or very difficult to see, as the human eye is insensitive to changes in the relative phase of light (Becker et al. 2003: 6). A thin section of cellular tissue such as an epithelial cheek cell is a good example. Under plain brightfield light, it is very hard to see the edges of the cell or its internal features; it has almost no contrast with standard mounting media. In order to see phase specimens, it is necessary to change the amount of contrast between the sample and its background (which may be the mounting media on a slide) or the surface of an artifact under reflected light. Common methods of enhancing contrast are discussed below: staining (see Box 7.2); darkfield light; differential interference contrast (DIC); or phase contrast illumination.

The most common and effective method of viewing starch is under transmitted brightfield light. It is under these conditions that most of the diagnostic features of starch discussed in this book have been observed. For practical application to a residue/starch study, lenses with magnifications above x10 (e.g., x20, x40 or x50, and x80 or x100) are needed. Cross-polarising filters are also necessary to induce specimen contrast, in both transmitted and reflected light, and to observe the diagnostic extinction cross of native, unmodified starch granules (Plate 10).

To ensure that the sample is well illuminated, glare free, and with good image resolution and contrast, the microscope must be properly set up for Köhler illumination. There are many published books (e.g., Bradbury and Bracegirdle 1998: 99–100; Ruzin 1999: 14) or websites (http://www.olympusmicro.com; http://www.microscopyu.com) that will assist you. The function of Köhler illumination is to ensure that the specimen is illuminated with a uniform field of light, something that must be achieved for each objective used. This is particularly important for observing starch and other objects that often display low contrast in the mounting media. Poor microscope setup may prevent you from resolving internal features such as lamellae or the hilum (cf. Chapters 2, 7), or in some cases the exterior shape of the granule.

Reflected Light

An important aspect of starch research is the analysis of the surface of stone tools in order to compare use-wear traces with the nature and distribution of starch residues accumulated through use of the tool (cf. Chapter 9). A reflected light microscope (commonly called metallographic or metallurgical microscopes) with photographic attachments is essential for the study of tool surfaces (Figure 3.1). Reflected light refers to microscopes that illuminate the specimen by passing vertical incident light down through the objective where it is reflected from the specimen surface back up the objective to the eyepiece (and usually split for binocular viewing). These microscopes were designed to look at solid, opaque specimens such as rock, metal, and mineral thin sections. For effective use with artifacts, these microscopes should be equipped with brightfield and darkfield illumination, polarising filters, and long working distance objectives. Although more expensive than regular objectives, long working distance objectives are considered essential at very high magnifications (i.e., above x200) because normal objectives are designed for polished specimens and require a very short physical distance between the tool and objective (also referred to as the working distance) to maximise resolution. Stone artifact surfaces are usually so rough and irregular that only the high points are in focus. Long working distance objectives are therefore required to focus on the higher points as well as down into adjacent depressions, without damaging the lens by direct contact with the hard tool surface.

An advantage of examining artifacts under reflected light is that the starch residues can be studied *in situ*, in the context of use-wear and other artifact features. Weaknesses are that the starch granules may be difficult to detect and cannot be observed as clearly as is possible under transmitted light. Starch granules can be observed under reflected light, but for comprehensive viewing and for reliable quantitative measurements, it is recommended that a slide mount be prepared and viewed with transmitted light (see Boxes 6.2 and 9.6)

IMAGING TECHNIQUES

Viewing of specimens can be enhanced by employing a number of special techniques, including specimen staining with chemical-specific dyes and illumination setup, such as darkfield, polarisation, and phase contrast techniques.

Staining

Staining biological specimens is one way of inducing contrast in low-contrast specimens such as cellular tissue and starch granules. The wide range of stains that has been previously assessed for starch research is listed in Table 7.1. For example, as noted in Chapter 2, relatively transparent starches will turn blue-purple-black in the presence of iodine-potassium-iodide, depending on starch structure (see Table 7.3 and Box 7.2), whereas cellulose will turn yellow. However, staining can render other features of the granule invisible, such as its internal structure and the polarisation cross.

Darkfield Light

Darkfield light is a simple technique for enhancing contrast in specimens. It can be used in both transmitted and reflected light modes, though it is more commonly found on reflected light microscopes. In both illuminating modes, the principles of darkfield or darkground light are effectively the same. A stopper or 'patch stop' impedes the central core of direct light, thereby creating a hollow cone of diffracted light that is directed at the specimen (Bradbury and Bracegirdle 1998: 95). The image is then formed by light that is subsequently refracted, diffracted, or scattered by the object (Goldstein 1999: 71). In transmitted light with darkfield illumination, transparent specimens otherwise difficult to see because they lack contrast with the mounting media, stand out brightly illuminated against a dark background (Bradbury and Bracegirdle 1998: 103; Goldstein 1999: 74). In reflected light, darkfield is indispensable for viewing residues on the surface of stone artifacts because reflected brightfield light produces glare, which obscures surface relief (Plates 25, 26). In darkfield, residues are easier to see, internal structural details may be resolved, and colour is usually enhanced. On the other hand, brightfield reflected light is preferred by many microwear analysts for viewing use-polish. It is therefore handy to have both these complementary modes of lighting.

Polarised Light and Birefringence

A common and quick way to check for the presence of starch granules on a slide mount (viewed in transmitted light) is to scan the specimen under cross-polarised light. Under this lighting, they appear as bright white (birefringent) objects with a strong, black extinction cross and the mounting media appears very dark. Birefringence is a complex optical property that may be used to deduce a great deal about the structure and composition of certain biological and mineralogical specimens. For practical purposes here, birefringence is defined as the property of certain objects to affect the path (its direction and velocity) of plane polarised light (Bradbury and Bracegirdle 1998: 8). Starch possesses this property. Sunlight and white or clear light bulbs consist of light waves vibrating towards the observer in all possible orientations. A *polarising filter* restricts the orientation of light waves to a single plane; such light is referred to as being *plane polarised* (Bradbury and Bracegirdle 1998:8). A pair of sunglasses is an everyday application of a polarising filter, reducing the unwanted effects of scattered light, decreasing the amount of light reaching the eye, and darkening the overall image.

A transmitted-light microscope has two polarising filters. The first of these lies underneath the slide-mounted specimen and the second filter (also referred to as an *analyser*) is mounted between the microscope objective and the eyepiece. When one of these filters is rotated so that its polarising axis is at right angles to the other filter, a normally bright image appears completely dark as nearly all light is prevented from reaching the eye. Light passing through the first filter is now plane polarised before it reaches the slide-mounted specimen. If this light were to continue onwards through the specimen without its direction being altered, the second polariser would block it. Only light that has changed vibration direction as it passes through the specimen can now pass through the second filter. These objects now appear bright against the dark background of cross-polarised light.

The intensity of brightness varies between different objects and may be used analytically (Goldstein 1999). Attempts have been made to use variations in brightness under cross polarisation to differentiate between starch granules (Badenhuizen 1965; Seidemann 1964). Besides starch granules, common biological materials that are strongly birefringent under polarised light are cellulose and faecal spherulites (cf. Box 7.3).

Polarisation Cross

Another characteristic feature of starch granules under polarised light is that they will display two dark bands, which form a cross in various orientations across their surface. This pattern is referred to as a *polarisation cross* or any of the following names: extinction cross, maltese cross, interference cross, spherite cross, or cross-wires. Some biological objects are described as being *radially symmetrical* with respect to the way in which light passes through the object (Goldstein 1999: 131). The molecules in these objects (e.g., starch granules) are often arranged perpendicular to their surface. As a result of this, when the polarising filters of a polarising light microscope are crossed, i.e., at 90° to the other, these objects typically produce a dark polarisation cross. The property of starch responsible for causing birefringence and the polarisation cross is the highly ordered molecular structure of amylose and amylopectin layers within granules (cf. Chapter 2).

The cross occurs because amylopectin molecules aligned parallel with the axes of the polarising filter appear dark, while molecules at more or less 45° to the polarisation axes appear bright (Goldstein 1999: 131). The net optical effect is to produce a dark cross over the bright birefringent background. In the case of starch, but not all objects, the cross appears to rotate in synchrony as the polarising filters are turned or rotated in their mounts. In some cases this optical property of starch can be useful and help distinguish starch granules from other objects that also produce a polarisation cross in polarised light. For example, the regular crystal structure of faecal spherulites appears very similar to starch when light is polarised, but the cross formed does not rotate when one of the polarising filters is rotated (Canti 1998: 435).

Phase Contrast

Phase contrast microscopy improves specimen contrast in transmitted light by exploiting differences in the thickness and refractive index of the sample (Becker et al. 2003: 6). When light rays pass through a specimen, their velocity may be affected by the physical properties of the material. Such changes in relative velocity (the phase of light) are not normally visible to the human eye, so phase contrast microscopes convert these alterations into variations in specimen brightness (Becker et al. 2003: 6). Images in phase contrast can be said to be either 'positive' or 'negative' depending upon microscope setup. In 'positive' phase contrast, regions of the material that are thick (with longer optical paths) will appear dark, while thinner areas or areas with a low refractive index will appear brighter (Goldstein 1999: 47). For images produced in 'negative' phase contrast, contrast is reversed and thick zones appear bright (Goldstein 1999: 47).

A downside of this technique is that materials with a higher refractive index than the mounting medium are encircled with a bright fringe or halo of light that may obscure specimen detail (Goldstein 1999: 53). Variations in microscope setup can alleviate this problem and the technique remains widely used for the analysis of living, unstained specimens (Becker et al. 2003: 7). Unlike differential interference contrast (cf. below), phase contrast is not affected by birefringence and so internal structural details may be resolved more clearly. Donald Ugent (Box 1.4; Plates 5, 22) has had excellent results in portraying starch granules with phase contrast.

Differential Interference Contrast

Differential interference contrast (DIC) is a more sensitive and expensive technique used to enhance contrast in unstained specimens. It is particularly effective with very thin specimens and may resolve very fine structural details (cf. Plates 27, 28). DIC resembles phase contrast in principle, but recombines light rays in a different manner (Becker et al. 2003: 7). DIC requires a polarising microscope and a pair of Wollaston prisms (the Nomarski method) that split and recombine the light beams that have changed in phase after passing through the sample specimen. The resulting image appears to be one of pseudo three-dimensional relief shading (which relates to optical properties of internal structure, not actual surface topography) of the specimen and is characteristic of the technique (Goldstein 1999: 66). Contrast of the specimen may be altered by shifting the lateral position of the Wollaston prism or by changing the intensity of light through rotation of the analyser or adjustments to the condenser diaphragm (Goldstein 1999: 67). It may be necessary to experiment with different settings to find the best image. One downside of DIC is that in its requirement for cross-polarised light, birefringence induced by the specimen can affect image quality. Typically, the formation of the extinction cross may obscure the internal structure of starch granules and some granules may take on a 'pyramidal' appearance (Plate 21). However, details of granule shape and visibility of granules of low relief may be greatly enhanced (Plates 27, 28). Working in partial rather than full cross-polarised light can overcome some of the image degradation induced by specimen birefringence.

SCANNING ELECTRON MICROSCOPES

The scanning electron microscope (SEM) is an additional method of analysis that can provide more information about samples than can be obtained from light microscopy (Figure 3.2). Magnification can range from about x20 to well over x100,000, though in practice, maximum magnifications of about x2,000 to x3,000 are sufficient for starch analysis. Depending on the machine and the settings used, the intensity of the electron beam for magnifications at or above this level may destroy starch granules. Techniques for preparing modern organic material for the SEM are often lengthy, and some require the sample to be attached to a metal stub and coated with carbon or gold (see Box 10.5). However, newer machines and techniques such as environmental and variable pressure scanning electron microscopy do not require immobilisation or prior preparation of the sample. As well as being nondestructive and noninvasive, the environmental and variable pressure SEMs are more flexible because more of the sample surface can be examined.

Figure 3.2. An environmental scanning electron microscope. *Photo by Gail Robertson*

The technique has been usefully applied by Robertson (1994, 1996; Box 4.5; Plates 33–36) to the analysis of modern starch residues formed on pounders during experiments, in Ugent et al.'s (1982) studies of ancient starch preserved in desiccated tissues (Box 1.4; Plates 6, 7, 8), charred residues (Hather 1993; Paz 2001; Willcox 2001), and ancient desiccated food products containing starch. Samuel's (1996a, 2000) analyses of Egyptian bread and beer manufacture made effective use of the SEM to identify starch granules modified by enzymatic attack that was characteristic of fermentation (Plate 60), something that would not have been possible using light microscopy.

SUMMARY

The basic techniques required to produce good images of starch granules necessary for identification and photography have been reviewed. In practice, even the most basic light microscopes are suitable for imaging starch granules; however, research-grade microscopes will generally be fitted with better optics and automatic 35 mm or digital camera systems. As unstained starch is largely transparent and of low contrast, correct microscope setup is essential to view surface features (e.g., facets and contours) and internal features (e.g., the hilum, lamellae, and fissures) of granules, which will aid in identification (see Chapter 7). Correct choice of mounting media (see discussion in Chapter 6 and list in Table 6.4) is also important to help induce contrast between the specimen (starch granule) and the mounting medium. Other repeatable techniques for inducing contrast in transparent specimens, including phase contrast and differential interference contrast, have been assessed. As noted, use of stains can also induce contrast, but is nonreversible and, especially when using the iodine stain, may obscure other features that may aid in characterising starch. If starch is partially gelatinised or damaged, selective use of stains can be invaluable (see Table 7.3). The use of the SEM remains widely used in the food sciences for imaging starch granules, and has advantages when using macroscopic food residues (Chapter 10), but is more difficult to use with the small residue samples prepared from artifacts and soils. The selection of techniques will largely depend on the questions asked. However, for most routine analysis of starch granules, a standard compound light microscope will be sufficient.

CHAPTER 4

STARCH PATHWAYS

Wendy Beck and Robin Torrence

Understanding the natural processes responsible for the formation of starch within plants and how they lead to differences in the starch granules, as described in Chapter 2, establishes a useful baseline for archaeological research. The next step is to build up knowledge about how and in what form starch might have entered the archaeological record in order to select appropriate sampling strategies and efficient methods for retrieving starch (e.g., Rossen et al. 1996; Spicer 1991). Furthermore, the archaeological interpretation of ancient starch relies on an understanding of the potential relationships between human plant use and starch granules.

Archaeologists use knowledge about the relationships between people and plants in the present in order to reconstruct past processes. In our case, it is essential to consider the cultural factors that could have affected the quality, quantity, location, and spatial patterning of ancient starch. Relevant behaviours to consider include factors such as the following: (1) the modification of vegetation through burning, clearing, tilling, ploughing, and cropping; (2) the selection of starch food sources or raw materials; and (3) methods of treatment, such as the processing and consumption of foods or the manufacture, use, and discard of tools. The need to understand cultural formation processes poses a serious methodological challenge that is just beginning to be met by relevant research. This chapter identifies the types of cultural factors that are important for starch research, highlights relevant research, and identifies priorities for future work.

DEFINING PATHWAYS

The generalised model for starch pathways depicted in Figure 4.1 provides a very useful analytical tool for making explicit the possible routes in the life history of a starch granule from its formation in the plant until it enters the paleoenvironmental or archaeological context. This simplified depiction of the myriad potential paths illustrates that the chain of inference from ancient starch granules to human behaviour is unlikely to be direct and straightforward. Although the distinction is not always unproblematic, it is useful to differentiate between the cultural and natural formation processes (Schiffer 1976, 1987) that determine the nature of the starch in the archaeological record. This chapter will describe general principles and relevant research needed to study the *cultural* formation processes that lead to

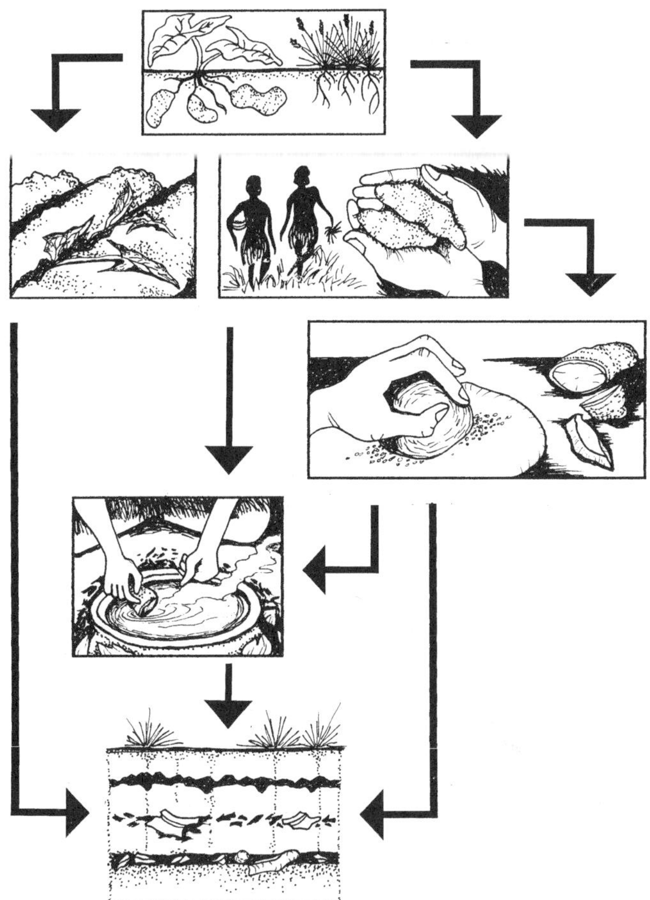

Figure 4.1. A model of potential starch pathways. *Illustration by Rebecca Conway*

the incorporation of starch in the archaeological record. It focuses on the life history of the starch in which humans are directly involved, beginning with the harvesting of the plant and ending with the archaeological record. The first of the *natural* formation processes, i.e., the production of starch in the plant, has already been described in Chapter 2. The second group of natural processes, those occurring after starch is deposited in an archaeological context, will be discussed in Chapter 5.

The starch pathways portrayed in Figure 4.1 represent the major routes that might have been followed in the life history of starch from its creation within a plant until it was discarded. The most direct path is from a plant which decays leaving starch granules in the soil. The type, quantity, and spatial patterning of plants may be heavily determined by human processes of plant management (e.g., firing, weeding, harvesting) and agricultural practices. The next path involves the human selection of a plant followed by consumption. It is likely that many food plants will be prepared in some manner as, for example, by grinding or peeling as shown in Figure 4.1. Often cooking follows. As a consequence of these tasks, the tools used are likely to retain residues when they are discarded. Starchy plants may also be converted into artifacts themselves (e.g., wooden hafts on stone tools, containers, adhesives) which will also eventually enter the archaeological record. If discarded within a midden, tools may come in contact with plant remains and may acquire additional residues. Although not depicted on this model, the starch could be later disturbed, re-enter new cultural pathways, and then be redeposited into a new context, e.g., through agricultural or building activities.

RELEVANT APPROACHES AND DATA SOURCES

Building up a model of starch pathways relevant for a particular research setting involves making sensible inferences about potential human behaviours and their effects on the deposited starch. As demonstrated by the case studies presented in this chapter, useful sources of data for finding links between human behaviour and starch deposition are very diverse. The fields of ecology, ethnobotany, history, ethnography, and ethnoarchaeology have important roles to play, as do specific experiments carried out to examine relationships between human activities and the resulting starch granule form and spatial patterning. Multidisciplinary approaches are often required (cf. Cotton 1996: Table 4.6).

Ethnobotany, History, and Ethnography

Ethnobotany is the study of relationships between people and plants from all time periods (Pearsall 2000: 2; cf. Balick and Cox 1996; Martin 1995), but it is particularly concerned with botanical knowledge and plant use in traditional societies (cf. Cotton 1996: 90 for a review of strengths and weaknesses). Indigenous people hold a vast amount of traditional botanical knowledge about the spatial and temporal distribution of starchy wild food resources as well as their management, processing, and relationship to animal species. Additional information can be gained from oral history, linguistics, and art history. For example, the historical importance and distribution of some plant species have been studied using data from 350 South American languages (Berlin 1992).

Gott (1989) assesses the value of historic sources such as scientific journals, floras, herbarium sheets, and word lists from traditional societies for ethnobotanical research. As the only primary records for how some plants were used in the past, they can play a special role in research. On the other hand, they should always be carefully evaluated and all references checked because some may be based on a misunderstanding of botany or use obsolete scientific names and incorrect botanical identifications. Gott (1989) also suggests that field checking by someone with a good knowledge of plant collection methods and taxonomy is important for most historical projects.

Excellent examples of how a range of historical sources can be used to reconstruct past uses of starchy plants are Gott's (1997, 2000) studies of Australian Aboriginal plant uses in southeastern Australia. Her exhaustive research illustrates that a very large range of starchy plants were used as food, raw materials, and medicines. Furthermore, her studies help predict which species might be found associated with particular stone tools, such as grinding stones or sharp-edged flakes. With the aid of these extensive databases, archaeologists

Box 4.1. Australian Aboriginal Food-Processing Technology

Wendy Beck

This case study illustrates how lists of utilised plants derived from ethnohistoric, ethnographic, and phytochemical data can help generate broad-scale generalisations concerning why and how particular plant foods might have been eaten and what their relative importance might have been in the past. Complex food processing is often associated with starchy foods, because raw starch is indigestible and not readily broken down in human intestines (Milton 1984: 161; Stahl 1984: 164). Stahl (1984: Table 1) notes that if uncooked plant parts are arranged in a scale of 'edibility,' based on cellulose and starch content, then fleshy fruits and flowers are at the high end whereas woody plant parts are least edible. Seeds and underground storage organs fall somewhere in the middle of the scale. My research in this area has investigated the importance of food processing within the Australian continent (Beck 1985).

I selected the most botanically reliable, detailed, and geographically comprehensive primary and secondary records of ethnographic and ethnohistoric records of plants. Descriptions were coded according to processing type, part eaten, and whether the food was recorded as a 'staple.' Next, data were gathered from the phytochemical literature: firsthand observations of palatability, data on toxicity to animals and people, and toxic components (Figure 4.2). Obsolete plant names used in historical documents had to be standardized before tabulation. Ethnobotanical records for 691 Australian food species and 902 cases of plant processing were analysed.

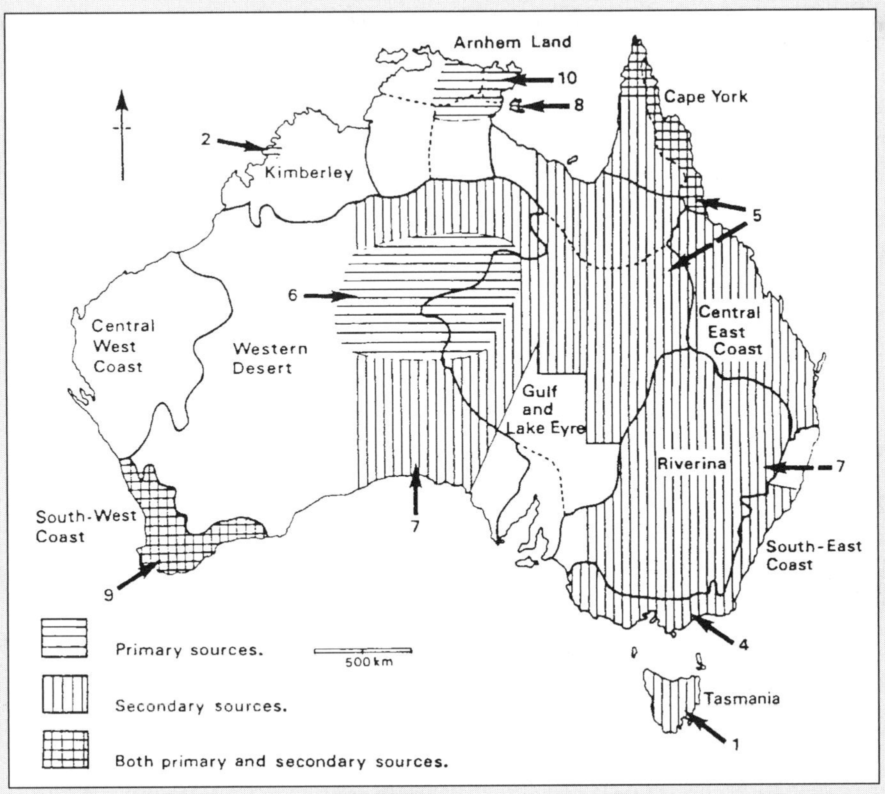

Figure 4.2. Spatial coverage of ethnobotanical and phytochemical sources relating to starch plant use by Australian Aborigines.

Box 4.1 (continued)

My results indicate a moderately strong and significant relationship between processing type and plant organ. I found that processing effectively doubled the number of plant foods available over those that are edible in a raw state, because food-processing technology improved palatability and digestibility and removed harmful substances. Complex processing techniques (2 percent of all cases) and multi-step processing (16 percent of cases) were defined as those that require more than one step and are relatively time-consuming. Examples include grinding and making into cakes or bread, water leaching, and/or prolonged roasting. More than the expected number of seed foods (102) and root foods (44) required complex and multi-step processing, whereas the more edible parts (e.g., fruits) required only simple processing.

This pattern can be explained by the physical and chemical barriers to digestibility often present in raw seeds and roots. An example of a chemical barrier occurs in cheeky yam tubers *(Dioscorea bulbifera* var. *bulbifera)* formerly a staple food in parts of northern Australia (see Box 4.2). These are inedible if eaten raw because they contain very bitter substances known as disobulbins, which, although not toxic, need to be removed by leaching overnight in running water (Webster et al. 1984). An example of a physical barrier is the hard coat on the outer surface of many grass seeds, which prevent digestion in the human gut unless broken down.

The literature review also identified a strong relationship between temperature and amount and season of rainfall as monitored by latitude and the frequency of seed and root foods in the diet. In the tropical regions of Arnhem Land and Cape York, fruits and roots are most commonly eaten; in arid central Australia, the main foods are seeds and fruits; and in temperate Victoria, roots and seeds figure most prominently (Figure 4.3). It seems likely that Aboriginal exploitation of plant organs partly reflects their natural occurrence in the environment. For instance, underground storage organs are least frequent in regions with low, mainly summer rainfall, such as Central Australia (Pate and Dixson 1982: 229).

Surprisingly, it was found that there was no association between toxicity and processing type. This result suggests that there are a variety of ways in which Aboriginal people can overcome toxicity. For example, the toxin present in seeds of *Cycas*, an important food in northern Australia, can be avoided by focusing on naturally aged seeds because the toxin decays over time (Beck 1992). Complex processing of plants can also be used for other reasons, such as preparation for long-term storage, for trade, or in response to scheduling needs. Overall, then, this study provides the basis for predicting which economic plants would be most relevant to the study of archaeological starches in Australia and their likely geographical distribution.

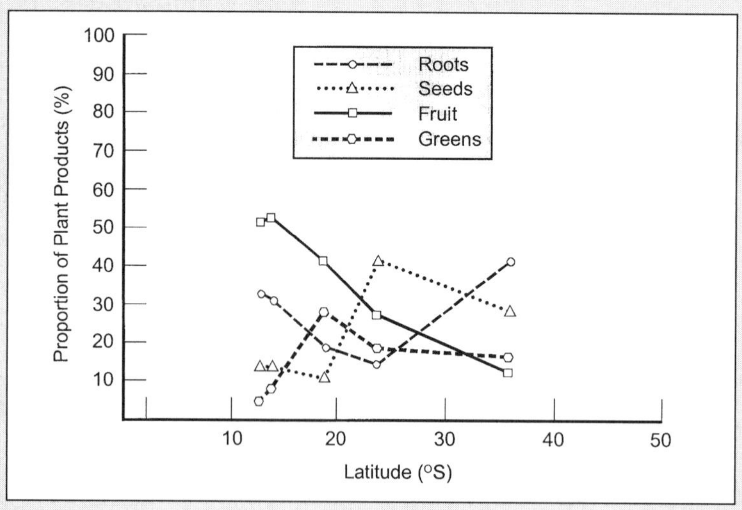

Figure 4.3. The frequency of plant parts eaten in different areas of Australia.

working in this region can then target their search for ancient starch towards important economic species (cf. Box 6.4). Beck's (Box 4.1; cf. Beck 1985) study of starchy plant use in Australia also uses historical data to understand where particular plants were favoured. In this case emphasis was placed on the processing and use of toxic plants. Studies like those by Gott and Beck can be invaluable for predicting which starches might be present at an archaeological site or associated with specific types of tools.

Ethnoarchaeology

One of the most valuable sources of information about cultural formation processes is ethnoarchaeology, defined here as the study of ongoing modern behaviour deemed relevant to archaeological research. By observing the choice, use, and consumption of starchy plants within an actualistic context, one monitors where, how much, and in what condition starch is deposited on artifacts and within potential archaeological contexts. Discussion with people who utilise starchy plants themselves is also useful for understanding why certain plants were chosen and why particular processing or manufacturing techniques were used. Ethnoarchaeological studies are also valuable because they often raise many additional questions not previously considered. Boxes 4.3 and 4.4 demonstrate the considerable potential value of this approach for ancient starch research (cf. Box 6.1; Chapter 9).

Experimental Studies

Experimental studies are another way to acquire useful data for interpreting cultural formation processes (cf. Boxes 4.5, 4.6). For example, Chikwendu and Okezie (1989) attempted to simulate the relevant ecological conditions under which domestication of yams (various *Dioscorea* sp.) might have occurred. It might be very useful to reappraise the large number of published experimental studies currently aimed at interpreting macroscopic plant remains in order to identify generalizations that could be relevant to archaeological starch, since many of these involve processing, deposition, and decomposition of food plant remains (e.g., Dennell 1976; Jones 1987; Beck 1989; Hastorf 1999). The role and importance of experimental approaches is considered in more depth within Chapter 9, which considers the deposition of starchy residues on stone tools, and is further illustrated in Chapter 10.

Biology

Biological data can be a useful source of information relevant to cultural formation processes, as demonstrated by studies that have examined factors affecting the production and distribution of pollen and phytoliths (e.g., Head 1989; Bowdery 1989). In terms of starchy foods, Harlan (1989) has examined the origin and spread of domesticated sorghum through studying the botany and geography of wild sorghum races. The distribution of wild-type taro may also assist our understanding of the process of domestication, as discussed by Matthews in Box 1.2 (cf. Matthews 2003). Ecological studies of animal behaviour, particularly how wild primates collect and process starchy plants, are also relevant sources for archaeologists (e.g., McGrew 1992).

Multiple Sources of Information

Beck's study of Australian Aboriginal use of cheeky yam (*Dioscorea bulbifera* var. *bulbifera*) summarised in Box 4.2 provides a good illustration of the value of using multiple lines of evidence from botany, ethnobotany, ethnoarchaeological observations, and replication to answer questions about cultural formation processes. Although cheeky yam was an important food for Aboriginal people within parts of northern Australia in the recent past, macroscopic remains of this USO have not been found in archaeological contexts. In order to predict whether ancient starch granules might be used to monitor prehistoric exploitation of this plant resource, the research project developed a set of predictions about how, why, and where cheeky yam starch may be deposited in archaeological contexts.

First, an examination of botanical and ethnobotanical documentary sources showed that cheeky yams produce large underground tubers seasonally, that the tubers are relatively high in protein and energy, and that various methods of processing are complex and involve either prolonged roasting or

Box 4.2: Processing Cheeky Yam in Northern Australia

Wendy Beck

This case study describes the processing and discard sequence used by Aboriginal people in northeastern Arnhem Land (Australia) for cheeky yam (*Dioscorea bulbifera* var. *bulbifera*). *Dioscorea bulbifera* is the only species of Dioscoreaceae that occurs in both Africa and Asia in both wild and cultivated forms suggesting that it may be one of the earliest tuber crops brought into cultivation. This case study aims to demonstrate how botanic, ethnographic, archaeological, and experimental sources of information can all contribute to answering questions about how starch finds it way into the archaeological record and how we might detect it archaeologically. The recovery of starch granules is important because macroscopic remains of this yam have not yet been found in Australian archaeological contexts.

Botany

Botanic data can answer important questions relevant to understanding where and why people used cheeky yam in Australia. What is it like? Where does it grow? Is it of economic value? How long has it been in Australia? *Dioscorea bulbifera* var. *bulbifera* is an indigenous Australian yam. It is a twining annual vine which occurs in vine scrubs in northern Australia. The underground storage organs of the plant consist of a short, fleshy structure with tuberous outgrowths which are renewed at the start of each dry season (Figure 4.4). The above-ground part of the plant simultaneously dies back in the dry season. These yams reproduce both sexually and vegetatively from starchy bulbils in the leaf axils. It is known in northern Australia as cheeky yam because of its bitterness (Webster et al. 1984). It is the only *Dioscorea* in Australia that requires complex processing; the other three (*D. transversa, D. hastifolia, D. bulbifera* var. *elongata*) are palatable. The harvested tubers of this variety of *Dioscorea bulbifera* can weigh up to 1.2 kg, but average about 100 g. They are relatively high in protein (1.0–1.6g/100g) and energy (127–253kj/100g prepared) (Beck 1985). This bitter variety may tend to be less deeply buried in the soil than nonbitter yams, as it may be better protected from natural predators. The botanical research shows that cheeky yams could be a useful starchy food plant in the dry season.

Ethnography

What is its economic value? Is it likely to be archaeologically visible? Cheeky yams were recorded by ethnographic researchers as being staple foods in parts of Cape York and Arnhem Land. It was also one of the few foods that was stored (Thomson 1939: 217). Cheeky yam processing also played an important part in initiation rituals. After preliminary roasting, the complex yam processing takes three regional forms: prolonged roasting (Kimberley), leaching in water (Arnhem Land), and mashing through sieves (Cape York). We can conclude from the ethnographic data that the cheeky yam was an important food in northern Australia for both economic and social reasons and that its processing is likely to leave large amounts of starch residues on processing tools and at processing sites.

Ethnoarchaeology

How is it processed? What are the likely starch outputs at each stage? Here is where ethnoarchaeological research can be very useful. In mid-1982 I was fortunate to be able to observe Aboriginal people processing cheeky yams by leaching at both coastal and inland locations in northeast Arnhem Land (Kopanga and Donydji, respectively) (cf. White 1978;

Box 4.2 (continued)

Meehan 1982). Predictions for how the starch might enter the archaeological record can be drawn from my ethnoarchaeological fieldwork.

Harvesting

Once a yam is located, the surrounding ground is cleared and the yams are dug out of the ground using metal or wooden digging sticks or the metal blades of bush knives (Figure 4.4). Sometimes, yam tops are replanted back into the soil. The remains of yam roots, stems, and leaves on the surface of the ground and spread around the holes left from yam digging are present at harvesting sites and could lead to the archaeological association of ancient starch with yam holes and vegetation clearing. The digging implements could also acquire small amounts of starch and soil on them from contact with the tubers, although ethnographic specimens have not yet provided this type of evidence (Fullagar et al. 1992: 41).

Preparation

The first step in yam preparation was cooking. Due to the need for twelve hours of processing, it is likely that this activity occurred at an overnight campsite. Cooking was done in two different ways: either by prolonged roasting (up to 20 hours) in an earth oven (Figure 4.5) or by boiling in a billy (metal pot) over an open fire. This would lead to changes in the shape and size of the starch granules but few residues would be expected until the next stage of processing as the yams are still whole at this point. After cooling, the tubers were peeled and sliced. At Kopanga a particular technique was used, whereby the yams were steadied on a digging stick which had been staked into the ground and were sliced with a snailshell grater

Figure 4.4. Frank Gurrmanamana digging up cheeky yams, Arnhem Land, Australia. *Photo by Wendy Beck*

Figure 4.5. Nancy Bandeiyama placing cheeky yams in the earth oven, Arnhem Land, Australia. *Photo by Wendy Beck*

Box 4.2 (continued)

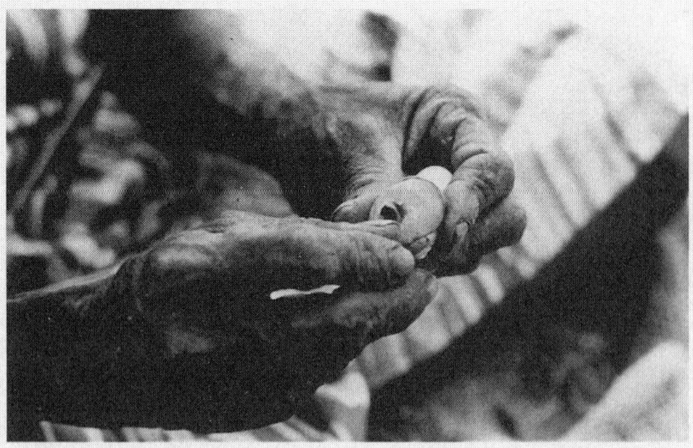

Figure 4.6. Snail shell grater being used by Anbarra woman Nancy Bandeiyama (now deceased) at Aupanga, Blyth River, Australia, in 1982. *From the Meehan and Jones Collection*

(Figure 4.6). Starch residues became attached to the grater (Plates 40, 41). The slices were collected onto a sheet of paperbark and placed in a pandanus fibre bag. The bag and contents were rinsed, taken to a nearby water hole, and left overnight. The final product was left to drain for 30 minutes before being eaten. At Donydji, a similar slicing process was used, but the tool was a wallaby bone knife (Figure 4.7) rather than a snail shell grater and the slices were leached overnight in a string bag.

Consumption and discard

Since the tuber slices were entirely consumed, the only possible residue from this stage could be starch granules preserved in coprolites.

Experiments

In order to further investigate the effect of processing on starch granule morphology, I replicated the observed processing techniques in the laboratory and sampled and photographed thin sections of the yam (Beck 1985). When compared with the uncooked material (Plate 29), the gelatinisation of the starch granules in the cooked and leached specimens is immediately obvious (Plate 41). In the experimental specimen most of the cell walls were still intact after leaching.

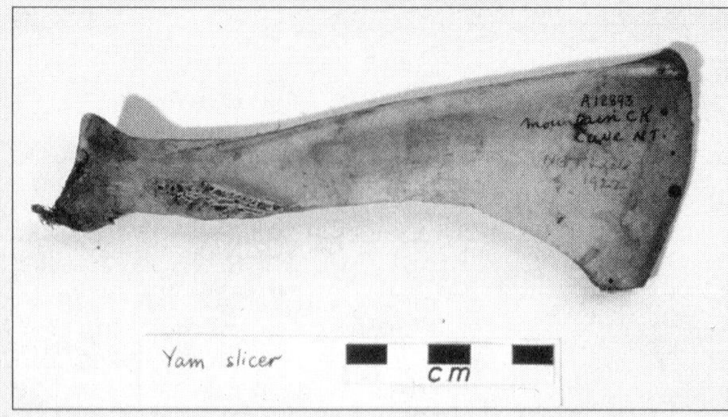

Figure 4.7. Wallaby scapula bone yam grater from Northern Territory, Australia. *Photo by Wendy Beck, courtesy of the South Australian Museum*

Archaeology

What are the implications of this research for archaeological recognition of cheeky yam starch residues? First, cheeky yams may have been an important food plant in the past, and because of the complex processing required, may be more visible archaeologically than other yam species. Second, the highest frequency of starch residues is likely to occur on slicing tools (especially shell graters and bone knives), inside the leaching bags, and in coprolites. Smaller amounts should be preserved in the soils at the locations where yams were processed with the smallest quantity deposited at the cooking site, harvesting site, and on digging tools. Cheeky yam starch may be found in association with yam digging holes, with hearths or cooking pits, and with water sources. Most of the starch granules will be in a cooked, rather than raw form.

Conclusions

This case study demonstrates the value of using multiple lines of evidence to explain how, why, and where starch might have been deposited into archaeological contexts. These useful expectations can now be examined archaeologically. Most importantly, I can demonstrate that the history of *Dioscorea* processing may now be traceable, whereas without the development of starch analysis, this plant would continue to be invisible in the archaeological record.

prolonged leaching in water. Next, the firsthand observation of yam processing among Aboriginal people indicated the most likely depositional contexts for starch. Finally, a number of simulations of yam processing were conducted in the laboratory to obtain more precise observations on the effects of processing on starch. Using these several research strategies, Beck found that the best places to collect starch relevant for understanding the history of yam utilisation in Australia would be on the processing artifacts, such as knives and graters, and in the sediments where food preparation took place. She also noted that due to changes in the granule, cooked yams will be much harder to detect than uncooked yams.

USES FOR STARCHY PLANTS

Having identified a set of methods that can be used to study cultural formation processes, we can now use the general model of starch pathways to identify the areas where research would pay off. The beginning of the starch pathway for any society is the choice of plants. Ethnobotanical studies of plant uses in modern or recent communities have identified four basic categories: (1) wild plants; (2) cultivated plants; (3) plants as material culture; and (4) plants used for their chemical properties.

Starchy Plants as Food

Scholars have identified a number of generalisations about human diets that could be useful in predicting the occurrence of ancient starch. For example, studies by Cordain et al. (2000) have suggested that latitude and population density are the major factors affecting the dietary mix of wild plants vs. animals among hunter-gatherers and that the concentration on plant foods is highest in latitudes between 11 and 40 degrees from the equator. In these cases the mix of plants generally includes 40 percent fruit, 25 percent seeds and nuts, and 25 percent underground storage organs (Cordain et al. 2000: 684).

In another study of this kind focused on Aboriginal Australia, Beck (Box 4.1) analysed the relationships between plant foods and food processing of 691 plant food species. She found that seeds and roots were usually treated with more time-consuming methods such as grinding, pounding, and leaching than nonstarchy foods such as fruits. The processing costs of nuts and seeds, which were nearly always borne by women, were particularly high. Plant processing had high benefits. Food-processing techniques removed harmful substances and improved palatability and digestibility to such an extent that the number of plant foods available for consumption nearly doubled, compared to the number that could be eaten raw. She also noted that fruits predominate in the diet in low latitudes where the fewest number of plant species are used (i.e., Arnhem Land), whereas seed foods were most common where the range of exploited plant species is greatest (Central Australia). This generalization might also have implications for the occurrence of ancient starch in other parts of the world. The main point here, however, is that predictions about food choice, which are important at an early stage of starch research, may be assisted by cross-cultural research.

Starchy Plants as Material Culture

The use of plants in material culture among traditional societies is still extremely important. Cotton (1996: 191) reports that between 18 and 61 percent of recorded plant uses in recent ethnobotanical studies were for material culture. All parts of plants may be used, including both timber and nontimber products (such as leaves, stems, petioles, bark, roots, flowers, seeds, fibres, exudates, and pigments). Starch can enter the archaeological record as part of the raw material for artifacts, building materials, canoes, masks, clothing, etc.

In Box 4.3 Gott describes how starch is likely to be incorporated within the archaeological record as part of tool production involving tubers from several species of bulrush *(Typha* sp.) These plants are used both as a source of food and for making fibres. She notes that the starch shed by the tubers will be present in the final string product as well as occur within residues on the shell scrapers used to process the fibre.

Starchy Plants as Chemicals

The diverse chemical properties of plants provide another important use of starchy plant parts. Starch is a soluble macromolecule that is highly viscous, adhesive, and is a useful surface coating. It is not surprising, then, that Lillford and Morrison (1997:

Box 4.3. How *Typha* Starch Enters The Archaeological Record

Beth Gott

A detailed description of how plant fibres from the starchy underground parts of *Typha* plants were processed, based on the historical, ethnographic, and biological data on the anatomy and autecology of the plant, enables us to understand how starch might enter the archaeological record. The two native species in Australia (*Typha domingensis* and *Typha orientalis*) provided starchy food staples for Aborigines along the main river and lake systems of southern Australia (Gott 1999). As with other bulrushes in many parts of the world, they were also an important source of fibre for the making of string.

The plants occur as dense stands on the margins of water bodies (Figure 4.8). They grow during spring and summer and die back in autumn and winter. The importance of *Typha* as a source of food and fibre on the Murray River is indicated by the numerous terms for different plant parts in the Wemba Wemba vocabulary collected by Stone (1911) at Lake Boga (Table 4.1).

Figure 4.8. Bulrushes (*Typha* sp.) growing in south-eastern Australia. *Photo by Beth Gott*

Table 4.1. Wemba Wemba words relating to the different states of the *Typha* plant (Stone 1911).

Wemba Wemba word	Meaning from Stone (1911)
GANG	rush
GUMBUNG	rush root
WANGULL	rush root, old after water has gone
BOORTITCH	rush root, old after cooked
JEERK	rush root, old, after fibre

The underground horizontal stems (rhizomes) were harvested preferentially in autumn/winter, after burning off the dry leaves, but would have been available year-round. The central part of the rhizome is rich in starch and contains strong fibre bundles (Figure 4.9). Digging sticks and even stone hatchets would have been used to dig out the rhizomes which were cut into lengths about a half-meter long (Figure 4.10). They were usually steamed in earth ovens. For large gatherings of people very big ovens were constructed to hold large quantities of rhizomes ('half a ton' is cited in Mathew 1899: 91). An alternative method of cooking was merely to lay the roots in the ashes of a fire.

Peter Beveridge (1865, 1883, 1889), an early European settler on the Murray River, gave a detailed account of the cooking and processing by the Wati Wati tribe. A hole 3 feet in diameter and 18 inches deep was dug by the women. Pieces of clay about the size of cricket balls were placed to one side. The hole was swept out with grass or boughs, and filled with

Box 4.3 (continued)

Figure 4.9. *Typha* rhizome from southeastern Australia. *Photo by Beth Gott*

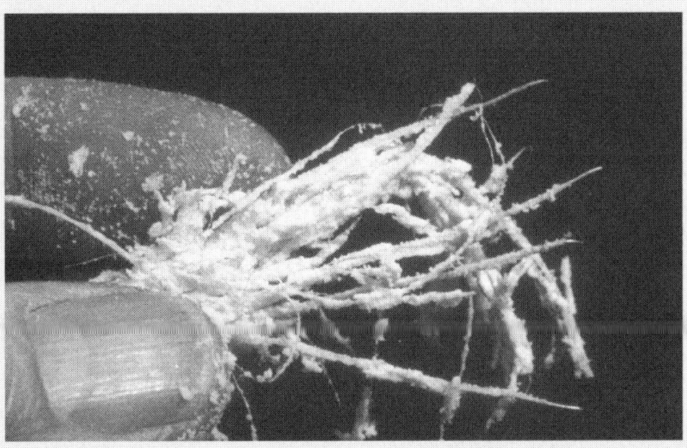

Figure 4.10. Starchy fibre bundles within the *Typha* rhizome were used as food and as a source for string by Australian Aborigines. *Photo by Beth Gott*

firewood, which was then set alight with the clay balls on top. When the fire had died down, the clay balls were removed with two sticks used as tongs, and the ashes were swept out. The hole was lined with moistened grass on which the rhizomes were placed, more moistened grass was used as a cover, and the hot, baked clay nodules placed on top (Beveridge 1889: 19). Mounds still remaining on the floodplain are full of these baked clay balls (e.g., Coutts et al. 1979: 60, 69).

After cooking, the cortex of the rhizome, which is poor in starch, was peeled off, and the central part containing much starch and fibre was twisted into a knot and crammed into the mouth. After the starch was chewed off, the remaining fibre was packed into bags which were temporarily used as pillows. When required for string-making, the knots were soaked overnight in water, teased out, and scraped with shells of the fresh-water mussel until they were clean, after which they were stored in small, neat hanks ready to be made into two-ply string. The resultant string was made into fishing lines, fishing nets, and duck nets (up to 60 m long). *Typha* string was very strong and long-lasting (Beveridge 1865: 15) (Figure 4.11).

In southwestern Australia the rhizomes were cleaned and pounded between flat stones into a paste, presumably to break up the fibre, and then made into a cake and baked (Grey 1841, II: 291; Moore 1884: 81). The Maori of New Zealand ate the pollen as well as the rhizomes (Crowe 1981: 107–8).

At all stages of harvesting and processing, the rhizomes might be expected to shed starch granules and these might even persist in the string artifacts made from the fibre. By contrast, young soft shoots of *Typha* were eaten raw in spring, so would probably leave no trace in the form of starch. Figure 4.12 is a flowchart showing the possible pathways of *Typha* starch granules. The starch would be deposited in all the contexts outside the boxes in the figure. It shows that there are many possible fates for starchy granules to enter the archaeological record. These range from granules deposited directly into the soil at gathering sites, to the fibres in the finished bags and baskets containing residual starch.

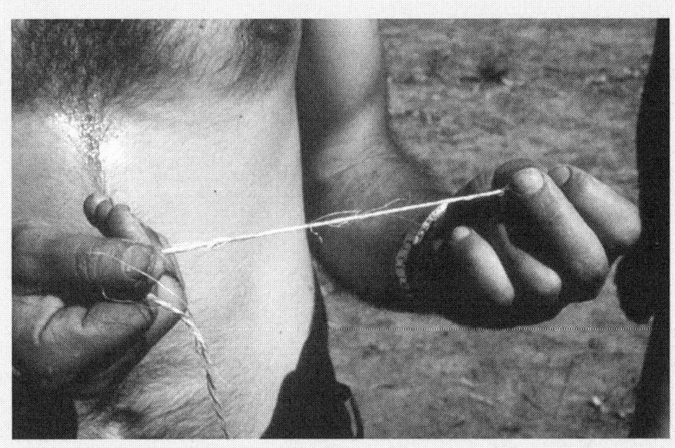

Figure 4.11. String made from the starchy *Typha* rhizome. *Photo by Beth Gott*

Box 4.3 (continued)

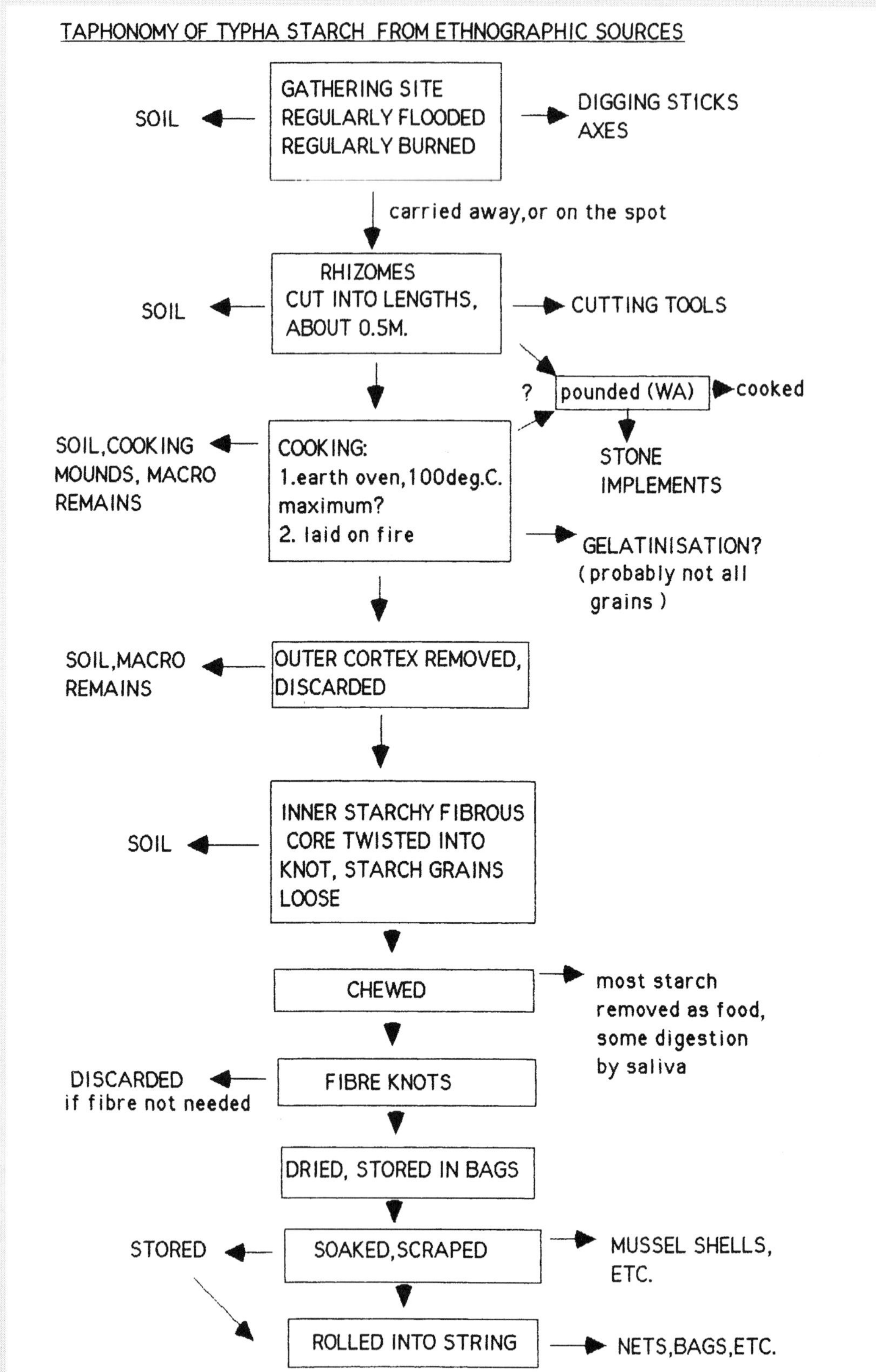

Figure 4.12. Flow chart based on ethnohistorical data showing potential cultural pathways by which starch granules of *Typha* might enter the archaeological record.

3) suggest that 70 percent of the current starch production in Europe and the United States is used for industrial purposes, rather than directly as food. The historical record provides a wide range of uses for starch. For example, the first-century Greek physician Dioscorides noted that wheat starch could be valuable in the treatment of ulcers, sores, and eye inflammations.

> Amylum is effectual against fluxes of the eyes, hollow ulcers and pustules. It stops the casting up of blood when consumed orally, and can prevent the occurrence of certain types of problems with your arteries. It is also mixed with milk and sauces. There is still another type of starch made from cereals in which the grain is steeped and macerated for one or two days, following which it is kneaded in the hands as dough, and dried in a very hot sun. However, this kind is not very useful for medicinal purposes.
>
> (Gunther 1959: 134)

Another medicinal usage for starch was described in 975 A.D. by Abu Mansur, an Arabian teacher and pharmacologist. He reports a treatment for wounds made by mixing starch with saliva.

In a further case, the Egyptians used starch paste to cement strips of papyrus stems together for use as writing paper as early as 4,000 B.C. (cf. Leach and Tait 2000; Fleider et al. 2001). A procedure for the extraction of starch for papermaking and other purposes was given in a Roman treatise by Cato in the year 184 B.C. Another great Roman authority, Pliny the Elder (23–74 A.D.), mentions the use of wheat starch for the making of papyrus documents, as well as for the whitening and stiffening of cloth and the powdering of the hair. In ancient times starch was also used as a sizing to provide paper with some resistance to ink penetration. Chinese documents of about 312 A.D. indicate that the most common species used for that purpose were rice, wheat, and barley. It is also worth mentioning that possibly as early as the fourteenth century starch was used in northern Europe for stiffening linen, both colored and uncolored starches were used as cosmetics, and uncolored starch was a hair powder (cf. Whistler et al. 1984; Radley 1968). Within traditional societies starch is also widely employed for its chemical properties, for example, as sizing for textiles, as gelling agents, and as weak adhesives (e.g., Cotton 1996: Table 8.1, 218). The starchy parts of plants also provide useful poisons, pigments, adhesives, fragrances, and drugs.

POTENTIAL PATHWAYS

After making predictions about the kinds of starchy plants that might have been used in the past, the next step is to model the potential pathways leading into the archaeological record. The generalised model presented in Figure 4.1 illustrates three major cultural routes for starch: (1) directly from the plant to the soil through disposal or abandonment; (2) by being consumed directly or altered by people and then discarded by them; and (3) by being incorporated into starchy residues on processing tools made from lithic, ceramic, or other materials which are then discarded or lost.

From Plant to Soil

The most direct pathway for the incorporation of starch into the archaeological record is direct from a plant that decays within the soil. This usually occurs close to or directly in the growing position of the plant. Environmental reconstruction is based on the assumption of a relatively straightforward link between starch in soils and its antecedent plant community. In many cases, however, the nature of the plant community may have been affected to various degrees by cultural processes, such as foraging strategies, selective weeding, firing, and cultivation. Analysts may use an understanding of cultural formation processes to reconstruct past landscape uses and management practices. Currently, this pathway is very poorly researched so that relevant actualistic studies are seriously required before this highly significant pathway can be used productively in archaeological inferences. For example, pollen studies have identified characteristic assemblages derived from current mowing, grazing, manuring, or burning practices that provide models for interpreting such practices in the past (e.g., Galliard et al. 1992; Vorren 1986). Research utilising the plant to soil pathway are described in more length in Chapter 8.

Box 4.4. Damage On Starch From Processing Andean Food Plants

Maria del Pilar Babot

Processing technologies are an important part of how humans have utilised and managed plant resources. The interpretation of damaged starch granules recovered from artifacts and sediments may enable archaeologists to reconstruct past methods of plant processing. This case study, focusing on the diverse techniques known from the Andean region, highlights the potential of an experimental approach.

Many of the cultivated plant species in the world originated in the Americas, the Andean region being one of the most prolific regions. For thousands of years Andean farmers have been domesticating and further developing foods based on roots, tubers, fruits, vegetables, grains, nuts, and beans. Among the food crops were tubers, cereals, pseudocereals, and legumes, such as the highly significant crops white potato *(Solanum tuberosum* L.*)* and maize *(Zea mays* L.*)* as well as the less known quinoa *(Chenopodium quinoa* Willd.*)*, amaranths *(Amaranthus mantegazzianus* Passer. and *A. caudatus* L.*)*, and the common bean *(Phaseolus vulgaris* L.). Their edible parts have an important starch content and are still basic staples in parts of this region (Parodi 1991).

Procedures for preserving, processing, and cooking, which are necessary to make these Andean starchy organs edible, may be tracked into the distant past. Tuber freezing and air dehydration of seeds are among the most common preservative strategies. The inhabitants of the High Andes store their tuber crops in a desiccated form called *chuño*: tubers spread out on ice-cold soils are repeatedly freeze-dried at night and unfrozen in the sun during the day. Then they are trod on to eliminate the remnant water content. *Tunta* is a variant of the above-mentioned process in which frozen tubers are buried in a wet hole and then covered up with straw to be sun-dried after some time (Cortella and Pochettino 1995; Parodi 1991). Preservation by air dehydration takes place by storing seeds after their maturation.

Coarse and fine milling to make stored dry starchy organs edible by obtaining flour or other byproducts (e.g., *chuño*, *tunta*, *ancua*, and rosettes of quinoa and amaranth) from seeds or tubers are among the most important forms of processing (Figure 4.13). Several processes, such as washing, rehydration and peeling, and roasting and popping of seeds, may take place before milling (Babot 1999). Washing quinoa seeds, termed *desaponification*, is necessary to eliminate their bitter saponins (Hunziker 1952; Martínez Ungria 1989). Slight rehydration and peeling are required to eliminate bran and husk from some seeds. Roasting is a common cooking strategy. Dry kernels of maize may be roasted in hot ashes to make *tostado* for further milling or to obtain popcorn called *ancua* or *aunca*. Quinoa and amaranth seeds may also be roasted to prepare flour or to make edible rosettes (Hunziker 1952; Ochoa de Masramón 1977). Finally, by charring quinoa grains, leaves, stems, and parts of other plants, *llipta* or *llucta* is obtained, which is an alkaline additive to coca leaves *(Erythroxylum coca* var.

Figure 4.13. *Chuño,* a dehydrated form of the potato, is displayed next to the fresh product in this market scene from the city of Huancaya, Peru. *Photo by Donald Ugent*

From Plant to Person to Soil

Several pathways fit under the general grouping of plant to person to soil. First, some of the starch consumed is resistant to digestion by humans and animals and will enter the record through defecation (see discussion in Chapter 5). Designing ways to exploit this pathway is quite important since it could provide direct information about human diet and animal forage. Second, starch might be incorporated in plants that are discarded incidentally to human activities, as for example, fuel by-products. Third, the creation and deliberate disposal of waste products from food preparation or material culture production is a very important pathway involving both native and modified forms of starch, depending on how aggressive the employed techniques were. Tools often used in these activities will attract residues, but are considered in a further pathway discussed below.

The processing of starchy wild plant foods is likely to be one of the most common ways in which starch enters the archaeological record. Food processing creates starchy residues at many stages, beginning with the initial discard of inedible portions of the plant at the harvesting site and continuing to the peeling, cooking, slicing, and more complex food preparation techniques. A comprehensive listing of plant food processing techniques can be found in Stahl (1989). Ethnographic and ethnohistoric studies, such as those concerning grain harvesting (e.g., Harlan 1989), cereal crop processing (e.g., Hillman 1984; Jones 1984), nut and root productivity (Koyama 1981), wild grass and forb seed management (Harris 1984), and seed deposition (Pearsall 2000), present useful data on procurement, processing, and discard behaviour which could be used to predict the distribution and abundance of ancient starch in archaeological contexts. Additional useful studies include Gremillion (1997), Hastorf and Popper (1988), Schultes and von Reis (1995), and the comprehensive review in Hastorf (1999). Pearsall (1989: 169) echoes the observation provided by Beck's cheeky yam study in Box 4.2 that the paucity of root and tuber macroremains (often high in starch content) in archaeological sites is probably closely related to how these resources were processed and used. Often the only part of a root or tuber that remains after use is the peel; for example, it is not uncommon to recover dried *achira* and *jicama* peelings in early coastal Peruvian deposits (cf. Box 1.4).

Box 4.4 (continued)

coca) that helps chemical reactions in the traditional practice of *coqueo* (Martínez Ungria 1989).

An experimental program was conducted to examine the impact of food processing techniques on starch granules (Babot 1999, 2003). The single or combined effects of dehydration, freezing, roasting, charring, and milling created diagnostic patterns of damage on starch granules. The summary of the experiments presented in Figure 4.16 shows that Andean food processing procedures generate physicochemical changes in starchy materials that modify the crystalline structure and related properties, completeness, and appearance of some starch grains. Dehydration caused by the action of air, heat, or cold generates structural damage in granules that alters and reduces birefringence. Mechanical damage, such as milling and treading as well as the effects of ice crystal formation, disrupts the tissue in the granules and promotes bursting and disjoining of the compound structures (Plate 31). Heating of almost dry seeds produces effects similar to gelatinisation (Plate 32). Clearly, some of the effects on the granules will affect their preservation in the archaeological record, although damaged granules may survive in some circumstances (cf. Chapters 5, 10).

Now that the cultural transformations of the granules have been identified, it should be possible for archaeologists working in the Andean region to use the diagnostic properties of modified starch granules as the basis for inferring the presence of ancient crop-processing techniques. The study of starch granules could therefore make a significant contribution to the history of plant utilisation in this and other regions.

Box 4.5. Crushed Starch Granules And Electron Microscopy

Gail Robertson

A study was conducted to investigate the effects of plant processing with stone tools on the morphology of starch granules (Robertson 1994, 1996). Three plant species (*Blechnum indicum, Alocasia macrorrhiza*, and *Canavalia rosea*) from southeast Queensland were selected because they are frequently mentioned in ethnographic and ethnohistoric records as an Aboriginal source of plant food which required pounding with a stone implement. In the first stage of the research a comparative reference collection of starches was assembled from plants positively identified by Phillip Cameron, Technical Officer, Botanic Gardens, Mt. Coot-tha, Brisbane.

Next, using an ElectroScan ESEM, images of starch granules and crystalline raphides (cf. Box 7.7) were examined. Notable differences among the sizes and morphologies of the starch granules from the three species were observed. The simple starch granules of *C. rosea* are large, generally round to oval (Plate 33). No raphides were observed. The majority of the starch granules of *B. indicum* are simple, but due to crowding, some aggregates of two granules can be observed, especially at higher magnifications (Plate 34). Granule shape ranges from elongated oval to round, although this shape is truncated in the crowded granules by a single flat facet. No raphides were observed. Finally, the separate granula from the rounded compound starch granules of *A. macrorrhiza* are multifaceted and often wedge- or square-shaped (Figure 4.14). Double pointed raphides which were elongated in shape can be observed singly and in bundles within some, but not all, cells. The distinctive characteristics of the three species were the very small size of *A. macrorrhiza* granules (<10 mm); the large numbers of multifaceted single granulas in the sample of *A. macrorrhiza* in combination with the round compound granules; the elongated or ovoid shapes of *B. indicum* granules, combined with the presence of some obvious compound and several truncated single granula; and the generally larger, round to oval granules of *C. rosea*.

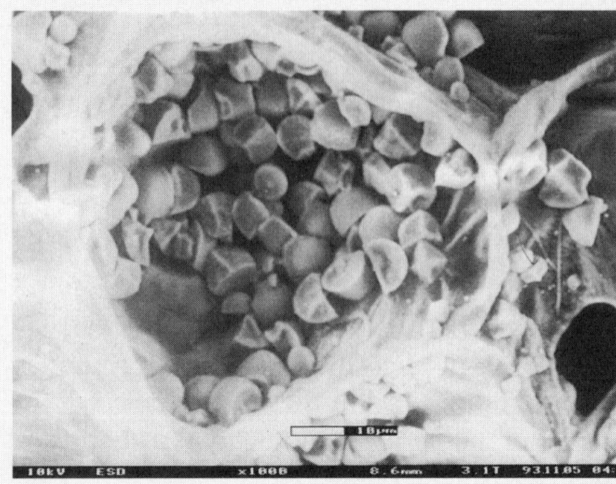

Figure 4.14. ESEM micrograph of compound starch granules and individual granulae in a fresh sample of *Alocasia macrorrhiza* taken at 1,000 diameters. *Photo by Gail Robertson*

The next step involved using silcrete river cobbles to pound and scrape samples from each of the three plant types. Care was taken to keep the different plants separated from each other during this activity. The experimental tools were then set aside.

Six months later residues were removed from the pounders. The morphology of starch granules of *C. rosea* (Plate 35) was essentially unchanged. Imaged granules are single, round to ovoid, with smaller granules more common. No raphides were detected. Starch granules in the experimental residue from *B. indicum* processing (Plate 36) also showed no obvious morphological changes. Granules were elongated to ovoid together with some truncated granules. Again, no raphides were observed. Experimental starch residues of *A. macrorrhiza* (Figure 4.15) exhibited an increase in the number of single, multifaceted granula. These

Processing starch for consumption or within tool manufacture can modify the morphology and physico-chemical characteristics of starch granules, which can affect their preservation (cf. Chapter 5) or make them more difficult to identify. In contrast, such changes may provide important clues about ancient technology. For example, pounding and grinding can crack and break granules, with dry milling having a greater effect than wet milling. The nature of mechanical damage varies according to the manner and duration of treatment. For example, starch granules placed in a modern rolling ball-mill showed evidence of surface roughening and cracking after five hours of treatment. After eighty hours of continuous milling, the amylopectin regions of the granules had lost their crystalline structure and would no longer display birefringence under cross-polarised light (Yamada et al. 1997). Dry heating starch, which makes it more digestible, causes little morphological change. In contrast, heating starch in the presence of water may lead to gelatinisation depending on the amount of water present (Chapter 3; cf. Plate 58). The cooling of heated starch causes retrogradation, a process in which birefringence, lost during heating, can return to a limited extent. Very little is known about the effects on starch granules of burning and charring of plants or products, although they are known to have survived (Hather 1991: 671–72; see examples in Box 1.4).

Babot (1999, 2003) has conducted experiments to investigate the effects of a wide range of processing techniques on starch granules. In Box 4.4 she notes that many different methods were used in the Andes to preserve and prepare starchy plants. The resulting effects on the granules are summarised in Figure 4.16. Further experimental studies looking at the effects of particular methods for preparing foods would be very valuable to starch studies. As Babot notes (Box 4.4), it could be possible to use characteristic damage on granules to make inferences about past food-processing techniques. Although experiments are useful because conditions are carefully controlled, ideally one would combine them with ethnoarchaeological studies, such as those described in Boxes 4.2, 4.3, and 4.6, since these are conducted in more naturalistic settings and reflect potential variability in behaviour.

Box 4.5 (continued)

were imaged at 1,000 diameters for uniformity of analysis, but also examined at higher magnification for more precise identification. A coating of debris in some experimental samples made observation of starch granules marginally more difficult than it was for fresh plant specimens.

In conclusion, the morphology of starch granules of both *Blechnum indicum* and *Canavalia rosea* remained essentially unchanged by short-term aging and processing. In contrast, processed *A. macrorrhiza* granules exhibited reduced average area, length, and width because the physical breakup of the large compound granules created a greater number of smaller, multifaceted, single granula. These results illustrate that while comparative reference collections are essential for the accurate identification of ancient starch granules, the possibility of changes in some of their physical properties due to various processing methods and/or taphonomic processes should be considered.

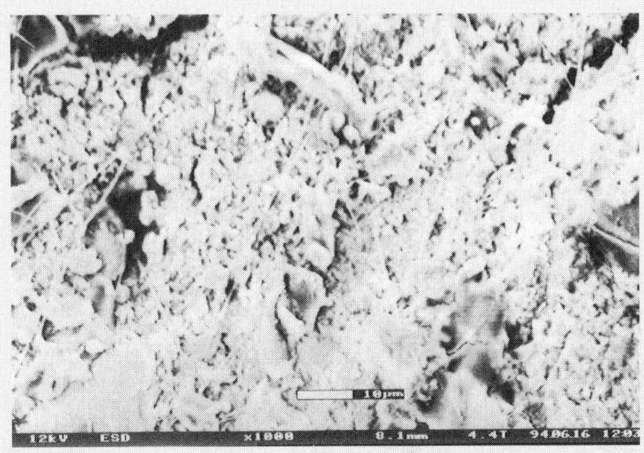

Figure 4.15. ESEM micrograph of starch granules and raphides in residues of *A. macrorrhiza* on an experimental pounder, taken at 1,000 diameters. *Photo by Gail Robertson*

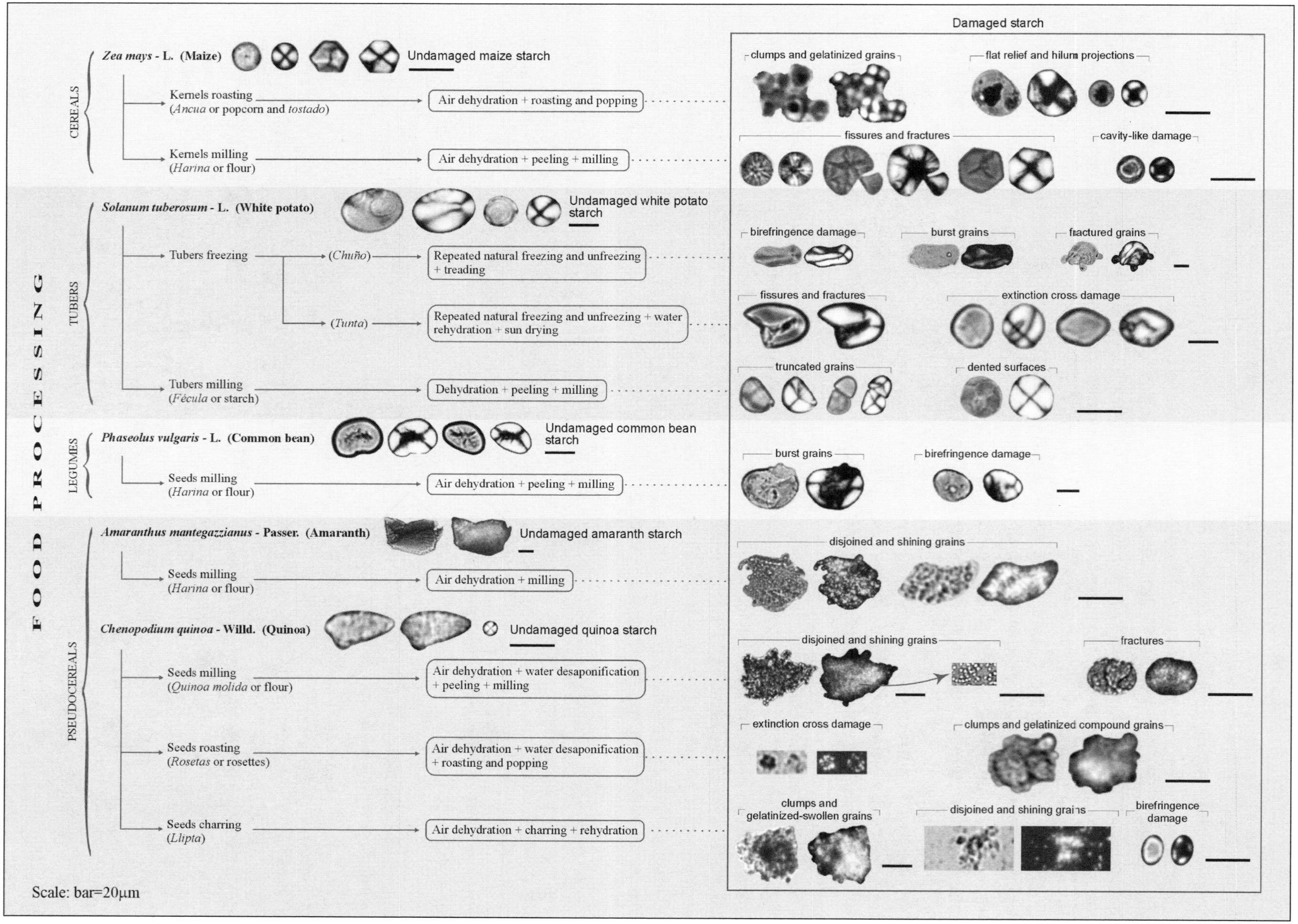

Figure 4.16. Effects on starch structure and morphology of food-processing techniques in the Andes region. Granules are compared before and after processing. Starch granules are presented as normal (left) and under polarised light (right). *Prepared by Maria Babot*

Box 4.6. Giant Water Lily Processing In Northern Australia

Jennifer Atchison

An ethnoarchaeological study of the processing sequence for water lily damper made by Aboriginal women in northern Australia illustrates how starch granules might be modified during plant processing as well as a number of routes by which starch could enter the archaeological record. The giant water lily, (*Nymphaea gigantea* Hook) is found in ephemeral pools, billabongs, and lagoons across the Kimberley region of Western Australia, the Northern Territory, Queensland, and down into northern New South Wales. Flowering is recorded for February to July and fruits for February (Wheeler et al. 1992). The Murinpatha, Kadgerong, and Jaminjung people from the eastern Kimberley today maintain traditional knowledge of this species and it is still eaten in various forms.

Both vegetative and reproductive parts of the species are eaten. The succulent green stalks are commonly chewed, and the seeds contained within a large bulbous fruit are also eaten raw. These seeds may also be processed and cooked into a damper. The sequence for this water lily damper production was demonstrated by informants Polly Wandanga (a senior Jaminjung woman), Biddy Simon (a senior Murinpatha woman), and other members of the Marralam community in July 1994 (Atchison 1994). Starch residues were collected at each processing stage for inclusion within a reference collection.

The women collected a 10-liter bucket filled with fruits. The fruits were placed in a small roasting fire for approximately 10 minutes (Figure 4.17). They were then removed from the fire and cooled before separating out the seeds which were roasted in the sun for three to four days (Figure 4.18) and turned regularly to ensure adequate drying. A large grinding slab and a hand-grinding stone were used to grind handfuls of the seeds into a paste with the addition of water (Figure 4.19). After grinding, the paste was wrapped up in succulent, fresh water lily leaves and baked in an ash fire for approximately 2 hours. The resulting damper was broken into pieces ready for eating (Figure 4.20). Processing the seeds in this manner resulted in a storable product.

Figure 4.17. In the first stage of food preparation, giant water lily (*Nymphaea gigantea*) bulbs are roasted in a small fire in northern Australia. *Photo by Jennifer Atchison*

Figure 4.18. Seeds are removed from the giant water lily bulbs (*Nymphaea gigantea*) and dried in the sun for three to four days. *Photo by Jennifer Atchison*

Samples of the water lily seeds were collected for starch analysis prior to roasting, during the grinding procedure, from the tool surfaces, and from the resulting damper. The average granule size for *Nymphaea gigantea*

From Plant to Person to Tool to Soil

This pathway involves three separate processes. In the first case starch enters the archaeological record through the decay of a tool made from a starchy plant, as, for example, the string described by Gott in Box 4.3. Her case study also highlights the second pathway in this group, which is the deposition of starch on tool surfaces during the manufacture of a starchy product. The third mechanism is the primary cultural transformation employed so far in archaeological studies of ancient starch—the deposition of starch on tool surfaces during food processing (cf. Chapters 9 and 10).

Two studies have used experimental and ethnoarchaeological methods to understand if, how, and in what condition starch might become attached to tools during food preparation. First, Robertson (Box 4.5) replicated prehistoric stone pounders and used them to process three plant species noted in ethnographic and ethnohistoric

Box 4.6 (continued)

Figure 4.19. Sun-dried giant water lily seeds (*Nymphaea gigantea*) are ground into a paste on a grinding slab with a grinding stone. A small handful of seeds are processed at a time with the addition of water. *Photo by Jennifer Atchison*

was less than 5 μm and their shape classified as subround (Atchison 1994). It was observed that after the damper was baked, the starch granules showed considerable deterioration, no doubt due to the long cooking process which gelatinises starch. When viewed under the microscope, the number of intact granules was reduced and most appeared as clusters rather than as individual granules (Plate 42).

Figure 4.20. The Giant Water Lily seed (*Nymphaea gigantea*) paste is wrapped in lily leaves and cooked in the ashes of a slow fire to form a damper or cake. *Photo by Jennifer Atchison*

The case study yields two important observations for starch research. First, physical changes were observed in the nature of the starch granules as a consequence of cooking during processing. It is possible that both raw and cooked forms of granules might be recovered from an archaeological context and would need to be distinguished. Second, the types of tools used in the sequence were identified and the process by which starch was deposited onto their surfaces was observed. As a consequence of water lily processing, starch would be a likely residue on the tools involved and could assist researchers to determine past functions of stone tools. This study has already proved useful in interpreting the presence of starchy residues on stone pounders from the Jinmium site, which is located in the same area (Atchison and Fullagar 1998).

sources. When she analysed the tools six months later, she found that the basic morphology of granules had not altered significantly, but many of the compound granules had been broken into their individual constituents.

Second, Jennifer Atchison conducted an ethnoarchaeological study of water lily processing by Aboriginal people in northern Australia (Box 4.6). Her results identify the most likely places for starch granules to be deposited during tool use as well as their likely form and frequency. She also notes several physical changes in the starch granules which occurred with cooking and processing the flour into damper. If distinguishable in archaeological analyses, these could provide information about food-processing techniques. Similarly, Beck's (Box 4.2) study of cheeky yam processing within an Aboriginal Australian group concluded that starch attached to tools was probably the only way that it would survive in the archaeological record, because cooking the prepared yams altered the granules in such a way that they would be unlikely to survive.

ADDRESSING RESEARCH GAPS

Even this necessarily brief discussion of starch pathways makes it clear that there are a multitude of processes that need to be recognized and understood in order to make inferences about past human behaviour. Reality is potentially even more complicated since starch granules may be repeatedly affected by different processes during their lifetime. For example, granules that became attached to a bone knife during harvesting 3,000 years ago could be transferred several times through recycling, reuse, and redeposition before ending in the archaeologist's laboratory.

Describing the major starch pathways also points out where there are large gaps in our knowledge. While there is plentiful anecdotal information and ad hoc research, systematic experimentation and observation are now required. The precise nature of the research on cultural formation processes will be determined by the specifics of the case, but as Isaac pointed out:

> The eager archaeologist frequently finds himself tempted or even obliged to invoke evidence from site features for which the determining processes have never been investigated. In this situation archaeologists must either run the risk of dependence on *a priori* reasoning and untested assumptions, or they must undertake the investigations of relevant processes by experiment or systematic observation. (Isaac 1967: 31)

The study of food preparation and related activities, such as pounding, grinding, roasting, or charring, is an area of great potential because these activities are likely to have had major effects on patterns of starch morphology, spatial distribution, and preservation. While we understand the general effects of heating and fragmentation, there is no compendium relevant to a range of plant starches. Research conducted within the modern food industry provides useful information that can guide thinking, but ethnoarchaeological and experimental approaches are more appropriate for understanding ancient techniques since these were probably conducted on a much smaller scale with less aggressive technology and chemicals. Such information can then be applied to refining the identification methods for archaeological starches and for assessing (and perhaps predicting) their survival rate in particular behavioural contexts.

Since the greatest emphasis in archaeological research to date has been on starch preserved on artifacts, much less is known about how spatial distribution of starch granules preserved in soils relates to human activities such as defecation, or agricultural activities such as ploughing, plant processing, and trampling—to mention a few examples representative of several pathways. Clearly, systematic and extensive middle-range research will be required by all projects studying ancient starch in order to understand specific trajectories.

Studies at various geographical scales are also a priority for research. For instance, at the micro-level, there is a need to collect ethnographic or historic references to nonfood uses of starch, for example where starch-containing tissues incidentally or purposefully contact artifacts. At a more macro-level, one could usefully employ ethnobotanical sources for describing regional plant uses, thereby targeting priority plant species for reference collections and starch distribution patterns. There is also a need to study how land management and plant cultivation affects starch composition and movement in the landscape and in sediments.

Considering the role of formation processes will impact on research in at least three areas. To begin with, the necessity of accounting for the processes by which the archaeological record is formed should influence techniques for excavation, planning, and retrieval of sites, landscape, and artifact contexts. Being able to predict more certainly the likelihood of starch preservation will enable archaeologists to design specialized recovery techniques and more efficient data recovery. Second, identification techniques and the classification of economic plants may also be influenced by the study of cultural formation processes, whereby the likelihood that identifiable starch granules will survive particular cultural contexts can be assessed. Third, the design of sampling strategies to take differential preservation and presence of starch into account will enable archaeologists to understand the behavioural contexts more clearly.

DIVERSE APPROACHES

Understanding the natural and cultural formation processes that operate within the starch cycle is an essential part of ancient starch research. This chapter has presented a number of examples where archaeologists have successfully investigated cultural formation processes that might have affected ancient starch, often using interdisciplinary methods. We need further attempts to critically examine ethnographic and ecological sources of information and to construct theoretical models for formation processes. Most importantly, archaeologists should continue to design and carry out problem-oriented actualistic research, such as experiments and ethnoarchaeological observations, because these are the most direct ways of approaching the problems. These approaches will enable a sounder basis for constructing likely scenarios of starch forms, frequencies, and spatial relationships that archaeologists uncover. By implementing a diversity of approaches, our understanding of the cultural formation processes for starch will be strengthened.

CHAPTER 5

TAPHONOMY

Huw Barton and Peter J. Matthews

To date archaeologists have not been concerned with the taphonomy of starch in terms of its very long term preservation or how it might be moved around the landscape. In contrast, the specialist starch literature contains useful descriptions and discussions about the mechanisms that explain how starch granules degrade and transform (e.g., Buléon et al. 1998; Haslam 2004). We consider a study of taphonomy (e.g., Lyman 1994: 1) as the processes that might affect the condition or frequencies of biological remains, including starch, once they enter the archaeological record. This chapter is primarily concerned with postdepositional processes, as the other major transforming processes, those resulting from cultural activities prior to starch deposition, have been discussed in Chapter 4. Our discussion of starch taphonomy is primarily focused on two research areas: (1) the physical and chemical processes that might influence starch degradation and/or preservation; and (2) the postdepositional movement of granules. In our discussion of the latter, we review pioneering archaeological case studies that have tried to tackle the problem of starch survival and provide some much needed qualitative and quantitative data.

SURVIVAL, FORM, AND CONTEXT

After starch granules enter the environmental or archaeological deposits, they are subject to a wide range of taphonomic processes that influence how much and which types are preserved, in what ways and to what extent their form is altered, and whether they are transported from the initial point of entry to a different context. The remarkable and inescapable fact about starch is that it survives, albeit sometimes in small quantities, within a wide variety of environmental and archaeological contexts and on a variety of substrates. The broad global distribution and very wide variety of settings where ancient starch has been documented are illustrated by the selected cases presented in Table 1.1. For example, archaeological sites containing starch include the arid areas of Africa, Australia, and North and South America, as well as tropical areas in Southeast Asia, Oceania, and South America. Sites located in sheltered conditions, such as caves, as well as open contexts have produced ancient starch. Furthermore, starch has been recovered in a variety of contexts: e.g., on flaked and ground stone tools (Atchison and Fullagar 1998; Barton and White 1993; Fullagar and David 1997; Field and Fullagar 1998; Furby et al. 1993; Loy et al. 1992; Piperno et al. 2000; Wallis and O'Connor 1998), attached to pottery fragments (Samuel 1996b; Juan-Tresserras 1998; Crowther 2001, 2005; Horrocks and Bedford 2005), and as isolated granules within sediments and storage pits (Tuross and Dillehay 1995; Barton et al. 1998; Barton 2005; Fullagar et al. 1998; Therin et al. 1999; Horrocks et al. 2002, 2004a, 2004b). Occasionally starch granules have been found entombed within dried cellular tissues (Rossen et al. 1996; Samuel 1996a, 1996b; Ugent et al. 1981, 1982, 1984, 1986, 1987; Ugent and Peterson 1988) or embedded within other organic products such as plant resins (Parr 2003b) and ancient glue (Flieder et al. 2001). The vast proportion of starch has been recovered as isolated, intact granules seemingly unaltered by the passage of time and, when taxonomic identification is possible, identical to modern reference material. One of the surprises is the very long time over which starch has been able to persist. It has been recovered from sites dating to the Middle Stone Age in Africa that are at least 180,000 years old (Van Peer 2003) and there is currently no reason to suppose that it could not survive for even greater periods. Such longevity of preservation is particularly surprising because starch is an energy-rich food for a myriad of large and small organisms. Despite the many factors that could have prevented its survival, the simple fact remains that very ancient starch does still persist and that

Box 5.1. Human Digestion Of Starch

Delwen Samuel

The most important factors in understanding taphonomy of starch through human digestion are the nature of the starch (native versus modified; cf. Chapter 3), its association with other food components, and the different functions of the digestive tract. The digestive process begins in the mouth, where chewing breaks food into smaller particles, thereby increasing the surface area available to enzymatic attack (Würsch 1989: 220). At this point it is acted on by amylase (an enzyme specifically associated with the conversion of starch into sugar) present in saliva (Dreher et al. 1984: 48; Sivak and Preiss 1998c: 157). Little starch digestion occurs in the stomach. When the stomach contents are released into the small intestine, pancreatic amylases resume starch breakdown. Pancreatic enzymes are produced in proportion to the amount of starch ingested. Cooked starch is rapidly broken down in the small intestine (Sivak and Preiss 1998c: 158). The final conversion of starch-derived, short glucose chains into simple glucose is accomplished by enzymes attached to membranes of the cells lining the small intestine. Glucose is then transported across the cell membranes and enters the bloodstream (Sivak and Preiss 1998c: 158–59).

Starch Digestibility

Englyst et al. (1992) classify the digestibility of starch into rapidly digestible starch (RDS), slowly digestible starch (SDS), and resistant starch (RS). This is a refinement of a previous scheme using readily digestible, partially resistant, and resistant starch (Englyst and Cummings 1987: 224, 230). Rapidly digestible starch refers to fully gelatinised and dispersed starch which is generally available immediately to enzymes for breakdown into glucose. Slowly digestible starch includes raw starchy foods (e.g., muesli) that are also completely broken down in the human small intestine, but more slowly than cooked starch (Englyst et al. 1992; Cummings and Englyst 1995).

Resistant starch has been further divided into three groups by Englyst et al. (1992). Resistant starch 1 (RS1) is physically inaccessible to digestive enzymes because it is sequestered in food fragments that have been incompletely broken down, for example in foods that are not thoroughly chewed or within partially milled grains and seeds. Resistant starch 2 (RS2) is inherently resistant to enyzme attack. In general, starches showing the B- or C-type X-ray diffraction pattern, and thus having B-type crystallinity, tend to be more resistant to pancreatic amylase attack (Englyst et al. 1992: S34). Plant foods that are normally eaten raw (e.g., slightly unripe bananas) or are prepared so that the starch is not fully gelatinised (e.g., biscuits) may contain significant quantities of RS2. The third type of resistant starch (RS3, retrograded starch) comes from cooked foods that are not consumed absolutely fresh, such as bread and cooled cooked potatoes. Brown (1996: 116) describes a fourth type of resistant starch produced by extensive modern industrial modification. Resistant starch is not digested in the small intestine, but work over the past two decades has shown that it plays an important role in the large intestine, or colon, where bacteria are able to ferment a proportion of it into physiologically important, short chain fatty acids (Topping and Clifton 2001). Englyst and Macfarlane (1986: 699) stress that resistant starch has always been a normal component of cooked starch foods, but the proportion of the different types vary according to the botanical source of the starch and the processing methods used to prepare the foodstuffs.

Archaeological Recovery

The industrially generated resistant starch described by Brown (1996) is not a component

property is a great boon to archaeologists and paleo-environmentalists.

Given the paucity of research directed to the identification and understanding of taphonomic processes for starch, it is not surprising that there is no clear explanation for its long-term survival. We adopt a pragmatic strategy for developing the field of starch taphonomy. First we identify the most likely biological and physical processes that might act on starch after it has left the cultural pathways described in Chapter 4, and then we examine their potential effects. This chapter concentrates on processes relevant to understanding (1) the preservation of starch, (2) the condition of preserved granules, and (3) the potential movement of granules within deposits.

STARCH PRESERVATION

Before discussing the postdepositional alterations of starch we must consider an important biological process embedded within cultural activity that will directly affect starch preservation: the process of human digestion. The activity of various fungal and microbial enzymes in the environment is expected to be the main pathway via which starch granules are degraded (see below). Human digestion is partly mechanical (chewing) and also a

Box 5.1 (continued)

of traditionally prepared foods and so is highly unlikely to be found in ancient remains. Cooled cooked resistant starch (RS3) is a very small component of prepared starchy foodstuffs. For example, wholemeal bread has 0.8 percent RS3 and spaghetti has 0.2 percent RS3 (Englyst and Cummings 1987: 225), although cooked cooled foods such as cold potato have higher levels. Human colonic bacteria is able to break down this type of starch irrespective of source (Englyst and Macfarlane 1986). Therefore, traces of retrograded starch may be undetectable in coprolites. Starch inherently resistant to enzyme hydrolysis (RS2) may be detectable in coprolites (see Box 5.2), but if present, may be found in small amounts. For example, banana starch consumed in its native state is passed more or less intact through the small intestine, but is almost completely fermented in the large intestine by colonic bacteria (Faisant et al. 1995a). In one study, only about 1 to 2 percent of the original intake of starch granules ended up in the faeces of normal human subjects in a recognisable form (Faisant et al. 1995b).

Physically inaccessible or sequestered starch in unruptured cell walls (RS1), which may pass into the faeces of healthy humans in some quantity, is the most likely type of ingested starch to be detected archaeologically. In an experiment in which patients were fed sweet corn, peas, or beans, Englyst and Cummings (1987: 231) found that up to 10 percent of faecal solids were starch protected within recognisable fragments of food particles. The presence of sequestered starch depends on processing techniques, such as milling and cooking, extent of chewing, and the nature of the starchy food ingredients themselves. For example, legumes have fibrous thick-walled cells which, although separated from each other, are not disrupted during cooking. The cell walls prevent complete starch gelatinisation and hinder access to starch-degrading enzymes. Starch digestibility of legumes varies according to species and variety (Würsch et al. 1986). Less efficient ancient technology was likely to produce more sequestered starch in food particles compared to modern industrial processing techniques.

In summary, the presence of starch in ancient coprolites depends on the botanical source of the starch, ancient processing technology, individual chewing habits, and starch detection methods. For starch in ancient gut contents, the timing and contents of the final meal as well as the digestive health of the individual are additional key factors. The recovery of ancient ingested starch has the potential to provide information about consumed foodstuffs, the nature of ancient food processing, and the nutritional status of ancient individuals.

process of enzymatic (saliva) and chemical breakdown (stomach and intestine) of food, including starch. This is an area of starch taphonomy that has already stimulated a lot of interest in the nutritional and health sciences, as the manner in which starch is or is not digested by humans is thought to have a major impact on health (see Box 5.1). What this research has also shown is that some starch is able to resist digestion completely, survive the attack of amylase in the mouth and by stomach acids, and then pass right through both the small and large intestines to be recovered from faecal material.

Such persistence clearly has implications for the longer-term survival of granules within the archaeological record. For example, Juan-Tresserras et al. (1997) have recovered starch granules (along with phytoliths and pollen) from dental calculus

Box 5.2. Starch Residues In Coprolites

Mark Horrocks

Archaeological excavations conducted at the Polynesian sites of Kohika (Irwin 2004) and Great Barrier Island (Horrocks et al. 2002) on the North Island of New Zealand, recovered prehistoric coprolites containing starch residues (starch granules and xylem cells) and other plant microfossils such as pollen and phytoliths. Coprolite samples from Great Barrier Island identified as either human or dog were recovered from prehistoric living surfaces buried by sand dunes. A radiocarbon date from sediments directly above the coprolites is dated at 467±67 uncal. BP (Horrocks et al. 2002). Samples from Kohika, in the Bay of Plenty, are those of dog derived from waterlogged deposits and date from A.D. 1700 to 1750 (Irwin 2004).

Starch residues were extracted from the coprolites following the procedure given in Horrocks (2005), using a method similar to that of the heavy liquid separation technique in Box 8.4. Analysis of the coprolite residues revealed starch granules in various stages of decomposition, fragmented xylem cells, and largely degraded cellulose tissue (Horrocks et al. 2004a). Two distinct plant species were represented in the samples. Coprolites from Great Barrier Island solely contained starch residues from the introduced crop sweet potato (*Ipomoea batatas*) (Plate 37), and those from Kohika had only starch residues from the indigenous bracken fern (*Pteridium esculentum*) rhizome (Horrocks et al. 2004a: 255). Both plants were reported as being under cultivation at the time of early ethnographic recording. In the latter example, the starch analysis of the coprolites provided the earliest direct evidence for the use of bracken fern as a carbohydrate food resource in New Zealand.

Many starch residues had retained their original morphologies allowing identification to the level of species. Some granules had retained the properties of birefringence (Plate 38), although the extinction cross was lost in many. Horrocks et al. (2004a: 257) noted that loss of the cross was a result of decomposition or heating during cooking. However, enzymatic damage to granules during time in the gut may also affect their internal structure and birefringence. Many granules had also changed colour and were recorded as amber or brown, as were many of the xylem cells. Horrocks et al. (2004a) implied that the colour change is caused by decomposition, although dry heat is also a possibility. Baker and Hobson (1952: 611) observed that starch baked in air changes colour from yellow to brown and finally to black; at the final blackened stage the granule structure is heavily altered or completely destroyed.

The survival of starch residues in prehistoric coprolites indicates that starch from wild plant species may be as resistant to enzymatic breakdown as that from many modern cultivated species. They can survive the acids and enzymes of digestion and then benefit from the protected setting of a preserving stool.

of skeletons dating to the Upper Paleolithic, Neolithic, Bronze Age, Roman, and Medieval periods from sites in the Iberian Peninsula and the Balearic Islands (cf. Scott Cummings and Magennis 1997). Furthermore, the technique used to remove the plant fossils from the dental calculus was quite aggressive: soaking in a 33 percent hydrogen peroxide solution for 12 to 24 hours and immersion in glacial acetic acid. Despite the various attacks on the starch granules, a wide range of plants, mainly representing common foods in the Mediterranean region (grasses, acorns, legumes) were represented as having been a part of the diet of the individuals on the basis of distinctive starch granule morphologies.

In another study demonstrating the persistence of starch to digestion, Horrocks (Box 5.2) has recovered starch from ancient coprolites in New Zealand, where the nature of granule survival was so good that he was also able to make important inferences about the composition and nature of their prehistoric diet by identifying the plants represented. Clearly starch granules are far more resistant to decay than one might have suspected solely on the basis of their chemical properties.

Once starch has entered the archaeological or environmental record, there are a host of factors that could potentially affect the degree and nature of its preservation. Currently, archaeologists generally report instances in which starch was recovered, rather than those in which it was not detected despite searching. Consequently, we lack a coherent understanding of the environmental conditions under which starch is expected to survive versus where preservation is unlikely. One approach to testing how and why starch might survive is to conduct experiments in real or simulated conditions. As described in Box 5.3, Lu placed a number of grinding stones used to process several species of plant in several environmental conditions (exposed, partially buried, and sheltered) for a period of just over two months. She found that starch was very poorly preserved when exposed to the elements, but that it survived quite well in the protected settings. Additional well-controlled experiments of this type would be very useful for testing which environmental parameters are conducive to starch preservation.

Regardless of the paucity of empirical evidence, it is useful to make predictions about likely taphonomic processes that would affect the degradation of starch following deposition. We do this based on our understanding of what is known about the nature of soils and the decay of organic material. It must be kept in mind, however, that since starch granules appear to survive a wide range of environmental conditions, they may be a special case of organelle that is not subject to the normal processes of organic decay.

The main factor leading to the removal of starch from archaeological deposits is likely to be the activity of microorganisms. To understand these we begin by looking at the physical and chemical conditions of the soil that determine the activity of microbes and the persistence of organic matter in soils. These include the degree of sediment compaction causing restriction of gross physical movement, soil moisture, soil pH, temperature, and oxygen content (Eglinton and Logan 1991: 320).

Soil Depth and Sediment Compaction

Many bacteria, fungi, and other organisms which produce starch-degrading enzymes (Table 5.2) seek out this high-energy food source. At a macroscale the major decomposing organisms will be earthworms, ants, and other burrowing insects, whereas at the microscale, fungi and bacteria are the dominant decomposers of organic matter in soil. These organisms are estimated to be responsible for at least 95 to 99 percent of all the decomposition and molecular breakdown of all organic matter in soils (Andrews et al. 1974; Jorgensen 1983). While organic breakdown is known to be rapid and extensive within soil (Haslam 2004: 1721), the observed persistence of some starch in ancient sediments and on artifact surfaces must be understood in terms of the biological and physical environment—and microenvironment—in which decay takes place. Several studies have shown that decay of organic material follows an asymptotic curve so that the majority of material is lost within the first three days following its incorporation into a soil (Haslam 2004: 1721). However, conclusions drawn from these analyses are based on soil chemistry and indirect measures of organic/starch content; visual observations of the soil for the presence of microscopic traces of organics such as starch are rarely made. When this has occurred and organics detected, new theories of organic decomposition involving particulate organic matter within

Box 5.3. The Survival of Starch Residues In A Subtropical Environment

Tracey Lu

Experimental Design

Starch residues have often been found on the surface of grinding tools recovered from archaeological excavations in several parts of the world (e.g., Field and Fullagar 1998; Piperno et al. 2000; 2004; Pearsall et al. 2004). To investigate whether starch can also be expected to survive on similar artifacts recovered from a range of different settings in subtropical China, I conducted a set of simulation experiments in the spring of 1999 (Lu 2003).

Materials and Methods

Due to the difficulties of examining the surface of large grinding slabs under ordinary microscopes, small pieces of stone were used as the test samples instead of exact replicas of archaeological artifacts. Sandstone and granite were selected because they were commonly used for grindstones in southern China. The surfaces of the sample grindstones were washed thoroughly under running water using a toothbrush. After cleaning, each stone sample was given a unique sample number, and two observation circles (e.g., Taro 1, Taro 2) were randomly marked on each stone surface (Table 5.1).

Each stone was then used to grind material from one of four plant species for approximately five minutes: foxtail millet (*Setaria sativa*); rice (*Oryza sativa*); yam (*Dioscorea* sp.), and taro (*Colocasia* sp.). Three samples broke during use. The remaining 13 samples were observed to have starch granules covering the entire area within the observation circles. The total quantity of starch present on each stone sample was estimated by counting starch granules visible within a grid of 100 squares placed on photographs taken at x400 within the observation circles. A count of starch on each stone piece was undertaken at the beginning and completion of the experiment and the results compared.

The samples were divided into three groups each of which was placed in one of three different environmental conditions: (1) shallow burial; (2) surface exposure; and (3) partial shelter from the elements. The first group was buried about 5 cm beneath the surface of a red sandy soil (pH 7), where they would be protected from wind erosion but exposed to the

Table 5.1. Survival rates of starch for various plant types in an experimental study (based on Lu 2003).

Plant Type			**Yam**			**Rice**		**Millet**	
Conditions	**Observation circle**	**Survival rate %**	**Observation circle**	**Survival rate %**	**Sample number**	**Observation circle**	**Survival rate %**	**Observation circle**	**Survival rate %**
Open	Taro 4.1	28.9	Yam 7.1	1.6	13	Rice 13.1	36.8	Millet 16.1	49.4
	Taro 4.2	64.6	Yam 7.2	17.9		Rice 13.2	47.1		
Mean		*46.8*		*9.8*			*42.0*		
Buried	Taro 2.1	78.8	Yam 5.1	70.4	3	Rice 3.1	84.9	Millet 15.1	88.0
	Taro 2.2	83.0	Yam 5.2	60.4		Rice 3.2	75.8	Millet 15.2	65.2
	Taro 14.1	53.7							
	Taro 14.2	87.2							
Mean		*75.7*		*65.4*			*80.4*		*76.6*
Sheltered	Taro 1	98.6	Yam 10.1	81.4	11	Rice 11.1	80.4	Millet 9.1	68.1
	Taro 2	83.9	Yam 10.2	75.3		Rice 11.2	79.8	Millet 9.2	74.5
Mean		*91.3*		*78.4*			*80.1*		*71.3*

Box 5.3 (continued)

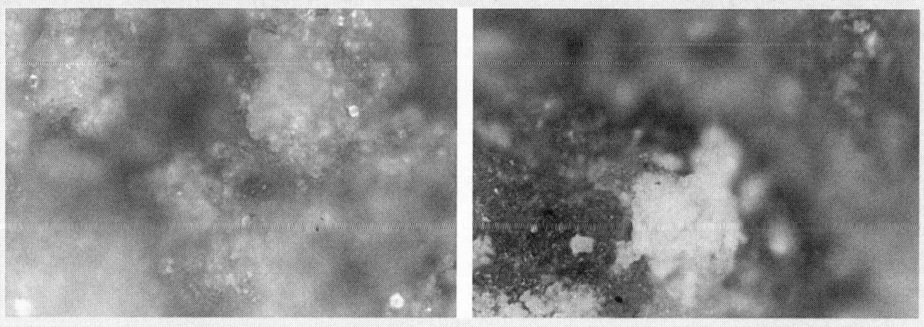

Figure 5.1. Observation circle No. 2 on sample No. 7 with yam which was placed in an open condition: (left) starch observed on 18 March 1999 (x400 magnification) completely covers the stone; (right) observation on 26 May 1999 shows that the area of starchy residue has reduced as indicated by the dark surface of the stone (x400 magnification). Images reproduced with permission of the author and Pandanus Books, Australian National University. *Photos by Tracey Lu*

effects of rain and changes in surface temperature and biological decomposition via insects and microorganisms. The second group was left on the ground surface with no cover. The third group was positioned under the shelter of a building which provided some protection from the elements simulating conditions found in rock shelters or caves which have some exposure to wind, rain, and biological activity. During the 71 days that the experiment was conducted (18 March to 27 May 1999), ambient temperatures ranged from 12.7 to 31.5° C, the humidity from 31 to 98 percent, and winds of force 5 to 6 were recorded on three days. There were a total of 32 days with rain and the total precipitation was 343.3 mm.

Results

The results summarised in Table 5.1 show that starch residues on tools left in open conditions fared worst, with less than 50 percent of the original residue remaining after 71 days. The pair illustrated in Figure 5.1 shows the loss of starch over time. The mean survival rate of all starch in buried sediments was 74.8±3.8 percent and for tools in sheltered conditions, 80.2±3.2 percent. Not surprisingly, sample grindstones that had some shelter from the elements or were buried in sediment were in conditions that are more conducive to starch granule survival.

There was less variation between plant types in buried and sheltered conditions than in the open setting, but overall starch from taro and rice fared slightly better than yam and millet. Samples with yam starches were most seriously affected with less than 10 percent of the original sample remaining. Within one observation circle, yam starch survival was less than 2 percent.

Conclusion

These experimental results indicate that exposure to strong wind and heavy rains has a significant effect on the survival of starch residues on grinding slabs, but over relatively short periods have a limited impact on starch on artifacts buried in the soil or deposited in sheltered conditions. It is possible that the rate of water flow, which is thought to be the major force removing starch grains from the tool surface, was significantly reduced in the buried and sheltered samples and this led to a higher rate of survival for starch granules. These preliminary results require further testing over much longer periods of time. The results clearly show that artifacts found in open sites should be examined for possible starch residues, although survival is likely to be poor, especially if the artifacts had been exposed for a long time. In contrast, artifacts recovered from sheltered sites such as caves and rock shelters could possibly contain starch residues.

Table 5.2. Starch-degrading enzymes and optimal conditions for activity (after White and Kennedy 1988).

Enzyme	Source	Optimal conditions for enzymatic activity	
		pH	**Temperature °C**
α-Amylase	Animal, plant, microorganisms, e.g., fungi (*Apsergillus* spp.), bacteria (*Bacillus* spp.)	4.8–6.5	wide range
ß-Amylase	Plants, some microorganisms, e.g., bacteria (*Bacillus* spp.)	plant 5.0–6.0, microbe 6.0–7.0	c. 45°
α-D-Glucosidase	Microorganisms, e.g., fungi, bacteria, and yeast (*Saccharomyces*)	fungi 3.5–6.0, bacteria 4.0–7.0, yeast 7.0-7.5	unstable to heat, some species 60–75°, unstable to heat
exo-(1-4)-α-D-Glucanase	Bacteria	6.5–10.5	47°
Cyclomaltodextrin	Bacteria (e.g., *Bacillus macerans*)	5.4–6.2	c. 60°
Pullulanase	Bacteria	5.0–7.0	45–50°
Isoamylase	Bacteria	5.0–6.4	40–50°
Amylo-1, 6-D-glucosidase	Yeast, mammal tissue	No data	No data

soils have been proposed (e.g., Waters and Oades 1991; see also Guggenberger et al. 1999; Haslam 2004: 1725–26).

A soil is a highly heterogeneous, three-dimensional landscape comprised of mineral aggregates and voids and characterised by chemical and moisture gradients. Consequently, the various locations of higher and lower microbiological activity must have a direct impact on the nature of starch preservation. The rate of starch hydrolysis by microbes will depend upon the optimal conditions for enzyme activity (Table 5.2), the physical condition of the starch granule under attack, and the physical and chemical condition of the soil. In most soils, the upper layers will be organic-rich and well aerated. Aerobes such as worms, fungi, and some bacteria will be responsible for most of the removal of organic material in this layer. As depth increases, reduced levels of oxygen will adversely affect the activity of aerobic decomposers but will favour anaerobic organisms such as bacteria and some chemical processes. For example, the rapid sealing of an occupation layer and the anoxic conditions of the buried sediments was argued to be largely responsible for the high levels of organic preservation, including starch, at the site of Monte Verde (Tuross and Dillehay 1995).

Depending on the degree of soil compaction, the porosity of soils may also play an important role in restricting the activity of microbes. The size of soil pores, which varies according to the type of soil and in relation to the mineral composition of sediments, will in turn have an effect on the movement of water and microbes as well as the capacity of microbes to acquire food. For example, Van Veen and Kuikman (1990) calculated that 95 percent of the pore spaces in a silt loam should be too small to be accessible to bacteria. Other soil types may similarly display pore spaces difficult for microorganisms to remain freely mobile (Golchin et al. 1998: 261).

As soil compaction increases with depth, microbial activities and the nature and rate of organic decomposition can be expected to decrease (Haslam 2004; Imshenetsky 1968; Taylor et al. 2002). For example, sampling for cellulose-decomposing microorganisms in various soil horizons failed to recover any microbes at depths below 60 to 80 cm (Imshenetsky 1968: 263). A more recent study by Taylor et al. (2002: 391) also noted that microbial biomass was reduced by 70 percent at a depth of 1 m and no fungal activity at all was detected at 2.7 to 3 m depth.

Studies of compost and humus formation have shown that when decomposition reaches a certain point, it slows down and a proportion of organic material remains intact. Given these basic principles, once starch granules pass through the upper layers of a deposit, it seems that the chances of preservation and long-term persistence are greatly increased. In addition, if additional organic material is not added to a soil, the activity of microorganisms will be reduced to a minimum (Rodale 1974: 883). While the deep soils—i.e., those 3 m—studied by Taylor et al. (2002) were still metabolically active, it was also shown that there existed a strong positive correlation between bacterial abundance, enzyme activity, and organic matter content (Taylor et al. 2002: 399). Any reduction in organic matter will cause a corresponding decrease in microbe numbers and rates of enzymatic decay.

Soil Moisture

The availability of water to permeate soil pores and hydrate the starch may be an important factor affecting starch degradation because the absence of water will restrict the activity of some microbes (Eglinton and Logan 1991: 320). Water will also dissolve many soil components that affect soil pH and mobilise minerals and other compounds that are cofactors in the interactions between enzymes, food substrates, and microorganisms. For example, in very dry soils or soils that are very compact and can exclude moisture, enzyme-catalysed hydrolysis will become inhibited due to the lack of water. In this case sugars cannot be extracted and starch may remain unavailable as a food source even if microorganisms are present.

Starch has been found in a wide range of conditions, so at this stage there does not appear to be a regular relationship between soil moisture and starch survival. As illustrated in Table 1.1, well-preserved granules have been discovered on tools buried in dry sediments in the sheltered conditions of arid-zone caves such as Chevelon Canyon, Arizona (Briuer 1976), and Hinds Cave, Texas (Shafer and Holloway 1979), and in tropical caves in New Ireland (Barton and White 1993), at Kilu Cave, Bougainville (Loy et al. 1992), and at Aguadulce, Panama (Piperno and Holst 1998), as well as from dry cave sediments at Niah Cave, Sarawak (Barton 2005). Starch has also been recovered from sediments and from artifact surfaces at a number of open sites in both arid and tropical conditions (Table 1.1).

Temperature and pH

There is little information available on the effects of temperature and pH on starch in archaeological sites. Starch from some taxa will begin to gelatinise in the presence of water when temperatures rise above 50° C (Table 2.1), but it is unlikely that ambient temperatures of many sites would rise so high. Since few sites have published details on sediment pH, general patterns cannot yet be drawn. At the site of Peztkes Cave, a sandstone shelter in New South Wales, Australia, Balme and Beck (2002) report starch survival from sediments that ranged from acidic (pH 3) to alkaline (pH 9). They also determined that the spatial variation in starch density was not correlated to variations in pH or moisture (Balme and Beck 2002: 161; cf. Box 8.9). At the site of Monte Verde, South America, organic material, including starch, was sealed within an anoxic (oxygen-free) and slightly acidic environment (Tuross and Dillehay 1995). Relatively well-preserved starch has also been found at the site of Zaña Valley in Peru, in conditions of moderate to extremely alkaline soil with pH ranges of 8.11 to 8.57 (Rossen et al. 1996).

Protected Settings

Some starch may persist over long time periods because it has been protected or shielded from bacterial and fungal activity, perhaps because it was encapsulated in fragments of plant tissue that have incompletely broken up during plant decay, food processing, and human food consumption (Englyst et al. 1992). Once in the soil, the hardy plant cellulose, itself relatively resistant to attack and digestion by enzymes, may be enough to prevent microbes from reaching the granules within. Other examples of starch protected from microbes by shielding might be the starch granules found in plant-based glue (Flieder et al. 2001) and plant resins used as a hafting or binding agent for artifacts (Parr 2003b).

A good example of prevention from decay is the desiccated and partially charred potato tubers from sites in the Casma Valley, Peru, where the rainfall is only 5 mm per year. Well-preserved starch granules were recovered from many dried and some partially dried potato (*Solanum tuberosum*), sweet potato (*Ipomoea batatas*), and achira or canna (*Canna edulis*) (Ugent et al. 1981, 1986, 1987; Ugent and Peterson 1988) (Plates 5–9; cf. Box 1.4). Associated evidence indicated that root crops had been baked next to wood fires. Well-preserved starch granules were recovered from the interior parts of many dried and some partially charred tuber or rhizome skins. Most granules were birefringent under cross-polarised light and reacted normally to the stain iodine-potassium-iodide (see Box 7.2). Although some granules had deep fissures or were cracked right through, most had retained their structural and chemical properties.

Experiments conducted by Hather (1993) to produce charred tubers comparable with those found in many archaeological sites demonstrate the extent to which the cellular structure of tubers may shield granules from degradation by heat. Tubers were dried between 24 and 96 hours at 70° C before charring at a temperature of 250° C for 2.5 hours. Although these conditions were probably more extreme than those associated with Ugent's finds, clearly identifiable starch granules were still present in some samples (Hather 1993: 77).

Protection by Artifacts

Since the major context from which starch has been recovered in archaeological studies is from the surface of an artifact (Chapter 9), it seems likely that the object itself affords some protection against organic decay. This observation has been made by most researchers investigating organic residues and starch preservation (e.g., Fullagar et al. 1998; Haslam 2004; Pearsall et al. 2004: 437; Piperno and Holst 1998; Piperno et al. 2000). Starch granules may escape the attention of fungi and other microorganisms by becoming trapped in pits, cracks, and crevices on coarse-grained materials such as those used for grindstones (Haslam 2004: 1726). At Monte Verde in southern Chile, abundant plant remains were recovered from an open air, wetland context associated with clay-lined pits, from cracks in a wooden basin, and from the edge of a stone tool (Ugent 1997; Ugent et al. 1987). Piperno et al. (2000) also recovered numerous starch granules from the surfaces of milling stones at the Aguadulce Shelter in Panama, with sample sizes reaching to 54 granules on one edge-ground cobble. By contrast, the sediments from the site were nearly devoid of starch.

Other organics such as pollen may also be afforded some protection from degradation by simple association with artifacts. In a study of pollen from a seventeenth-century refuse pit in Virginia, U.S.A., Kelso et al. (1995) found that at depth only pollen found directly associated with an artifact had preserved in the sediment section, whereas pollen in the soils had decayed. The artifact-pollen association appears to have provided some kind of lasting barrier against degradation, perhaps simply acting as an impermeable shelter against microbes that are mobilised by water movement within the soil.

Though the mechanism of artifact-residue preservation remains a mystery, well-preserved starch has been repeatedly recovered from many artifact surfaces including shell, bone, pottery, organics, and from highly homogenous smooth surfaces such as obsidian (Chapter 9) which offer few hiding places for starch granules and other organic residues. Whatever the precise nature of the preserving mechanism, it does not appear to require special conditions for organic persistence, because starch has been recovered from artifacts, associations with artifacts, and directly from buried sediments.

Particulate Organic Matter

One explanation for the survival of starch in a wide range of archaeological contexts may be the protection afforded by aggregates within the soil. Golchin et al. (1998) have discussed the existence of water-stable soil aggregates comprised of a core of plant material surrounded by mineral particles. Macro-aggregates that are more than 250 μm in diameter often consist of soil particles bound by fungal hyphae or rootlets and may be coated with further inorganic soil particles. Micro-aggregates in the size range between 250 and 20 μm contain a nucleus of plant material encrusted with inorganic components. Once encrusted with mineral particles, these micro-aggregates are extremely stable. Consequently, the cellular materials may become physically isolated from decomposer organisms (Golchin et al. 1998: 261; Guggenberger et al. 1999). Distinct cellular anatomy was visible under a scanning electron microscope when these aggregates were broken open (Golchin et al. 1998; Waters and Oades 1991: 168–71). The binding agents of soil micro-aggregates include persistent organic materials, crystalline oxides, and highly disordered aluminosilicates. Rapid wetting or disturbances from agricultural activities do not destroy the aggregates (Waters and Oades 1991) and the presence of aluminium may act as a barrier to microbial decomposition (Russell 1968: 91). Individual starch granules lie within the size range of micro-aggregates, so they may also be afforded protection by inclusion within them (cf. Haslam 2004).

Organic breakdown of soil micro-aggregates may also lead to the generation of new substances that inhibit decomposition (Hunt 1977). One of the most significant of these may be the tannic acids or tannins derived from decomposing plants. Since tannic acid has been shown to inhibit the digestion of starch by human salivary enzymes (amylases) *in vitro* (Thompson and Yoon 1984), it might inhibit similar starch-degrading enzymes within the soil.

Studies concerning bone taphonomy might also be useful for understanding starch preservation. Persistent organic materials such as humic substances have been implicated in the unexpected preservation of bone. For example, Nicholson (1998) undertook experiments designed to degrade bone by burial in a well-oxygenated environment with abundant decaying organic matter. After seven years, the bone was very well preserved, completely contrary to expectations. Nicholson's favoured explanation was that bone collagen had cross-linked with humic substances in the soil, forming stable molecular bonds and retarding dissolution of the bone (Nicholson 1998: 402). It has also been suggested that polysaccharides (which includes starch) may be shielded from enzyme attack by associations with other components of soils such as clays, hydroxides, oxides, and humic substances (Clapp et al. 1993: 56). If able to form molecular associations with other soil components, then starch granules might also be afforded additional resistance to degradation by bacteria and fungi.

CONDITION OF PRESERVED STARCH

Many of the granules recovered from archaeological contexts, regardless of age or particular context of preservation, retain their original morphology. This resistance to deformation or shape change has also been noted in some rare examples of fossilised starch (e.g., Baxter 1964; Wilkinson 1983). The best examples are three-dimensional casts and moulds of granules from mangrove hypocotyls replaced by iron pyrites dating to the Eocene, which in all morphological respects resemble modern examples (Wilkinson 1983). It is this property more than any other that permits the identification of starch from the archaeological record. In this respect starch appears similar to pollen in its physical toughness and persistence in soils and on artifacts. The ability of some types of starch to resist deformation and cracking even after extended physical treatment is quite remarkable. For example, even after grinding for one week, samples of maize starch still displayed rounded polygonal contours, polarisation crosses, and birefringence (Baker and Hobson 1952: 610).

In contrast, altered granules (cracked and split) have been recovered from archaeological contexts, as, for example, the desiccated potato starch granules identified by Ugent et al. (1981) or milled granules from grinding stone (Juan-Tressarras 1998). Given the increased susceptibility of these to enzymatic decay, however, it seems likely that few altered granules will survive unless microbes are excluded soon after their deposition.

Physico-chemical Properties

Physico-chemical properties of starch include granule birefringence under cross-polarised light, the production of an extinction cross, and reaction to biological stains. Starch granules recovered from nearly all archaeological and environmental deposits and within dried plant remains display the properties of birefringence under cross-polarised light and react normally to stains such as iodine-potassium-iodide (IKI), although there are some exceptions. For example, starch granules extracted from desiccated tubers at the site of Zaña Valley in Peru produced recognisable starch granules within preserved cellular tissue as observed by the SEM, but the granules failed to display birefringent properties when viewed with the light microscope (Rossen et al. 1996).

The retention of birefringence indicates that the well-ordered organisation of the polysaccharide polymers (amylose and amylopectin) in the crystalline regions of the granule has not been greatly affected by the passage of time, heat, ultraviolet light (Baker and Hobson 1952: 611), or by the digesting enzymes of microbes. The production of an extinction cross, related to the radial arrangement of ordered polysaccharide within the granule, is further support for the high level of internal structural integrity of archaeological starch.

Changes to the crystallite layers may be responsible for visible degradation of the extinction cross such as at the site of Balof 2 (Barton and White 1993), where a sample of starch granules displayed strong birefringence but the arms of the extinction cross appeared thickened and disjointed at the intersection with the hilum.

Reaction to Chemical Stains

The use of stains to test for the presence of starch has also yielded positive results for ancient starches, indicating that at a molecular level, the long chain carbohydrates have been preserved and not mineralised, replaced with other compounds, or substantially degraded. At Chevelon Canyon, Arizona, in strata dating from 8,700 years BP to dates within the last few millennia, Briuer (1976) found residues on stone tools and tested for starch granules and other plant materials using the IKI stain. The physical condition of the starch granules was not noted. He also demonstrated the persistence of intact cellulose, as adhering plant material, by staining residues with Sudan III (Briuer 1976). A wheat starch granule recovered from a grain silo at Fayum, Egypt, dating to 4441±180 B.C. had visible lamellae and also displayed the characteristic blue coloration when stained with iodine (Hallam and Osborne 1974, Plate 39). Starch granules on stone tools were also found to react to the iodine stain in levels dating between 9,420 and 14,240 at the tropical site of Balof 2, Papua New Guinea, although in most cases the granules turned a deep red rather than blue-black (Barton 1990: 59). Toluidine blue was also used to detect the presence of cellulose (Barton 1990: 59).

An early study using dyes to determine the structural changes of milling on starch (altered granules) from potato, wrinkled pea, and maize showed that some granules displayed increased susceptibility to dyes such as Congo red, Safranin, and Niagra blue that was related to disassociation of the polysaccharide crystallite within the granule (Baker and Hobson 1952). On exposure to alpha amylase in buffered solution, native granules attacked by this enzyme gradually lost their capacity to react to the iodine stain. They also found that the number of granules that took up a stain of Safranin and Niagra blue (indicating internal damage of the crystallites) was closely related to the number of granules that were eventually digested by the amylase (Baker and Hobson 1952: 611).

Starch baked in air changes in colour from yellow to brown and finally to black (Baker and Hobson 1952: 611). At the final blackened stage, the granule structure is heavily altered or completely destroyed. At lower temperatures, c. 200° C, starch granules turn a light-yellow colour, yet when mounted in nonpolar liquids still display intense birefringence and normal polarisation crosses (Baker and Hobson 1952: 611). This study also showed that within these heated granules polysaccharide breakdown had begun, as some dextrinization (conversion of starch to sugar) had occurred within the granule (Baker and Hobson 1952: 611).

Enzymatically Degraded Starch

Although clinical studies of the morphology of enzymatically degraded starch use a very wide range of different enzymes which will attack the starch in different ways, some generalisations about the process and its effect may be made. The early stages are particularly highly characteristic. The susceptibility of starch to enzyme attack and the morphological pattern of degraded starch varies according to the structure of the crystalline areas of the granules, the amylose content and other physico-chemical attributes, the enzymes used for starch breakdown, and the length of time that enzymes are active (Bhat et al. 1983: 261; Franco and Ciacco 1992; Gallant et al. 1992: S11). Granule size is also significant for the rate at which starch is hydrolised. Baker and Woo (1992) found that larger granules of rice starch (15–20 microns) were more than twice as susceptible to enzymatic attack as smaller granules (5–10 microns). Larger compound granules will be broken up by the enzymes into their component granulae; the individual smaller granulae are then slowly degraded by pitting (Gallant et al. 1982: 258).

Gelatinisation, as with mechanical damage, affects the structure and integrity of the starch granule and may promote further degradation. In one *in vitro* test and comparison of untreated and fully gelatinised wheat starch, the latter was 4,000 times more susceptible to alpha-amylase digestion (Gough and Pybus 1971). In the same study, incubating wheat starch in warm water (50° C) for 72 hours increased its susceptibility to alpha amylase by only four times. This might not seem a very significant increase, but it took place within the short time-span of a laboratory experiment. In contrast, under field conditions exposure to warm water and enzymes over long periods may lead to complete degradation. For the purposes of archaeological interpretation, experiments that somehow mimic very long-term or many cycles of exposure are needed.

During the earlier phases of amylase attack, the softer, more amorphous layers of the starch granule are preferentially attacked, exposing the harder, more crystalline shells. The X-ray diffraction, A-type starches have a more dense crystalline structure compared to B-type starches, but the latter are more resistant to enzymatic action (Gallant et al. 1992: S11). In all diffraction types, the crystalline layers are more resistant to enzymatic attack than the amorphous zones, which lead to a layered appearance as amylase action progresses.

With amylase-resistant potato and B-type starches, attack begins on the granule surface (exocorrosion) causing roughening, and the edges of the growth rings are exposed. As the enzyme reaction progresses, large pits sink into the interior, they widen in a conical pattern, and the edges of growth rings are detectable on the surface of the pits (French 1984: 194; Gallant et al. 1992: S11, fig. 7; Oates 1997: 381). In contrast to changes in surface structure, granules remain birefringent throughout until extensively degraded (Leach and Schoch 1961: 41).

Other starches are attacked by a pattern of endocorrosion, whereby pits appear on the surface of the granules from which channels develop into the interior of the granules. The precise pattern of corrosion varies according to botanical source. Some granules have specific zones of weakness susceptible to enzyme attack, such as a surface fissure or structural imperfection. For example, the large, lenticular granules found in the seeds of the *Triticeae* (barley, wheat, rye) are preferentially pitted along the equatorial groove, as well as on random areas elsewhere on the surface. Amylase penetrates towards the interior of the granule, where the centre of the granule is much less resistant. Once the surface is breached, amylases digest from the inside to the outside of the granule, randomly creating channels and canals in the softer shells of the granule. Starch granules from taxa such as maize, taro, rice, and some yams, as well as small granules of the *Triticeae* are randomly pitted all over the granule surface. The enzyme penetrates to the interior and then works outwards attacking the softer shells at every layer. The surface of granules affected by endocorrosion may look relatively intact, but the interior of the granule may be nearly completely digested away and they may lose birefringence in the hilum (French 1973: 1054; 1984: 194; Gallant et al. 1992: S11, fig. 7; Galliard and Bowler 1987: 63; Oates 1997: 381; Palmer 1989: 154; Leach and Schoch 1961: 41). Finally, the X-ray C-type starches are the most enzyme-resistant, a characteristic apparently related to the larger crystallite blocklets in the crystalline zone (Gallant et al. 1992: S11).

In an important paper, Juan-Tresserras (1998) summarises the results of an impressive range of case studies based primarily on the analysis of desiccated residues preserved on the walls of ceramic vessels from sites in the Iberian Peninsula dating from the Late Bronze Age through to the Iron Age. Like Samuel (1996a; also see Chapter 10), Juan-Tresserras (1998) has also been able to reconstruct various stages in beer-making by studying the visible condition of the preserved starch granules. The physical properties of the granule exterior helped distinguish whether the residues had been derived from whole grains, partially milled flour, or fermented products. The starch granules from the fermented stages were especially notable because of the presence of pits and channels on their exterior surfaces due to enzymatic attack. In many cases the lamellae had been accentuated by the selective removal of parts of the granule.

Large clusters of gelatinised granules lacking the extinction cross indicated that grains had been heated in a wet medium before fermentation. Within the solid, gelatinised mass some granules had been protected and were unaffected. The presence of a mix of granules both with and without extinction crosses within a distinctive mass of material is very similar to the residues resulting from cooking experiments reported by Crowther (2001, 2005, Plate 58). These studies show that starch granules can survive various types of food processing and, most importantly, that the nature of granule modification (see Chapter 10) provides an important source of evidence about the specific technology used in plant processing (cf. Babot 2003; Box 4.4).

POST-DEPOSITIONAL MOVEMENT

Many starch granules lie within the size range of pollen and phytoliths, and as a result we expect many of the same general processes that move these materials in the environment to apply to starches as well (e.g., Pearsall 2000: 258–63, 392–95). In this section we focus our discussion on several innovative experiments designed to study starch movement in situations that might influence behavioural interpretations of artifacts and studies of starch in archaeological sediments.

As discussed in Chapter 4, starch can enter the archaeological record from a wide range of natural and cultural pathways. Starch from underground organs may remain buried and other sources may become buried through a range of processes. Previous studies have shown that microfossils such as pollen and phytoliths may be mobilised within soils after deposition (Hart 2003; Kelso 1994). In addition, starch that remains on the surface, either trapped within decomposing plant remains or released from destroyed cells, could also conceivably be moved from the original site of deposition by the forces of wind and water. Experiments are needed to determine the full extent of post-depositional movement of starch in these contexts. While studies by Hart (2003) and Humphries et al. (2003) indicate that phytoliths may move in soil profiles, their interpretations are based on statistical interpretations of phytoliths within existing soils, but not controlled experiments under laboratory conditions (see Therin, Box 5.5, for such an experiment with starch granules). In addition to the potential effects on sediment sequences, at a very localised scale, starch preserved in soil might become attached to an artifact, leading to a false interpretation of tool function. Alternately, organic residues themselves might be local sources of starch that might then move from tools and into the surrounding environment.

Starch from underground sources (storage organs, roots) would not be expected to move laterally over great distances, unless exposed to wind and surface water through disturbance of the upper soil horizon. Starch from decaying litter may be slightly more mobile than its underground counterpart, but is unlikely to be transported much beyond its original source. Once below ground and within soils, vertical movement is more likely. Starch may be subject to processes of bioturbation, but this should be identifiable during routine analyses during excavation and prior to sediment sampling. It is useful to consider the post-depositional movement of starch at various scales from smallest to largest: (1) the artifact; (2) the local context; and (2) the environment at large.

Artifact Level

Beginning with the very localised environment of the artifact, it is important to know whether starch mobilised within a deposit can become attached to artifacts and potentially be treated falsely as a residue related to tool use. For this reason, Williamson (Box 5.4) conducted an experiment in which sediments containing stone tools and unrelated starch were irrigated over a two-year period. When the tools were analysed, she found that very little starch derived from the soil had become attached to them, suggesting little movement within the sediment. In contrast, she noted that some starch from foodstuffs placed adjacent to the tools was transferred to an artifact surface. Barton et al. (1998; Box 9.7) also found starch on archaeological tools that was not related to their use and was also deposited on a natural stone, but as in Williamson's case it is possible that the starch had not necessarily been mobile within the archaeological midden, but had originally been deposited in close proximity to the tools. Support for her findings comes from an archaeological examination of starch granules on tool surfaces and in associated sediments from the rock shelter site of Jinmium, Northern Australia (Atchison and Fullagar 1998: 121). Since starch granule density on tool surfaces was much higher than in the surrounding sediments and the morphology of the granules indicated that the starch in the sediments was derived from a different plant source, starch had remained relatively stable within the deposit.

Single Location

Moving to the scale of a single location or site, the prolific activity of microorganisms may be the best safeguard against mass quantities of starch being moved across land surfaces. They ought to ensure that any starch granules remaining exposed above

Box 5.4. Investigation Of Potential Contamination On Stone Tools

Bonny Williamson

A number of archaeologists have questioned whether the residues found on stone tools are solely due to prehistoric use or are at least partly the consequence of post-depositional contamination from material in the surrounding sediment (cf. Barton et al. 1998; Box 9.7). It is therefore important to find out if background starch within an archaeological context can adhere to stone tools. To test this possibility an experiment replicating stone tool use and burial conditions was established.

Three identical flowerpots (20 cm in diameter and 7 cm deep) were filled with alternating layers of soil with tools and soil without tools. Flakes of hornfels and quartzite were made for this experiment and used on one or a combination of different contact materials or included as clean, unused flakes. Contact materials included fresh meats (pork, lamb, chicken flesh and chicken bone, beef, beef bone marrow), dried beef, leather, and vegetables (potato, carrots). Some tools were buried wet, i.e., immediately following use or exposure to tap water, and some were allowed to dry after use and before they were buried to see if the condition of the tool determined degree of contamination. Finely ground maize meal was mixed with the soil in Flowerpot 2 in the proportion of 1 part maize meal to 15 parts soil prior to burial of the tools to simulate a worst case scenario. The sizes of the starch granules in the maize meal were in the order of 6 to 28 mm.

All three flowerpots were left outside and subjected to the elements (summer temperatures 16–30° C; winter temperatures 1–18° C; summer rainfall average c. 500 mm/year) for two years and four months. The contents of Flowerpot 2 and the upper half of Flowerpot 3 were then excavated, but Flowerpot 1 and

Table 5.3. Contamination on stone tools: experimental results.

Layer	Contact Material	Buried	Residues
Flowerpot 2			
1	Wet with tap water only (hornfels)	wet	No starch granules, only rootlets
1	Wet with tap water only (quartzite)	wet	Few starch granules (2–4 m), rootlets, mycohyphae and spores
2	Fresh chicken (quartzite)	wet	Animal tissue, red blood cells, few starch granules 4m, black film, rootlets, mycohyphae and spores
3	Clean, buried with potato (hornfels)	dry	Many starch granules (2–4 m), starchy deposit, rootlets
3	Clean, buried with potato (quartzite)	dry	Many starch granules (2–4 m), starchy deposit, rootlets
3	Clean (hornfels)	dry	No residues
4	Potato (quartzite)	wet	Few starch granules (2–4 μm),rootlets and large spores
Flowerpot 3			
Top	Fresh lamb only	wet	No starch granules, mycohyphae, rootlets or spores
Top	Lamb and potato	wet	Small starch granules, brown film, mycohyphae and rootlets

ground without some form of shielding or protection will be rapidly consumed. It would seem logical that the best chance any starch has of long-term persistence is to become incorporated and locked up within soils as rapidly as possible, thereby escaping the relentless activity of the microorganisms discussed previously. Given the previous discussion about starch preservation, it appears that the relationships between starch granule density, water flow rate, and sediment porosity are the most likely critical factors which affect the proportion of starch that moves downwards and escapes degradation.

In contrast, major disturbances to soils caused by human activity or other processes could lead to transportation but may also be quite detrimental to starch preservation. Any activity that floods a soil with fresh oxygen, moisture, and heat are likely to induce renewed biological activity from bacteria and fungi breaking down the remaining soil organics, including granules hitherto trapped within the soil matrix. It would seem likely that in order for starch granules to persist for a very long time, they require physical and chemical stability to escape breakdown and damage from microorganisms. Thus factors that continue to disturb

Box 5.4 (continued)

Table 5.4. Contamination on stone tools: results for the soil samples.

Layer	Starch	Hb	EDTA
Flowerpot 2			
Top	Few small starch granules (1–2 μm)	+2	0
Middle	Starch granules mostly small (1–2 μm), but a few larger (3–4 μm)	+3	0
Base	Starch granules mostly small (1–2 μm), but a few larger (3–4 μm)	+2	0
Flowerpot 3			
Top	Very few small granules (1–2 μm)	Trace/0	0
Middle	Very few small granules (1–2 μm)	Trace/0	0

the remainder of Flowerpot 3 have been left for further study. The artifacts as well as the soil samples from the top, middle, and bottom of Flowerpot 2 and from the top and middle of Flowerpot 3 were collected and analysed in terms of the detection of starch granules, pH determination, and the Hemastix test for the detection of haemoglobin/myoglobin (Williamson 2000).

The results are presented in Tables 5.3 and 5.4. From these it can be seen that starch within the soils does not arbitrarily adhere to artifact surfaces, except in very small quantities, but whole vegetables buried with the tools can be a source of contamination. Reasonable quantities of starch granules were only observed on the tools that had been buried in contact with a potato, whereas fewer were found on tools used to cut a potato. The starch found on the artifacts was much smaller than that in the maize meal, suggesting that it was from a different source. This is true even if the artifact surface was covered with a relatively sticky substance such as blood, as in the case of the tools used to cut fresh lamb and chicken. It therefore seems unlikely that contamination from soils with starch in the order of up to 7 percent will not be an important factor in the composition of residues on stone tools, but that tools buried in contact with vegetables may attract starch. In this case a comparison with use-wear would be required to distinguish use from contact (cf. Barton et al. 1998).

Further experiments of this type designed to test the level and nature of contamination within the depositional context would be very useful.

Box 5.5. Starch Movement In Sediment

Michael Therin

This experiment was designed to assess the rate of downward movement of starch granules of differing sizes through a sand-type deposit. The selection of the variables was an attempt to simulate conditions encountered in the tropics with a high annual rate of rainfall. Sand was selected as the medium for the observations as this could be supplied as sterile laboratory-grade material containing no unwanted starch or other organic material.

The downward movement of microfossils such as pollen and phytoliths (starch-sized objects) has been recorded, and thus post-depositional mobility needs to be considered when interpreting starch recovered from sediments and artifacts. The studies on pollen transport indicate very different rates of downward infiltration which appears to be strongly site dependent relating to site-specific micro-environmental condition, as might be expected. At the site of Matley Wood, Hampshire, Dimbleby (1985) estimated that pollen, with a known date of introduction, was moving downwards at a rate of one centimetre every 30 years. Studies of pollen distribution from a seventeenth-century refuse pit in Virginia, USA, by Kelso et al. (1995) estimated a much more rapid descent of up to one centimetre every four to five years. The movement of particles in soils is known to be dependent upon several factors including the size of the particles, porosity of the soil, and water flow rates. The nature and size of soil pores may also help to trap and immobilise particles which are known to occur with pollen (Dimbleby 1985), and also, if pores are small enough such as occur within silt-loam, may even immobilise microorganisms such as bacteria (Golchin et al. 1998: 261; Van Veen and Kuikman 1990). However, if soils are not compacted and pore spaces of sufficient size, downward movement of starch-sized particles may be anticipated.

The Experiment

The simulated movement of starch grains in sediment profile was measured using four columns of sand (Table 5.5) of different particle size (Table 5.6), irrigated at a controlled rate for a period of two months. The different sands used in the experiment were labelled by the laboratory suppliers as G1, G400, and Ground Sand. At the head of each column a 0.1 g sample comprised of an even mixture of rice grains (3–8 microns) and corn starch (5–25 microns) was added and sealed, then sealed under 2 cm of sterile sediment. A control sample was included in the experiment to ensure that no extraneous starch grain contamination of samples occurred. This consisted of an identical sand column run under the same conditions but without the addition of starch grains.

Table 5.5. Sand size class and irrigation rate used in post-deposition experiments.

Column No.	Sand Size Class	Irrigation (ml/day)
1	Sand G1	100
2	Sand G400	100
3	Ground sand	100
4	Sand G1	200

Table 5.6. Percentage of particle sizes in the different sands used in the experimental study of starch movement.

Particle size (µm)	G1 (%)	G400 (%)	Ground sand (%)
500	6.8	0.2	0
400	16.3	0.3	0
300	28	0.8	0
200	38.6	67.2	0
100	10.4	31.2	56.3
<100	0	0.3	43.7
Total	100	100	100

Box 5.5 (continued)

Sand G1 has the greatest distribution of particle sizes (Table 5.6) with 83 percent occurring between 200 and 400 microns maximum diameter. Sand G400 contained a more limited distribution of sizes with 98 percent falling between 100–300 μm, and the Ground Sand was all 100 microns. Irrig ation rates (Table 5.5) were selected to represent an average annual tropical rainfall of four metres of rain per year. The diameter of the experimental column was used in combination with the rainfall figures to calculate the volume of irrigation required. A rate of 100 ml per day over a two-month period produced a simulated annual rainfall of 4,777 ml, while column four, irrigated at 200 ml per day for two months, matched the high end of the tropical rainfall range (9,554 ml).

At the completion of the experiment, the columns were sliced into 11 one-centimetre sections. A total of four of these sections was then selected (Table 5.7) for detailed analysis, from which a four-gram sediment sample was removed for starch extraction. Starch was recovered from the sand following the heavy liquid separation technique outlined in Chapter 8. For further details on the method see Therin (1998a).

The Results

Table 5.7 shows the movement of starch down each of the four columns. The counts from the upper layer (0.0–0.5 cm) are estimates of the total based on a sample field of view then extrapolated for the entire slide, as is also the case for layer 2 (1.5–2.5 cm) in Column 4. All other counts represent the sum total of starches counted from each four-gram sediment sample.

The total starch count of the 0.1 g sample was never made, but is likely to be considerably higher than the count estimate of starch that travelled through the first sampled layer. The distribution of remaining starch downwards through the column indicates that only a very small percentage of starch granules moved further than one centimetre through the sand and that the sand particle size and irrigation levels have significantly influenced granule movement. Overall, less than 0.01 percent of starch counts estimated in layer one (0.5 cm) has moved below one centimetre in columns 1 through 3 and in column 4, 0.3 percent. Doubling the irrigation rate moves more starch through the column, at least initially, though it appears as if much of this starch becomes trapped at a depth of 1.5 to 2.5 cm. At a rate of 0.01 percent starch movement per annum, it would take roughly 7,800 years for the entire starch sample to travel to the base of a 10 cm deep column irrigated at the simulated annual average. This of course does not take into account field conditions such as the gradual accumulation of new sediment above the starch source, nor the steady degradation of the starch source by microorganisms. Starch movement is slow, even within a particularly porous substrate such as sand.

Therin (1998a) also calculated data on the variation of starch movement based on gran-

Table 5.7. Results of post-deposition experiments: starch granule counts by layer (based on Therin 1998a).

Depth (cm)	Sand G1 100ml/day	Sand G400 100 ml/day	Ground sand 100 ml/day	Sand G1 200 ml/day
0.0–0.5 cm	780,000	740,000	750,000	760,000
1.5–2.5 cm	71	51	41	2526
5.5–6.5 cm	94	71	54	35
9.5–10.5 cm	84	56	42	38

starch deposits appear more likely to facilitate the removal of starch from the environment rather than influence any possible post-depositional movement of granules.

Potentially one of the most important taphonomic factors is transport of granules through sediments by water. When water enters sedimentary deposits, horizontal, slanting, and vertical movements of starch granules may take place. Downward movements are most likely to be caused by rainfall onto a porous, well-drained soil that allows the rainwater to flow through it freely. The issue of downward infiltration has been discussed by Therin et al. (1999) in relation to the discovery of starch granules within volcanic airfall tephras that formed the B-horizon and were devoid of evidence of human activity. Counts of starch granules per gram of sediment were low overall (cf. Barton et al. 1998) indicating that starch granule movement was restricted, even under conditions of high tropical rainfall.

Therin (1998a; Box 5.5) attempted a series of controlled experiments to monitor the movement of starch granules downward in sediments. He found that only small quantities of starch granules of any size moved through his experimental column, though small starch granules (< 5 μm) were more mobile than larger granules. On this basis he suggested that starch in archaeological sediments would largely remain immobile after deposition. Therin (1998a: 71) did recognize, however, that the long period of time available at an archaeological site may allow a higher proportion of starch to move downward than was witnessed in his experiments. His conclusion is relevant for starch that has already been mobilised and has escaped biological activity in the upper layers, but due to the action of starch-digesting organisms, only a small proportion of starch will ever be preserved and available to be transported over a long time period.

An area that has not yet received attention in terms of starch deposition is the role of animals, including humans, as unwitting vectors of starch that is protected and stored for short periods within the digestive system. Starch that has resisted digestion in the gut will either be naturally highly resistant to enzymatic decay or trapped within intact fragments of plant tissue (see above). The quantity of starch within individual stools seems to be quite variable and relates to the efficiency of the gut and the type of starch involved. The impact of this starch on archaeological interpretation remains to be tested by an adventurous scholar. The

Box 5.5 (continued)

ule size. His results indicate that the smallest starches (0–5 μm) are far more mobile than larger granules (6–25 μm). These granules travel faster and more freely through the sediment than grains of larger size. The largest granules (20–25 μm)—though rare in the overall starch sample (c. 1.5 percent)—were represented by only a few examples in layers below 0.5 cm depth, indicating that they are less likely to become mobile by irrigation and when they do move, move very slowly down the sequence.

Conclusion

This experiment indicates that starch can be mobilised downwards through sediment when irrigated with a simulated rainfall. As anticipated, the size of the starch granules, particle size of the sediment (which may also be a proxy for pore size in this case), and the rate of water flow influence the starch movement in sediments. It is also apparent that starch moves initially within the first centimetre or so of sediment, but then its downward rate is significantly decreased. This may be due to increased sediment compaction and/or reduced flow rates of water that has lower kinetic energy down the column than at the irrigation head in this experiment. In regions with much lower rainfall than the tropical averages simulated here and sediments with higher clay and silt content (small pore spacing), far less starch may be expected to move downwards through the section. Further experimentation would be of benefit utilising a wider range of sediment types and degrees of sediment compaction, but could be easily achieved and is, of course, encouraged.

problem, like many others in studies of taphonomy, is one of scale. Could starch derived from animal waste be of such quantity and spread over such a wide area that the natural starch background or starch from a cultural deposit be so altered as to confuse interpretation? We suggest that this is unlikely to be the case; however, as with any unknown, it should be considered a possibility until demonstrated otherwise.

Whatever the true extent of the post-depositional movement of starch, it seems unlikely to be severe enough to seriously disrupt interpretations of human behaviour or paleoenvironments, since recent studies have been able to detect differences in starch assemblages not only in modern soils and on the surfaces of rock shelters, but also in ancient sediments (e.g., Boxes 8.1, 8.8, 8.9).

Environment at Large

In terms of the largest scale of the environment, there have been no objective tests of starch mobility. Starch is heavier than air and denser than water and will settle in either media if left undisturbed, but it also has the potential to become mobilised. A study of starch movement within a hospital showed that powders consisting of starch granules could be released into the air from packets of examination gloves and then remain airborne long enough to be dispersed around a room and enter into the air-conditioning system (Box 8.7). A condition known as 'baker's asthma' can be caused by exposure to large quantities of airborne particles associated with starch produced by commercial flour mills, bakeries, and other areas where starch flours are stored (Anto et al. 1989; Hinojosa 2000). Surveys with air filters to determine the extent of airborne particulates containing starch from industrial sites handling flours found that granules could be transported by wind across a city (Anto et al. 1989). Taphonomic studies are required to test the mobility of starch in different environmental contexts and to assess how this might affect archaeological and paleoenvironmental interpretations.

FURTHER RESEARCH

A great strength of ancient starch analysis is its apparent ubiquity in terms of the wide range of depositional conditions in which it has been recovered and the longevity of its survival. Despite its apparent persistence, very little is known about the effects of potential taphonomic processes on the quantities, spatial patterning, and forms that have survived. A current challenge is to conduct fundamental research in order to monitor the processes underlying this long-term survival of starch in order to understand why and in what contexts starch does and does not survive, and whether some conditions could cause bias in the recovery of particular plant types or morphologies. At this early stage of taphonomic research, it appears that starch is generally resistant to decay in the absence of starch-degrading microorganisms. In other contexts, having escaped the attention of starch-degrading enzymes, it remains stable over very long periods of time and does not appear to alter its physical shape or physico-chemical properties. This is good news for the future of ancient starch research, but additional research is necessary to confirm and enlarge on these findings.

A major research area is to determine the nature of ancient starch. From the experimental observations summarised above, it is suggested that the semicrystalline properties of the starch granule are largely responsible for its stability over very long time scales, but this remains an untested hypothesis. One way forward is to compare and contrast contexts in which starch has been preserved. It would be useful if scholars who recover starch recorded as much as possible about the preserving sediments and the conditions of the starch granules.

Further challenges for starch research exist in the undertaking of controlled experiments which simulate different preservation conditions in order to determine the effects of variables such as sediment type, soil porosity, pH, infrequent exposure to water, and the impact of prolonged surface exposure or rapid burial to starch survival. The single study of starch survival on stone artifacts left exposed to the elements suggests that some form of shielding against wind and rain dramatically increases the rate of starch survival (Lu 2003), but further studies are required. Hopefully, this chapter will spur interest in starch taphonomy so that future studies can resolve the unanswered questions posed here.

CHAPTER 6

REFERENCE COLLECTIONS

Judith Field

A reference collection is the major tool for identifying starch granules to the correct plant taxa. It is an essential resource for ancient starch analysis, particularly in the early stages of work, because it will be referred to frequently. Published material on starch granule morphology and atlases (see summary in Chapter 7 and especially Box 7.1) can be extremely useful, especially as a starting point, but these are not usually sufficient for most research because the granules are portrayed in two rather than three dimensions; polarised images are uncommon; the range of shapes presented is generally restricted; and the number of species that has been published is still limited, especially in terms of wild rather than domesticated or cultivated plants. For special topics, such as the effects of food preparation on granule form, morphological variation of granules within different parts of the plant, or variation due to seasonal or environmental factors, it is best to make special collections under well-monitored conditions. In addition, preparing a reference collection is also a useful task in itself because it helps to familiarise oneself with the material under investigation.

The aim of this chapter is to present a guide to the process of preparing a reference collection for use in ancient starch research. Compiling a reference collection involves the following stages: (1) designing the collection; (2) compiling the plant material; and (3) preparing voucher specimens and slides. Various approaches are outlined and important factors to be considered during this process are discussed. A range of protocols for the preparation of specimens is also provided.

DESIGNING THE REFERENCE COLLECTION

In designing a plant reference collection appropriate to ancient starch research, a number of points should be considered. The nature of the archaeological research questions is vital since this will determine the kinds and number of plant taxa required. Additional factors such as environmental variation may also be important.

Archaeological Questions

The content of the reference collection should be framed by the archaeological research questions. In some cases only one or a few plant types need to be targeted, whereas in others a large suite is needed. Studies of the domestication and/or use of a particular plant type—as described, for example, in Boxes 1.1, 1.2, and 1.3—require only one or a few plant species, although close relatives should also be included for comparison. The investigation of diet on the basis of residues on cooking vessels demands a collection of all likely economic plants within the region during the period when the archaeological record was formed. For example, as a model of past plant uses, Jenny Atchison collected plants that were known by local Aboriginal women in the Kimberley region of Australia to have been utilised in the recent past (Box 6.1). In contrast, a much more comprehensive approach was used in compiling the reference collections made by Lentfer and Therin (Box 6.2) in tropical Papua New Guinea and Field and Gott (Box 6.4) in a semi-arid part of southeastern Australia. Whereas Atchison and Lentfer and Therin relied heavily on indigenous knowledge of the local vegetation, Field and Gott also used historic sources such as explorers' journals (cf. Box 4.3) and paleoenvironmental data.

It is quite useful to enlist the help of a botanist at an early stage of the study because he or she may know what plant species are relevant for a particular archaeological research question and can often suggest useful sources of data.

Choice of Taxa

The required level of taxonomic identification should be defined at the start of the project since

Box 6.1. Indigenous Knowledge And Starch Reference Collections

Jennifer Atchison

The case study presented here describes the establishment of a starch reference collection from ethnoarchaeological research conducted in the Keep River region, Northern Territory (Atchison 1994). A reference collection was sought to assist the identification of starch granules recovered from prehistoric stone tools (Atchison and Fullagar 1998). Plant species that had been processed with stone tools by Aboriginal people in the Keep River region were selected as appropriate for the reference collection. This selection was influenced by

- loss of traditional knowledge about which plants were used or how they were processed;
- change over time in the species distribution in an area;
- limited access to remote areas; and
- seasonality of plant species.

Where possible, the selection of appropriate plant species for the reference collection involved collaboration with local, traditional Aboriginal custodians. Custodians played a significant role in identifying the relevant plants that might have been eaten and were also able to locate and prepare or process the plant. The edible portion was usually the focus of collection; however, a complete botanical specimen was always collected and photographed in the field for identification and future reference. Flowers were collected where available because they are an especially important aid in botanical identification.

Preservation and Transportation

Some plant species require processing to make the plant edible. It is important to collect and document the starch from these activities in case some stages of processing induce morphological changes in the starch granules (cf. Boxes 4.4, 4.5). For this purpose it is useful to have the traditional custodians demonstrate the sequence. Starch specimens can be collected at each stage of the process, photographed, preserved, and stored. In the Keep River region, for example, two yam species made a major contribution to carbohydrate in the local diet. *Dioscorea bulbifera* (Figure 6.1) requires extensive processing, including

Figure 6.1. In the Keep River region in northern Australia, the yam *Dioscorea bulbifera* (*mumunya*) is leached in running water to remove toxins and then roasted before it is eaten (cf. description in Box 4.2). *Photo by Jennifer Atchison*

Figure 6.2. The yam *Dioscorea transversa* (*dangar*) is commonly roasted by Aboriginal people in the Keep River region of northern Australia, but may also be eaten raw. *Photo by Jennifer Atchison*

Box 6.1 (continued)

Figure 6.3. Biddy Simon and Polly Wandanga harvesting the palm *Livistona* sp. B. in the Keep River region of northern Australia. Both the main stem and portions of the leaves are edible. *Photo by Jennifer Atchison*

Figure 6.4. The growing heart of the *Livistona* sp. B. palm is rich in starch. In the Keep River region in northern Australia, it is usually prepared by roasting in the fire. *Photo by Jennifer Atchison*

water leaching and cooking, to make it edible (cf. Box 4.4) whereas *Dioscorea transversa* (Figure 6.2) may be prepared by roasting it in the ashes although it can also be eaten raw.

In Figure 6.3, Polly Wandanga and Biddy Simon are shown harvesting the edible parts of the palm *Livistona* sp. B. Both the heart of the palm and the base of the leaves were eaten (Figure 6.4). About a teaspoon-sized sample was collected from each plant part that was thought to contain significant quantities of starch, placed into Eppendorf™ tubes, and covered with pure ethanol. Without a preserving liquid, samples would usually become covered in mould by the time they arrived back in the laboratory. It is preferable to store specimens under refrigeration but this is not always possible.

Processing and Recording

Once back in the laboratory a permanent record of starch granules was prepared.

- A small piece was cut from each sample and placed on a clean, dry petri dish. The rest was stored. It was especially important that some of the sample collected in the field was preserved should further preparations be required.
- Each petri dish with sample was placed in a fan-forced oven set to 25° C for approximately 6 hours, then covered and left for about a week.
- The samples were ground to a fine powder with mortar and pestle and transferred to labelled snap-cap micro centrifuge tubes (e.g., Eppendorf™).
- As small a quantity of powder as possible was placed into another tube, covered with 50 μl of distilled water, and vigorously mixed using a vortex mixer.
- A 10 μl sample was then placed onto a microscope slide and left to dry. Once dry, the sample was mounted in Euparol™.

All slides in the reference collection were labelled to facilitate cross-referencing of samples from the various parts of each plant.

it will assist in the selection of species types and number of individuals per species required for the reference collection. It is probably prudent to ensure that more than one species of a genus is sampled because the starch granules from the whole genus or even the family level may be similar. The rule of thumb is that more examples are always better than a few. If starch granules cannot be separated to species on the basis of morphological characteristics, then identification to genus or family may be the limiting factor on your analysis.

Additional Factors

The spatial scale (e.g., within or between sites) and nature of the environment can be significant in considering how many species are required for the reference collection. If the environment and landscape changed in the past, reference material will need to be drawn from a much wider field than the research area and possibly from a location that has a similar modern climate and vegetation to the prehistoric counterpart. Collecting a range of plants will have the added advantage of providing information about plant associations and the modern environment. If vegetation surveys have already been undertaken, then plant lists should be available and there should also be herbarium specimens which might be sampled. If detecting environmental differences is important or is suspected to significantly alter the character of the starch granules in the archaeological sample, then it may be important to collect specimens under a range of specific, defined conditions. Additional factors to be considered are whether starch morphology varies with season or growth or environmental conditions (e.g., Plates 17–20). If so, the sampling strategy will need to be adjusted accordingly. Depending on the research framework, the collection of wild varieties, if appropriate, may be preferable because granules within modern cultivars and other commercially grown plants may not be representative of ancient plants.

SOURCES OF REFERENCE MATERIAL

The major sources of reference material are listed in Table 6.1. It may be useful to combine specimens from a variety of them. In all cases accurate documentation should be maintained so that in the future it would be possible for someone to check identifications and/or repeat the study. As mentioned previously, published atlases such as those discussed in Chapter 7 may provide a useful substitute when reference material is not available.

Herbaria

Herbaria can be a good source of plant material, but one should keep a number of points in mind when approaching them for access to this material. The quantity of material in the collection may not be sufficient to provide an adequate sample for research purposes. The herbarium may not permit sampling of its specimens. One should check with herbarium staff that the identifications are current. As some species may have disappeared from your study region in more recent times, the herbarium records may provide useful historical information about past distributions.

Seed Banks

Government departments or privately funded research agencies have seed banks for different varieties of one or more species, especially for

Table 6.1. Potential sources of starch reference material.

Source	Example
Publications	Published atlases, journal articles
Herbaria	John Ray Herbarium, University of Sydney
Government Departments	Australian Winter Cereals Collection
Private Gardens/Botanic Gardens	Kew Gardens, London
Markets	Fruit and vegetable suppliers
Field Collections	Local (or similar) environs of the site under study

Box 6.2. Collecting Starch In Papua New Guinea

Carol Lentfer and Michael Therin

The analysis of starch granules (and other microfossils) in residues on stone tools and from sediments has become a major focus for monitoring changing patterns of subsistence, tool function, and landscape change for archaeological research in West New Britain province, Papua New Guinea (e.g., Fullagar 1992; Barton et al. 1998; Kealhofer et al. 1999; Therin et al. 1999; Lentfer et al. 2002). To further understand the floristics, ethnobotany, and cultivation practices of the region, it was necessary to undertake fieldwork. Compiling a starch reference collection was part of this research. Since palaeoenvironmental reconstruction is important for understanding the dynamic processes of human/environmental interaction and is necessary for archaeological interpretation, the focus of field collection was expanded beyond the economically important plants to encompass as many other ecologically significant starchy plants as possible. This approach was necessary to ensure that as many starch granules as possible obtained from archaeological samples could be identified.

Field Area

The major focus of archaeological research is along the north coast of West New Britain and offshore islands in the Talasea region. This is a typical, wet tropical environment once dominated by lowland tropical forests. The region presently consists of a mosaic of cultivated environments with village gardens, coconut and oil palm plantations, and various stages of regrowth forest. For this study, vegetation surveys and plant collection were conducted on Garua Island and the adjacent small, volcanic islands of Kaula and Garala, Walindi plantation on the north coast of the main island, and Nave River and Garu village environs on the southwestern coast of the Willaumez Peninsula. Ethnobotanical information was gathered from a variety of sources including firsthand sources (i.e., people living in the region), books and field reports on regional flora (e.g., Peekel 1984), and field notes in the herbarium at the Forestry Research Institute, Lae.

Field Methods

Plants were collected from a variety of vegetation communities, villages, new gardens, abandoned gardens, early regrowth forests, late regrowth forests, and logged forests. Field research assistants from the local area were employed to assist with the collections and ethnographic recording. Assistants with sound knowledge of the specific areas sampled in terms of plant use, cultivation history, and other activities (e.g., logging) which may have changed vegetation patterns were carefully selected from Garu village and Kimbe town.

To begin with, plants known ethnographically to have been used in the area were targeted. This included plants used for food, housing, artifacts, and ceremonies (Figure 6.5). The second collection strategy was to collect plants from the different ecological zones noted above. Third, random collection was also undertaken as new plants were encountered during the normal course of fieldwork. This included collections from roadsides, beaches, swamps, and forest tracts. Ethnographic information was recorded and photographs taken of live plants (Figure 6.6).

Where possible, leaves, stems, flowers, fruits, roots, wood, and bark were all collected. Plants were labelled and placed into polyethylene collection bags. To keep plant samples fresh, drops of water were added to the collection bags. As soon as possible after collection (usually within four to eight hours), plant samples were prepared for the West New Britain project herbarium (stored at Southern Cross University) using a method developed to enhance preservation in the humid tropics. First, samples were trimmed to a suitable size for herbarium sheets and dipped into a container with undiluted methylated spirits. When thoroughly

Box 6.2 (continued)

Figure 6.5. Making a reference collection in Papua New Guinea. Collecting the starchy heart of the palm (*Nypa fructicans*) with local informants. *Photo by Lisa Kealhofer*

Figure 6.6. Carol Lentfer and Michael Therin recording ethnobotanical information on Kaula Island (Papua New Guinea) with informant John Namuno. *Photo by Robin Torrence*

wet, samples were placed carefully between newspaper sheets and pressed in a wooden plant press. The press was then placed into a large polyethylene bag, additional methylated spirits was added to ensure uniform saturation of newspaper (approximately one to three cups), and the bag was sealed (see Womersley 1976).

Additionally, subsamples were set aside specifically for the starch reference collection. Leaves, stems, and flowers were soaked in 70 percent methylated spirits and pressed between newspaper sheets and stored in sealed polyethylene bags and portions of tubers, corms, and fruit were placed into labelled vials in the 70 percent methylated spirits solution. All samples were kept moist and stored out of the sunlight in natural air temperatures until starch extraction procedures in the laboratory commenced.

Herbarium

Pressed plant samples were transported to the herbarium at the Forestry Research Institute, transferred to clean newspaper and open-weave metal presses, oven-dried between 60° C and 70° C for 48 hours, and identified to species level with the assistance of resident specialists. A database was constructed listing species, locations, and ethnographic information. Voucher specimens of the West New Britain herbarium, representing more than 1,000 plant species, were lodged at the Southern Cross University, Lismore, Australia (Figure 6.7).

Starch Reference Collection

Once we returned to the lab, the samples taken specifically for studying starch were dried in an oven at 36° C. The wet samples (the tubers, corms, and fruit stored in vials with methylated spirits) were dissected. Only half of these samples were dried, the other half was retained for future reference; once dry, small samples were taken from each part of the dried plant, i.e., stem, leaf, root, and flower. The inner core, outer core, and skin of tubers, corms, and fruit were also sampled. All samples were ground with a glass mortar and pestle into a fine power. Permanent slides using a dusting of fine powder and mounted with Euparol™ were made and stored flat in wooden boxes (Figure 6.8). The remaining ground plant tissue was transferred to vials and stored for future reference. It is preferable to use temporary mounts during which the granules can be easily rotated. All microscope slides were examined using transmitted brightfield microscopy and starch grains were photographed under nonpolarised and polarised light at x200 to x500 magnification. The images were stored on a compact disc.

Box 6.2 (continued)

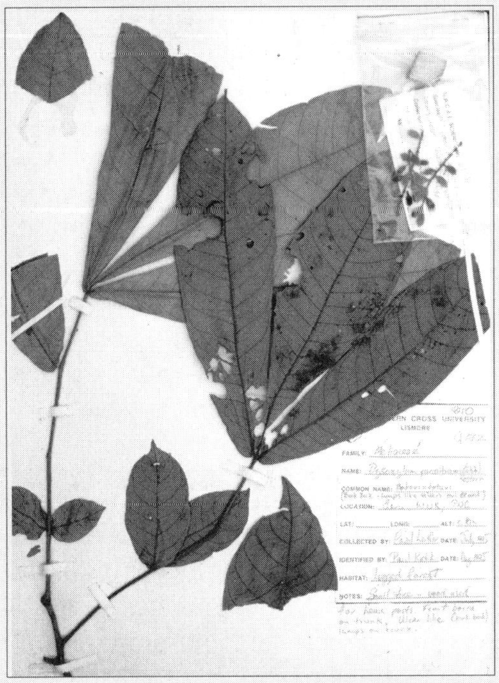

Figure 6.7. Voucher specimen in the West New Britain herbarium. *Photo by Carol Lentfer*

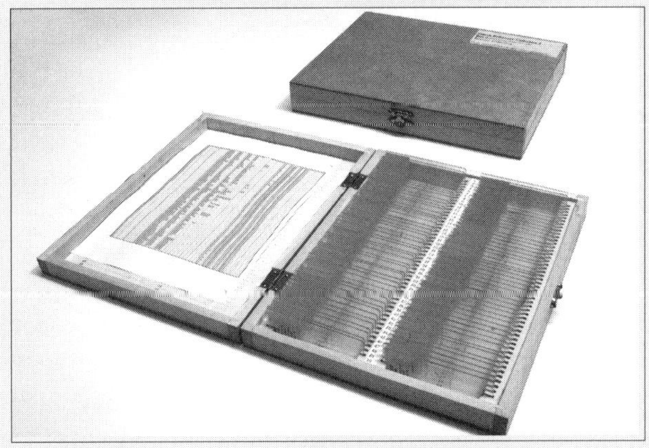

Figure 6.8. Permanent slide reference collection of starch granules at the Australian Museum, Sydney. *Photo by Stuart Humphreys, Australian Museum*

Digital images of individual granules were also made for a multivariate study of granule morphologies (Torrence et al. 2004).

Initial Investigation

The initial examination of the starch grains from the reference collection yielded promising results (Therin et al. 1997). Over 300 microscope slides of 172 starch-bearing species were prepared. From the analysed slides, starch granules were present in 96 percent, although only 27 percent of the sample had granules over 5 microns in size. Not surprisingly given the basic characteristics of plants (cf. Chapter 2), starch was abundant in underground storage organs (tubers and corms), seeds, and fruits but was also present in flowers, leaves, roots, and stems (Figure 6.9). The latter granules probably represent transient starch which is unlikely to be diagnostic to plant taxa (cf. Chapter 2; Haslam 2004).

The morphological variability between species was sufficient to enable the identification of individual species in most cases. Small starch grains (<5 microns) occurred in leaves and stems of most of the plant species. There was very little morphological variation under normal light microscope magnification and it was difficult to identify granules to species level on morphological grounds only. Of the slides with larger starch grains, nearly all were derived from food storage parts of plants (tubers, corms, nuts, and fruit).

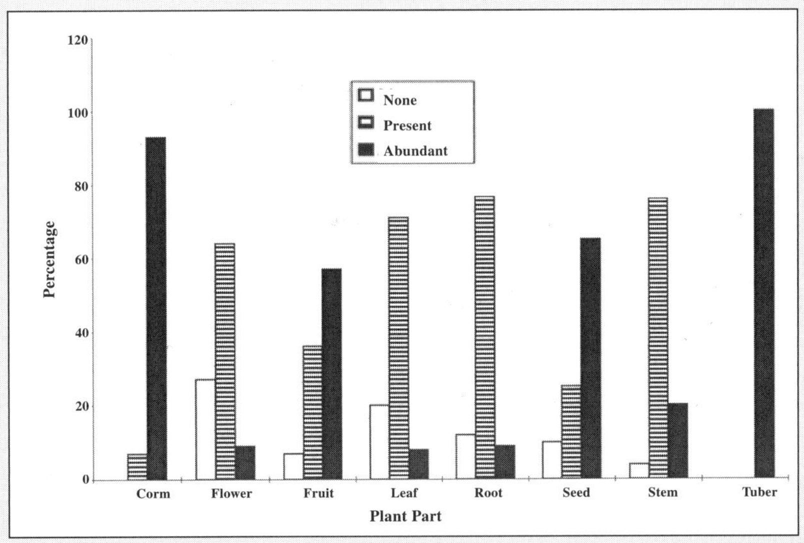

Figure 6.9. Summary of the abundance of starch granules in different plant parts represented in the Australian Museum reference collection. *Illustration by Michael Therin*

cultivars. Depending on the requirements of the reference collection, they may provide useful material. The advantage of this source is that species identification is generally assured and detailed information about the particular species and strain should be readily available.

Established Gardens

Established government botanic gardens and private gardens often propagate particular species of plants that are not commonly available. It may be possible to obtain fresh specimens of rare and uncommon plants that are not generally available or normally cultivated in a particular region. Material from these sources is grown under controlled conditions and identification can usually be confirmed. If environmental conditions are essential, the specialised growing conditions of a particular garden may not be suitable. Nonetheless, these plants will provide a foundation for exploring the range of variation in any particular species.

Another solution is to set up a living plant collection, as described by Peter J. Matthews (Box 6.3), although this demands an extremely high level of resources and commitment. It may be possible to collaborate with a botanic garden in this exercise, especially if only a limited number of species are required.

Markets

Fruit and vegetable markets can be an important source of reference material when other avenues are unproductive or beyond the project budget. In this case expertise with the identification of the relevant material is required. Since only part of the plant will be available, this is not a useful source for voucher specimens and so adequate records should be prepared in case validation is needed later. Photographic documentation, permanent thin sections of the plant parts sampled and additional information, such as vernacular name, source, uses, storage methods, date and place of collection, etc., are all important.

Field Collections

After other avenues of information retrieval have been exhausted, sampling the likely plant sources directly may be required. Both Atchison (Box 6.1) and Lentfer and Therin (Box 6.2) worked with indigenous groups who guided them to plants that they either still used or knew about. Lentfer and Therin also collected plants with no known local uses if they contained reasonable quantities of starch. These specimens provided a control sample for the species that were assumed to have been economically useful in the past. A different approach was taken by Gott and Field (Box 6.4) who used ethnobotanical information obtained from a range of historical sources in formulating their fieldwork. Finally, vegetation surveys will also provide important information concerning plant communities. An understanding of the biogeography of the region will allow, to some extent, the prediction of available plant types.

PREPARATION AND STORAGE

When preparing a reference collection, it is advisable to get help from a botanist or herbarium, many of whom produce brochures detailing the collection of plant specimens and preparation for storage (e.g., Albrecht 1993; Bedford and James 1995; see Table 6.3 for useful websites). Pearsall (1989: 129–36; 2000) also provides detailed information about preparing herbaria specimens (see also Womersley 1976, 1981; Bye 1986; Bridson and Forman 1992; Cotton 1996: 115). Equipment for field collecting is listed in Table 6.2.

Sampled species may range from trees down to small ferns or subterranean tubers but the amount of material to be collected will depend on research questions, the plant parts required or available for confident identification, and feasibility. For most plants, the minimum requirement will be leaves and stem apex, flowers, fruits, and seeds or sporocarps. Where bulky specimens have been collected, for example cones or large seeds, they should be cut lengthwise to produce a relatively thin section of the original cone or seed. It is recommended that the quantity of material collected should be enough, plant availability permitting, for at least two and preferably three specimens to be prepared.

Pressing and drying is the preferable method for temporary storage, but if this is not possible, then the specimens should be placed in plastic bags (but only if they can be refrigerated) until identification can be confirmed, usually back in a field laboratory. Fruit, seeds, and sporocarps should be stored in paper envelopes to allow for drying.

Box 6.3. Living Plant Collections As Reference Material

Peter J. Matthews

Modern starch reference collections can be made in many ways, beginning with plants growing in natural, disturbed, or cultivated habitats, or with living plant collections that have been assembled for starch sampling or other purposes. I have been involved with two living collections, comprised of taro and yam. These were originally assembled in order to learn about the evolution, domestication, and dispersal of these plant types. Both these collections have contributed to starch granule reference collections.

For morphological and genetic analyses, my colleagues and I needed the following: (1) examples of closely related species and genera; (2) broad geographical coverage of the main species of interest; and (3) examples from a variety of apparently natural or wild habitats, from disturbed or ruderal habitats, and from cultivations. We also desired multiple collections of identical genotypes, since the two plant types concerned are clonally propagated for cultivation. The cultivated species were not excluded because we wished to learn about variation in the crop in different regions and habitats. If our aims had been to collect diverse forms of the crop as efficiently as possible for a particular breeding program, the range of material would have been more restricted, and there would have been less tolerance of likely or obvious duplication.

Since the 1960s Doug Yen had managed to maintain a small number of taro and yam plants from various locations in Southeast Asia and the Pacific, hoping that these could someday be used for historical and agricultural research. Yen's collection at the Australian National University (ANU) was able to provide Tom Loy with starchy corm samples from a range of wild and cultivated forms of taro (*Colocasia esculenta*), closely related species in the same genus, and species in related genera (*Alocasia* sp. and *Xanthosoma* sp.) from the same plant family (the aroids, or Araceae). The samples were chosen in order to determine whether or not the starch from modern taro and other aroids could be distinguished from each other and from other cultivated plants. The material was invaluable for identifying plant residues on stone tools from Kilu Cave (Loy et al. 1992).

Some of the aroid material had been difficult to obtain and it was clear that not all of it could be preserved as living material after my departure and Yen's retirement. Parts of the living collection were subsequently maintained elsewhere within the ANU, at the University of Selangor in Malaysia, at the Royal Botanical Gardens in Sydney, and by the National Institute for Vegetables, Ornamental Plants and Tea (NIVOT) at Ano in Japan, where I later worked (see Matthews et al. 1992) . I continue to maintain a collection of frozen DNA samples at the National Museum of Ethnology. Sadly, the living taro collection could not be maintained anywhere in a complete form and those parts that were preserved have been either abandoned or lost, or face an uncertain future.

While Yen and I were working with collections of taro and yam in Canberra, Ryohei Terauchi (University of Kyoto in Japan) was independently collecting plants of *Dioscorea bulbifera* and other yam species from various parts of Eastern Asia, Africa, and Madagascar (cf. Matthews and Terauchi 1994). In 1995, Terauchi left Kyoto to work in Germany and was no longer able to maintain his living yam collection. Luckily, just before the collection was abandoned, I was still living in Japan and Terauchi kindly allowed me to remove a large number of tuber and bulbil samples. These are currently maintained in a freezer at -30° C at the National Museum of Ethnology, Osaka. Starch samples in the form of dried smears on glass slides were also prepared for analysis at

Table 6.2 Useful equipment for plant collecting

- *Plant press*—including plenty of newspaper and corrugated cardboard.
- *Plastic bags* for those plants that cannot be identified immediately.
- *Paper bags/envelopes* for collecting seeds, flowers, fruits or other relevant plant parts. Flowers can be used for pollen extraction and leaves/seeds for phytoliths if need be.
- *Labels and a notebook* for recording relevant collection information (e.g., date, location, etc.).
- *Hand lens* for observing flowers or other plant parts not readily visible with the naked eye.
- *Forceps* for handling small/delicate specimens.
- *Reference books* for identifying/keying out specimens.

Box 6.3 (continued)

the Australian Museum and a duplicate set of slides has been kept in Osaka.

For the initial survey of starch granules in modern *D. bulbifera* and related taxa, samples were chosen that represented (1) the main genetic lineages of *D. bulbifera* that had already been identified by Terauchi using DNA analysis; (2) other yam species that may or may not have starch that is similar to the starch of *D. bulbifera*; and (3) more distantly related species from a different plant order, the Liliales. By sampling within and between closely related taxa, and in distant or outgroup species, we will begin learning how easy or difficult it may be to recognise yams at the taxonomic levels of species or genus, in archaeological samples.

In both situations described here, starch samples were obtained opportunistically from living collections that had been established for other purposes and before it was known that starch samples might be needed. Fortunately, the collections were large enough and diverse enough to allow useful sample sets to be chosen for specific historical or archaeological purposes. Both collections were short-lived. Living plant collections are maintained by many different kinds of people and organisations in many parts of the world, so there are potentially many ready-made opportunities for obtaining materials for starch reference collections—at universities, agricultural research stations, public botanical gardens, and in the private collections of plant enthusiasts. When reliable identifications and adequate information about sources are available, it makes sense to use living collections that already exist.

Although it is very desirable for ancient starch researchers to share their materials and data, it is also useful for different researchers to develop their own starch reference collections, for the particular purposes of their own archaeological research. Living collections of starchy food plants are seldom maintained for more than a few years or decades but they can be an invaluable source of reference material.

If it seems desirable to establish a new living collection, be realistic about the logistical and financial challenges. Suitable facilities and services (land, protected growing areas, pots, fertiliser, water, labour, and so on) should be available. Living collections require continuous care and manual effort for watering, replanting, labelling, and propagation. There may be expensive but necessary quarantine restrictions on the import and export of living plants from foreign sources (including quarantine restrictions for processed starch samples). In the case of a plant species or group that is already well represented in one or more living collections, it may be easier and cheaper for someone to take starch samples from an existing collection rather than from original field sources. Ideally, reciprocal help should be offered to the people who are maintaining a living plant collection, in the form of labour, financial support, living plant samples from your own area, and/or research reports and publications that have benefited from the collaboration.

Box 6.4. Compiling A Reference Collection For Studying Pleistocene Grinding Stones

Judith Field and Beth Gott

The Late Pleistocene archaeological site of Cuddie Springs in southeastern Australia has yielded grinding stones dated to around 30,000 years ago (Fullagar and Field 1997; Field and Fullagar 1998) (Figure 6.10, Plates 61, 63). The early appearance of what may be seed-grinding stones has a bearing on a number of issues in Australian prehistory. The archaeological questions relate to the timing of the introduction of seed-grinding technology and how it correlates to the occupation of the arid zone by people during the Late Pleistocene (see Smith 1985, 1988; Balme 1991; Gorecki et al. 1984). Prior to the Cuddie Springs discovery, seed grinding was thought to have been a development of the Late Holocene period. It was important therefore to execute a functional study of the artifacts, i.e., to document morphology, use-wear, and residues (cf. Chapter 9). The residue study revealed the preservation of starch granules and phytoliths on the used surfaces of some grinding stones (see Box 9.4). Identifying the source of the starch and the phytoliths was important for building a case for seed-grinding use and in turn identification of the earliest known seed-grinding implements. Having identified starch granules on stone tools from the Cuddie Springs site, a reference collection was required for identification.

Species Lists

The compilation of a list with the likely economic plants for the region was essential in order to narrow down the range of material that was required for the collection. Publications such as *Plants of Western New South Wales* (Cunningham et al. 1992) and plant lists held by the National Herbarium in Sydney provided useful details on modern plant distributions. Research in various explorers' journals (e.g., Mitchell 1848) can help build up a list of some of the grasses utilised by Aborigines at the time of European contact. A comprehensive database compiled by Beth Gott (2000) from ethnographic sources, published literature, and consultation with Aboriginal people provided additional lists of plants used in the past.

Survey and Collection

Vegetation surveys were conducted at the site of Cuddie Springs and the surrounding environs during a number of seasons (Figure 6.11). Varying the time of collection was important because the annual rainfall can change dramatically from year to year. During high rainfall years many native grasses are found, while the drought years see little or no undergrowth (Figure 6.10). The presence of both summer and winter grasses also required staggered collecting trips. The region around Cuddie Springs was settled by Europeans in the mid- to late 1800s, and during the early part of the twentieth century the land was subject to intensive overstocking of cattle and sheep. With changes in ownership of surrounding

Figure 6.10. Excavations in process in the lake bed at Cuddie Springs, Australia, during the dry season. *Photo by Judith Field*

Box 6.4 (continued)

Figure 6.11. Beth Gott collecting plants during a vegetation survey at Cuddie Springs. Note the essential penknife and plastic bag. *Photo by Judith Field*

Figure 6.12. Nardoo (*Marsilea drummondii*) plant from Cuddie Springs collected for the herbarium. Note the presence of the sporocarps that are important for identification to species in addition to the stolons and leaves; scale in cm. *Photo by Beth Gott*

properties, the collapse of the wool industry in the 1990s, and subsequent changes in stocking rate and conservative land use, a number of native grasses have reappeared in recent years. Each season provides the possibility of collecting a range of plants that have not been seen for some time.

A number of soil types and associated vegetation are found within the study area. The lake floor supports a different vegetation community to the semi-arid vegetation of the surrounding red soil plains. As a result, plants were sampled over a wide area. Specimens from economically important plants from more arid regions were also sampled in response to information from a paleoenvironmental study that showed the area was much more arid during the Late Pleistocene (Figure 6.12).

Plants were placed in plastic bags (for later identification) or pressed between pieces of newspaper in a plant press. Seeds and other specimens that needed drying were placed in paper bags or envelopes. Information concerning the location, date, collector, etc., was recorded for each specimen.

Sample Preparation

Slides of starch granules from seeds were prepared according to Beth Gott's method described in Box 6.5. Documentation of the starch granules involved image capture using a digital camera linked to a computer.

A Cautionary Tale

An important finding of the study was the apparent changes noted in the relative sizes of starch granules from Nardoo (*Marsilea drummondii*) from different locations. Starch granules from the climatically stressed Simpson Desert in western Queensland appeared to be smaller than those from a permanently wet location in Melbourne, Victoria (cf. Plates 61, 62). The size variation in granules with respect to environmental variation demonstrates that collection strategies should cover a range of climatic conditions, especially if present-day conditions are not reflected in the prehistoric record.

Leaves, stems, and flowers should be laid out between a number of sheets of newspaper and every few specimens interleaved with corrugated cardboard or aluminium to ensure adequate ventilation. The specimens should then be placed in a plant press made of any sturdy material that will not buckle or warp easily (e.g., metal shelves from a refrigerator or marine plywood) and then secured with straps or bolts through the wood with wing nuts to allow easy access to the press. The paper in the press should be changed daily for a few days and the press stored in a well-ventilated area to allow for quick drying. If left too long between changes, fungal growth may destroy the specimen. In humid climates the paper in plant presses should be changed regularly and the application of ethanol to retard unwanted biological activity is also useful (cf. Box 6.2).

Dried material is then mounted onto herbarium sheets made from sturdy white paper. All the relevant details concerning collection will then be recorded on a data sheet, such as the one in Figure 6.13, and/or on a label stuck to the lower right-hand corner of the herbarium sheet, as in Figure 6.7. When collecting voucher specimens (see below), check to ascertain if the herbarium requires their own labels to be included when depositing specimens with them. Cotton (1996: 115) recommends arranging the plants to display the maximum information possible: e.g., flowers and leaves should be spread to avoid overlap, with at least one leaf turned to expose the underside of the leaf. Additional fruits and flowers can be attached in paper packets on extra sheets.

DETERMINED NAME: ____________________

FIELD ID: ____________ **FAMILY:** ____________

____

FORM: tree sm tree shrub herbaceous
grass other ____________

HABIT: climbing prostrate
caespitose rhizomatous other ____________

DESCRIPTIVE NOTES: ____________________

____

ABUNDANCE: dominant abundant frequent occasional
rare ____________

ASSOC. VEG. open woodland open shrubland other ____________

CHARACTERISTIC SPP.: ____________________

TOPOGRAPHY: ____________________

SOIL COLOUR: MUNSELL ____________________

TEXTURE: ____________________

MAP SHEET: ____________ **SCALE: 1:250000** ____________

LAT: ____________ **LONG:** ____________

LOCALITY: ____________ kilometres at ________ from ________

____

COLLECTOR: **DATE:** ____________

NUMBER: ____________

POLLEN SAMPLE COLLECTED: ____________

FRUIT SAMPLE

Figure 6.13. An example of a herbarium label used in the John Ray Herbarium at the University of Sydney (courtesy Dr. M. Henwood). The labels are stored with dried voucher specimens and provide detailed information about the specimen, time of collection, and environmental conditions, all of which may be critical to understanding the development of the plant and condition of associated starch granules.

VOUCHER SPECIMENS

Voucher specimens, the permanent records, are the backbone of every reference collection. They are essential for any reference collection since identification cannot be effectively validated without one. A voucher is a plant specimen (usually dried) that demonstrates diagnostic features, which in most cases means stems, leaves, flowers, and fruits, although the key factors will vary from species to species. For example, in taro the tuber and an intact lead and stem axis are necessary for a confident identification to species. Date, location, environmental setting, and collector should also be noted. Hosking et al. (1996) argue that voucher specimens are collected to ensure a comprehensive list of species for each region, validate current identifications, update name changes as they occur, and monitor name changes that are not on existing databases.

Once prepared, the voucher specimen should be forwarded to a herbarium or botanic garden for verification. In some cases plant identification is not straightforward so it is important that an appropriate specialist be engaged to undertake the identifications of voucher specimens (possibly involving a financial outlay). After identification, the voucher specimen should be lodged with an appropriate institution where long-term maintenance of the material is assured. Pearsall (1989: 133) recommends that a minimum of three voucher specimens be prepared: one to be lodged at the relevant herbarium; one for the laboratory/field herbarium; and a third used to produce specimens (e.g., starch granules) for the archaeological reference collection. Sources of additional information about collection and preparation of voucher specimens is provided in Table 6.3.

PREPARING STARCH REFERENCE MATERIAL

Estimating Quantity

The first task in preparing starch derived from the plant reference collection is to estimate the quantity that will satisfy the research goals. Samples should be large enough to cover the variability in types and morphology present in the plant, but some research programs may demand additional and possibly larger samples. For instance, if characterisation using quantitative measurement of attributes is planned, then the reference material needs to have a sample size of at least 100 to 500 granules (e.g., Piperno and Holst 1998; Torrence et al. 2004). Chemical stains (cf. summaries in Chapter 7) also require relatively large quantities of starch. Furthermore, if one is planning to use a range of techniques, and especially if these involve destructive protocols (e.g., chemical tests, scanning electron microscopy), then multiple samples will need to be prepared. Since different plant types produce varying amounts of starch, it is useful to make a preliminary wet mount to estimate the density of starch to assist with planning how much material needs to be processed.

Slide Preparations

The methods presented in Box 6.5 illustrate the wide range of methods currently used for preparing starch for a reference collection. Variations in the methods depend on the type of material involved, available equipment, and the length of time the slides will be used. In temporary mounts the starch is gradually drawn to the edges of the slide each time it is rehydrated and so they become less useful over time. Some permanent mounts, however, have the disadvantage that the granules cannot be moved around. Unless granules are frozen in different views, this can limit studies of three-dimensional morphology. It is useful if the permanent mounts are all prepared in the same manner. A range of equipment necessary for preparing slides either for starch granule reference material or for viewing residues extracted from artifacts (cf. Chapter 9) is illustrated in Figure 6.14.

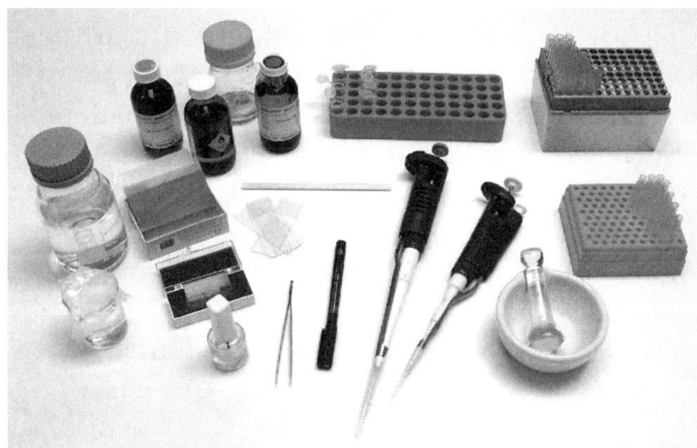

Figure 6.14. Equipment used to prepare slides for reference material or to view residues extracted from artifacts. Clockwise from bottom right: mortar and pestle; pipettes; pen for labelling slides; tweezers; clear nail polish; ultrapure water; slides; coverslips; mounting media; disposable stirrer; plastic vials; pipette tips. *Photo by Stuart Humphreys, Australian Museum*

Table 6.3. Information about preparing a herbarium

1. "Field Techniques used by the Missouri Botanical Garden" http://www.mobot.org/MOBOT/Research/Library/liesner/tpage.html
   The Missouri Botanical Garden Website provides a particularly comprehensive list of field gear.

2. "Preparation of Plant Specimens for Deposit as Herbarium Vouchers" University of Florida Herbarium http://www.flmnh.ufl.edu/natsci/herbarium/voucher.htm

3. "Preparing Herbarium Specimens of Vascular Plants," by C. Earle Smith, Research Notes, Ethnobotanical Leaflets http://www.siu.edu/~ebl/

Box 6.5. Methods For Preparing Reference Slides

Judith Field

I summarise a number of the methods most commonly used by a range of ancient starch analysts for mounting starch granules onto slides. Other protocols and mounting media are available and discussion with lab technicians or other researchers may provide other useful alternatives to those listed here. The choice of a technique depends on the need for a temporary or permanent mount, the type of material to be mounted, and the availability of equipment and mounting media.

Basic equipment required:

- razor blades or scalpel blades;
- metal spatula or blunt knife (e.g., bread and butter knife);
- glass microscope slides;
- coverslips (size 0);
- disposable pasteur pipettes and bulbs—and a sharps container for disposing of the waste pipettes;
- glass beakers or tubes to hold preparations as they are made;
- glass mortar and pestle for grinding up material, especially seeds (glass is non-porous so it can be cleaned effectively between uses, thereby avoiding contamination);
- sieves can be as simple as a plastic tea strainer made with a plastic mesh. (It is important to measure the diagonal width of the mesh opening to determine the mesh size. Otherwise an Endecott sieve of the appropriate size can be used. Determine the minimum size of the mesh required by looking at a wet preparation of your material and measuring the maximum size of starch grains. To exclude seed coats it may just be necessary to use a 100 μm mesh or larger);
- tap water is sufficient for the wet mounts. (Water should be checked first to ensure there is no contaminating starch; otherwise distilled water can be used);
- starch-free gloves are useful for handling archaeological material. (Care must be taken when buying gloves because often 'powderfree' gloves still contain starch. Gloves should be checked for starch by sampling them with water and mounting on a slide. See Box 8.7);
- mounting media (see Table 6.4);
- brightfield transmitted light microscope, ideally with a x40 objective lens, polarising filters, and an eyepiece micrometer to measure the starch grain diameters. (A 1 mm stage micrometer can be used to calibrate the units on the eyepiece micrometer. Access to a microscope with a trinocular head and camera system will allow documentation of the starch granules. See Chapter 3 for discussion of microscopy.)

Protocols for Seeds

Method A (Linda Scott Cummings)

1. Crush seeds in a glass mortar and pestle to a fine powder.
2. Mount a thin smear of material on a slide with glycerine jelly, as for root samples (see below).
3. Occasionally the seed coats need to be removed by wet sieving. It may be important to have some of the seed coat on the slide as these often appear in prehistoric samples.

Method B (Beth Gott)

1. Grind about one teaspoon of seeds into a coarse powder with a glass mortar and pestle.
2. Dry filter the ground material through a sieve (tea strainer with 0.4 mm mesh size) by pouring in the powder and lightly tapping the side of the sieve. Discard the residue in the sieve.

Box 6.5 (continued)

3. Grind again (to break up cells to obtain single granules and to eliminate seed coats). The result is a fine powder residue in the glass mortar and pestle.
4. Transfer a small amount to a slide and smear with a small metal spatula (or other flexible metal).
5. The aim now is to remove the larger particles. This can be achieved by holding the slide at an angle and tapping the slide on its side, and/or using a spatula with a gentle smearing movement. The smearing action also spreads the particles along the slide.
6. Mount with Karo™ syrup or Permount™—one large drop or enough that it will spread under the cover slip (size 0) to fill the space. Seal the coverslip with nail polish. We have found that the medium stays fluid for a number of years and the granules can still be tumbled for detailed examination.
7. Care is taken here to isolate the starch granules and eliminate any extraneous material. The aim is to provide a clean preparation to facilitate image analysis.

Protocols for Underground Storage Organs

Temporary wet mount

Method A (Peter J. Matthews, Gail Robertson, Tom Loy)

1. Sample freshly cut surfaces of the tuber by scraping with a toothpick.
2. Smear the residue onto three slides and allow to dry.
3. Place coverslip on slides and secure by placing a small drop of nail polish at four points around the edge.
4. Slowly add water to the mount by placing a pipette at the edge of the coverslip and adding water a drop at a time. Capillary action will draw the water under the coverslip.
5. For long-term storage of samples, Matthews recommends that small cubes of the tissue sample be cut and dried slowly in an open 1.5 ml snap-cap micro centrifuge tube (e.g., Eppendorf™) overnight in a ~36° C oven. These can then be stored frozen for long-term preservation as back-up samples.

Method B (Piperno and Holst 1998; Dolores Piperno)

1. Fresh samples are washed and peeled if necessary.
2. Tissue is macerated with a grater and placed in a large beaker (about 500 ml) of water.
3. Extract is filtered through cheesecloth with a mesh size large enough for the granules to pass through. (Larger portions of pulp are retained in the cloth.)
4. Extract is washed three times in water and then allowed to dry. Washing involves allowing the filtered extract to settle in a large beaker, then decanting the water. Particle size should be taken into consideration when allowing the sample to settle. A one-hour settling time will allow all particles >5 μm to settle. Care should also be taken at this stage to preclude atmospheric contamination with other starch particles. Perhaps covering the beaker with a watch glass or Parafilm™ may reduce contamination.
5. One drop is placed on a slide for analysis and then a coverslip is mounted with water. Water is used as the mounting media as it allows starch granules to be rotated by gentle pressure applied to the coverslip, which provides different views for measurement and recording purposes.
6. Coverslips are kept in place by clear nail polish.

Permanent mount

Method A (Beth Gott)

1. Using a sharp razor blade, cut a fresh surface on the tuber.

Box 6.5 (continued)

2. If tuber is large, take several cut surfaces across the section to be sampled.
3. Press the freshly cut surface onto a slide, allow to dry and mount in Karo™ or Permount™.
4. If mucilage is present, irrigate the slide with water and stir with a needle to mix starch with water. Dry in air at room temperature, preferably under a dust cover of some kind, then mount.

Method B (Gail Robertson, Tom Loy)

1. Sample tubers as for temporary wet mount (see above).
2. Allow to air dry on the slide (use dust cover).
3. Apply one drop of mountant (Permount™ or Euparol™ are recommended) onto slide and allow to spread.
4. Place coverslip on slide. The coverslip is applied by holding at an angle to the slide and allowing one edge to contact the mounting media. The coverslip is then slowly lowered onto the mounting media. This reduces the capture of air bubbles in the media. If air bubbles are trapped, then the coverslip can be gently tapped until the air bubbles move to the edge and escape.
5. Allow to dry overnight.

Method C (Huw Barton)

1. Scrape internal surfaces of tuber with a scalpel or razor blade.
2. Place residue into a snap-cap micro centrifuge tube (e.g., Eppendorf™) with 0.5 ml of (ultrapure) water.
3. Vortex for several seconds. Vortex machines violently agitate the tube to mix. If not available, then cover tube top with Parafilm™ and invert and shake the tube at the same time to thoroughly mix the sample.
4. Extract a 20 µl aliquot with a variable volume (e.g., Gilson™) pipette and place on a clean slide. If a variable volume pipette is unavailable, then use a pasteur pipette to draw up a very small volume.
5. Air dry and mount coverslip with Permount™ as described above or with glycerin (a semi-permanent mountant). When using glycerin the coverslip must be well sealed around the edges with clear nail polish and slides stored flat, as this media will continue to flow for some time. I have also seen good preparations made using clear nail polish as the mountant. However, once this dries, the sample will be immobile.

Method D (Linda Scott Cummings)

1. Slice the root with a razor blade or scalpel to expose the interior.
2. Take a blunt knife blade and scrape starches from the internal surfaces.
3. Mount on a microscope slide and add a couple of drops of glycerine jelly for a permanent mount. After approximately one month, seal the edges of the coverslip with clear nail polish. The slide has to be clean and dry so the polish will adhere to the surface. A thin line of polish around the edge of the coverslip is all that is required.

Method E (Michael Therin, Robin Torrence, and Richard Fullagar)

The protocol described here was devised for collecting and sampling plants in tropical climates.

1. In the field, plant samples are placed in 70 percent solution of methylated spirits after collection.
2. In the laboratory, samples are removed from the preserving solution and dried in an oven at or below 35° C.
3. Small samples are taken from each plant part collected.
4. Dried samples are ground with a glass mortar and pestle into a fine powder.
5. Glass microscope slides are then dusted with a fine coating of the ground sample.
6. Slides are mounted with Euparol™ and stored flat in wooden boxes (Figure 6.8).

Table 6.4. Mounting media and their refractive indices.

Mounting Media	Refractive Index	Notes
Water	1.333	Non toxic, temporary.
Karo™ (corn) syrup	1.484-1.486	Water based, non-toxic requires sealing; grains can be rotated; discolours Toluidine Blue stains.
Glycerol™, 50% water	1.397	Water soluble.
Aquamount™ (BDH Chemicals)	1.400 (other ri available)	Water based, nontoxic, tolerates some water in the sample.
Glycerine Jelly	1.440	Water based, nontoxic; slide requires sealing.
Glycerol™, pure	1.475	Water soluble.
Euparol™ (Carolina Biological Supply)	1.485	Solvent based, permanent; soluble in alcohol.
Numount (Brunel Micro)	1.500	Canada Balsam substitute.
DPX (ProSciTech)	1.500	Permanent, low toxicity, solvent based; soluble in xylene and toluene.
DEPEX (Electron Microscopy Sciences)	1.524	Replacement for Canada Balsam; soluble in xylene and toluene.
Permount™ (Fisher Scientific)	1.525	Toluene based, toxic, permanent; grains can be rotated for a few days before setting.
Wintergreen oil (methyl salicylate)	1.535	Natural product of *Gaultheria fragrantissima*; soluble in alcohol.
Canada Balsam (Fisher Scientific, Sigma)	1.541-1.547	Natural product from resin of *Abies balsame*; soluble in xylene and toluene. Rarely used now because of expense. Hardens and yellows with age.
Naphrax (Brunel Micro)	1.730	Diatom mountant; soluble in toluene.

Note: If a mountant takes more than a few hours to solidify (e.g., Permount™), then it is important to store the slides flat, rather than on their sides as the starch granules may sink to one edge of the slide, thereby making viewing and analysis problematic.

Box 6.6. Procedure For Making Hand Sections

Judith Field

1. Hold the sample in one hand. If it is very small, then place it between two pieces of polystyrene foam. The foam may also be cut during this process.
2. Draw the razor blade towards you across the area to be sectioned in a slow, smooth, sweeping action. Try to take the thinnest section possible.
3. Draw the razor blade plus section through a water-filled petri-dish or saucer to allow the section (plus foam) to float off.
4. Once you have accumulated a few sections, pick up the best and thinnest.
5. Use a small brush to lift the section(s) onto a bubble of water on a slide. Be careful because the section can get caught in the hairs of the brush and be damaged or lost. Some of the water can be drawn off using a piece of torn filter paper applied to the edge of the bubble. This will allow for quicker drying of the slide.
6. When the section has dried, mount with a coverslip either in the water or by adding a water-tolerant mounting media like Karo™ (corn) syrup.

It is very important to avoid methods that involve heating the slide. In laboratory experiments I have found that starch can be modified by heat on the slide (cf. Box 8.5). The starch granules may lose some of their distinctive surface morphology, the extinction cross may lose its clarity, and birefringence may also be affected (cf. Chapter 2).

Preparation of Sections

In addition to reference slides, thin sections are also useful because they provide clear evidence of how the starch granules are stored and reveal differences in nature and density of packing (as shown in Plates 19, 20). These may also be stored at the herbarium with the accompanying descriptions and collection information. Hand sections are usually prepared using a double-sided razor blade (broken in half). Injector razor blades are also very good as they provide a stable and very sharp edge to work with. If the amount of material available is small, then it can be wedged between two pieces of polystyrene foam to provide some stability during sectioning. Larger pieces can be hand held easily. A description of this technique is provided in Box 6.6. Preparing good hand sections is a skill, and it is advisable to practice beforehand with unimportant, readily available material. Sections end up being wedge shaped with the thinnest edges providing the information required for identification. The section must be as thin as possible (<1 mm) so they can be mounted on slides and coverslipped with a permanent mounting media.

Mounting Media and Refractive Index

For viewing starch granules, a mounting media that has a different refractive index from starch—i.e., higher or lower than 1.530 (Fredricksson 1998)—should be chosen. The range of different mounting media that have been used by researchers is listed in Table 6.4 (also cf. Coil et al. 2003: 1000–1). Selection of one of these will depend on personal preferences and availability.

DATABASE COMPILATION

Once the reference collection has been compiled, then relevant information about the starch can be documented. As needs and research questions will vary, the complexity with which the documentation is put together will also be different from project to project. A hard copy of the photos of the starch granules (with scale bars) may be a handy ready-reference when sitting at a microscope. In addition, the associated files can be sent electronically to colleagues when comparing notes.

CONCLUSIONS

A good reference collection is central to making confident identifications of starch recovered from archaeological contexts. Before constructing the collection, the size and scale should be defined by careful consideration of a range of factors. Foremost is the potential range of plant materials appropriate to the research questions being posed. While reference to published sources will provide a certain amount of information relevant to the study, collecting and preparing specialised reference material will facilitate firsthand observations of the variability of starch production and morphology within and between plant species.

CHAPTER 7

DESCRIPTION, CLASSIFICATION, AND IDENTIFICATION

Robin Torrence

The basic morphological features and chemical properties of starch granules have been studied by botanists and chemists for over a century, so it is somewhat surprising that the assignment of starch granules in archaeological contexts to their parent plant taxa is a relatively young technique. In developing methods for the description, classification, and identification of ancient starch, archaeologists benefit from the very extensive literature on starch granule morphology in economic and other plant taxa (e.g., Reichert 1913, 1919; Seidemann 1966; Jane et al. 1994) and the keys that are being drawn up for archaeological research (Piperno and Holst 1998, 2004; Piperno et al. 2000, 2004; Piperno in press; Pearsall et al. 2004; Perry 2001, 2002a, 2004, in press). Botanists have long realised that the morphology of starch granules can be diagnostic of individual plant genera and even species. Likewise, the early analytical approaches to the microscopic characterisation of powdered foods and drugs, which was a concern for chemists and pharmacologists during the rapid industrial growth of the early twentieth century, demonstrated that in combination with other cellular components starch granules could be used routinely to identify plant taxa (cf. Greenish and Collin 1904; Wallis 1957; cf. Flint 1994: 1–4). In many cases routine, day-to-day identifications can proceed with a light microscope equipped with cross-polarised light (cf. Chapter 3). The use of scanning electron microscopy and various chemical and physico-chemical techniques may add extra precision (e.g., Hall and Sayre 1973; Fitt and Snyder 1984; Huber and BeMiller 1997; Jane et al. 1994; Cortella and Pochettino 1994; Ugent 1994; Chapter 3). In Box 7.1 Ugent provides a useful survey of previous systems for starch classification. The keys he describes can be found at his website (www.siu.edu/~ebl/amylose.htm). In addition, detailed descriptions of many individual species have been published in the journal *Starch/Stärke*.

Before attempting identification, researchers need to compile a reference collection containing all the relevant species for their research question, as discussed in Chapter 6, and invest sufficient time to gain familiarity with the morphologies expressed. As part of this process one needs to be well acquainted with the various features of starch granule morphologies in general as well as the basic features of the starches derived from various organs of the plants in the comparative collection. Some research questions will require knowledge of more plant types than others. For example, artifact residue samples or soil samples from agricultural contexts may be dominated by starch granules from a few domesticated taxa, while environmental soil samples may be characterized by a wide array of wild plant starches.

Augmenting reference material with published descriptions of the morphological features of starch granules provides a powerful tool for confirming identification. Consistency is not a problem. Descriptions and measurements of starch granules derived from tropical plant taxa from Central America and northern South America are concordant between research collections recently described by archaeologists at the Smithsonian Tropical Institute and Southern Illinois University, Carbondale, and also with the same or related taxa recorded by other investigators (cf. Piperno in press; Piperno and Holst 1998, 2004; Piperno et al. 2000; Perry 2001, 2002a, 2004, in press; Pearsall et al. 2004; Reichert 1913: Appendix I). Because the plant specimens were collected from different regions of the globe over a span of ninety years for modern specimens and 8,000 years for archaeological samples, the evidence indicates that taxon-specific starch granule morphology for these species is stable across temporal, geographic, and environmental zones. This phenomenon also indicates that basic starch granule morphology is readily observed and described in the same ways

Box 7.1. History Of Starch Classification

Donald Ugent

Modern studies of ancient starch would benefit from studying previous approaches to classification before embarking on new keys. The great Dutch microscopist Anton Leeuwenhoek (1719) was probably the first person to publish illustrations of starch granules. Following his work, a number of investigators conducted experiments that were aimed primarily at revealing the internal composition of starch granules, but these physiologists were not concerned with the morphological differences between plant species. Fritzsche (1834) was possibly the first to consider the form of the starch granule as an aid to identification. He described the starches obtained from a variety of plants and noted variation among the granules obtained from different sources. In some cases the starch granule form was characteristic enough to determine the species of the plant, whereas in other cases only the genus or family was obtained.

While Fritzsche left us with some valuable illustrations and descriptions of the granules of different starch species, the work of producing the first taxonomic key employing starch granules as its basis fell to Schleiden (1849). Schleiden's key classifies all starch granules into three major categories: simple, compound, or amorphous. Simple and compound granules are further subdivided according to their shape (round, lenticular, elongated, etc.) and the position of their central cavity or hilum. All told, some 22 starch types are classified by Schleiden and over two dozen species and plant families are cited by him as examples.

An important monograph by Carl Nägeli (1858) dealt with the morphology, physiology, chemistry, and systematics of starch. Nägeli's key (Table 7.1) basically uses Schleiden's divisions, but has some significant changes. For example, he relegates the amorphous condition to a secondary rank consistent with that of starch granule shape and uses the ranks of simple versus semi-compound and compound as his primary key divisions. Moreover, in Nägeli's revision the number of recognizable starch types is reduced to 17. Due to its extensive coverage of starch plant species (both cultivated and wild), Nägeli's monograph could still be used as a basis for further studies.

The next important classification scheme, by Meyer (1895), was based upon the initial development of the starch granule rather than the division between 'simple' or 'compound.' Meyer rejects the earlier concept of compound granules on the basis that there are no known cases of granules being formed through the separation of an original single granule. Instead, he refers to these types as 'complex' starch granules. Although Meyer's physiological classification (based on the actual mode of development of the starch granule) would appear at first glance to give us a more legitimate basis for the recognition of different types of starch granules, his system proved, in reality, to be too difficult to use on a practical basis. For that reason, later scientists largely abandoned it in favor of the one devised by Nägeli.

Following Meyer, there were several other attempts to categorize starch granules, but none were as thorough or as sophisticated as their predecessors'. Muter, for example, who based his classification upon the histological and polariscopic peculiarities of starch (Leffman and Beam 1905: 88), presents us with a scheme in which each major group is designated by an important starch plant species. In contrast, Kraemer's (1907) classification into six major groups was based on whether the granules were simple or compound combined with their shape or texture, followed by size. He also noted that, as a rule, leaves, herbs, and flowers do not contain starch. Winton (1906) produced a typology that depended heavily upon the shape of the starch granule. He wrote that the forms of

Box 7.1 (continued)

Table 7.1. Nägeli's (1858) system of classification for starch granules.

A. Granules Simple.

I. Centric. Hilum in the mathematical center; lamellae always equal at two corresponding diametrically opposite points.

Type 1. Spherical. When the granule is free, both hilum and granule are spherical (*Zea mays*).

Type 2. Lenticular. When the granule is free, both hilum and granule are rounded; granules compressed; sometimes circular or ovoid; sometimes triangular or quadrangular (*Secale*).

Type 3. Oval. When the granule is free, both hilum and granule are oval to lanceolate-oval; occasionally kidney-shaped or somewhat curved; when on end they appear circular or somewhat compressed (*Phaseolus* and *Doliches*).

Type 4. Spindle-shaped. Granule linear or lanceolate, tapering towards the pointed ends, or of equal width with blunt ends; when on end they appear almost circular (*Euphorbia*).

Type 5. Bone-shaped. Granule elongated and compressed from the narrow aspect, but linear spindle-shaped from the broad aspect, with enlarged laminated ends (*Euphorbia*).

II. Eccentric. Hilum usually more or less removed from the mathematical center of the granule; lamellae coarsest and finest at opposite ends of the granule, respectively.

Type 6. Inverted cone-shaped. Granule on end almost circular; slender at the hilum end (*Solanum tuberosum*).

Type 7. Cone-shaped. Granules on end almost circular; decidedly thicker and broader at the hilum end (*Scilla peruviana*).

Type 8. Wedge-shaped or compressed. Granules flattened, of equal thickness throughout, or thicker but narrower at the hilum end than at the distal end (*Canna edulis*).

Type 9. Rod-shaped (*Iris florentina*, *I. pallida*).

III. Granules simple and structure obscure.

Type 10. Structure not fully developed, or not identified owing to diminutive size of the granules. Lamellae, hila, cavities, fissures, and clefts seldom observed (*Narcissus poeticus*).

B. Granules Semi-compound.

Type 11. Granules semi-compound. The component part-granules are enveloped partly or wholly by a common substance (*Hyacinthus orientalis*).

C. Granules Compound. The component part-granules are not enveloped by a common substance.

I. Composed of fused part-granules.

Type 12. Composed of fused part-granules. The part-granules are not separated by fissures, and even different granules may be fused with one another (*Zingiber officinale*).

II. Composed of separated part-granules. The part-granules separated by fissures.

Type 13. Granules in 1 or 2 rows. From 3 to 11 components arranged in 1 or 2 rows (*Polygonum fagopyrum*).

Type 14. Equally divided granules of few components. From 2 to 10 or more almost equal sized part-granules which when separated have 1 curved surface and 1 or more pressure facets (*Manihot* and *Batatas edulis*).

Type 15. Unequally divided granules of few components. From 2 to 10 or more unequal sized firmly united part-granules, which when separated have 1 curved surface and several flattened pressure facets (*Aconitum napellus*).

Type 16. Multiple granules. From 20 to many thousand firmly united part-granules which when separated are covered with pressure facets (*Oryza sativa*).

Type 17. Hollow spherical granules. The part-granules are arranged in a spherical layer, as if a globular shell had been divided radially (Algae).

Box 7.1 (continued)

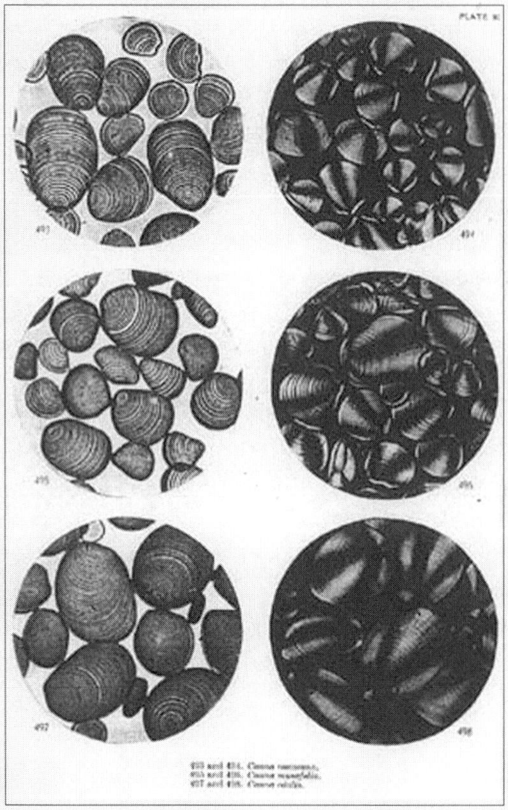

Figure 7.1. A sample plate of *Canna* sp. starch from Reichert's atlas of starch granules. *Reichert 1913: Plate 83*

the granules are so numerous, even within the same variety, that accurate classification was not possible, but six major shape categories were highlighted.

The next milestone in the history of starch taxonomy was the publication of Edward Tyson Reichert's (1913) monograph on 'The Differentiation and Specificity of Starches in Relation to Genera, Species, Etc.' Reichert offers a splendid review of the past literature in this field and presents a remarkable series of 100 photographic plates illustrating the starch of over 350 wild and cultivated species. He includes two photomicrographs of each plant type showing their appearance in ordinary and polarised light (Figure 7.1). Reichert also presented the results of a number of optical and chemical dye tests that were conducted on these species (Table 7.2), as well as whatever was known at the time of their morphometric properties. While Nägeli's system of classification is used throughout the text, Reichert also includes a chapter that details the essentials of the other important systems of starch classification.

Reichert was able to distinguish between the various starch varieties and species he tested through a combination of analytical tests. Although normally fewer than four tests are needed for the identification of any unknown sample of starch, he included eleven in his comprehensive work: (1) microscopic examination to determine the shape and size of the granules, position of the hila, characteristics of the lamellae, orientation, etc.; (2) degree of polarisation; (3) iodine reaction; (4) gentian violet reaction; (5) safranin reaction; (6) chloral hydrate-iodine reaction; (7) chromic acid reaction; (8) pyrogallic acid reaction; (9) ferric chloride reaction; (10) Purdy solution reaction; and (11) temperature of gelatinisation. Through the use of these methods, Reichert was able to ascertain the identity of an unknown sample of starch down to the species level, and, in some cases (as in maize), the technique could yield the specific botanical variety.

Reichert's (1913) work was reviewed and revised by James B. McNair (1930) who rearranged the data of Reichert's charts to best allow for the ready identification of unknown starch materials. The results of these tests for some 350 starch plants are given along with helpful information on the preparation of starch samples for analytical testing. (McNair's original article has been reprinted by the author as a chapter in his 1965 privately printed book, *Studies in Plant Chemistry*.)

Future attempts to make keys for identifying the botanical source of ancient starch granules would benefit from studying previous work since the starch granules from many plant types have been described in these classification schemes. Furthermore, these works present characteristics that could be used to discriminate among various species as well as useful information about chemical tests.

by multiple researchers working in different disciplines. Clearly, published descriptions and photographs in combination with data derived from a reference collection provide the basis for reliable identifications.

CHEMICAL AND PHYSICAL TESTS

There are two basic methodological approaches for identifying starch. These can be used separately or in combination. The first involves diagnostic tests of chemical and physical properties of starches, whereas the second uses the optical properties of the granules. By using a combination of histological studies with chemical and physical tests, Reichert (1913, 1919; cf. McNair 1930, 1965; Hall et al. 1989) was able to characterize a large number of plant species and in the case of maize recognize specific botanical varieties (cf. Box 7.1). Lamb and Loy (2005) discuss the use of Congo Red dye for identifying starch granules damaged by cooking or milling. A range of potential chemical and physiochemical tests are listed in Table 7.2. Industrial starch chemists typically use tests of the following characteristics, which are relevant to their commercial uses, but it is not clear that these are necessarily useful for the kinds of samples available to archaeologists: x-ray diffraction; amylose content; solubility; swelling power; susceptibility to α-amylose attack; water binding capacity; and gelatinisation temperature. Many of these techniques result in the destruction of the morphological integrity of the starch granules, thus rendering them much less useful for further study. A further limitation is that most of the modern tests used by researchers in the starch industry require very specialised and expensive equipment that is not widely accessible. A number of recent techniques developed within the field of chemosystematics, such as thin layer chromatography, may in the long run also be applicable to the classification and identification of starch granules (e.g., Ugent 1994).

Importantly, tests that measure specific chemical and physical properties that can be very useful in situations where there are large-size samples such as archaeobotanical macroremains (cf. Box 1.4) have much more restricted applications in paleoenvironmental research and archaeological studies of soil samples and artifact residues where the focus is on a smaller number of individual granules. Moreover, in many cases, these relatively small samples are quite adequate for concrete

Box 7.1 (continued)

Table 7.2. Diagnostic tests used by Reichert (1913, 1919).

Intensity of Polarisation

Iodine Reactions
- Lugol's Solution
- Iodine Vapor[1]
- Iodine-Chloral Hydrate[1]
- Iodine-Latic Acid[1]

Aniline Reactions
- Safranin
- Gentian Violet

Gelatinisation Temperatures

Media used include:

Water, Chloral Hydrate-Iodine, Chromic Acid, Pyrogallic Acid, Ferric Chloride, Purdy's Solution, Nitric acid, Sulphuric acid, Hydrochloric acid, Potassium hydroxide, Potassium iodide, Potassium sulphocyanate, Potassium sulphide, Sodium hydroxide, Sodium sulphide, Sodium salicylate, Calcium nitrate, Uranium nitrate, Strontium nitrate, Cobalt nitrate, Copper nitrate, Cupric chloride, Barium chloride, Mercuric chloride[2]

1. Previous results from these tests are reviewed by Reichert (1913: 170–71), but they were not used by him.
2. Results from the first five reagents are listed in Reichert (1913: 324–33; cf. McNair 1930) whereas results from the full set are provided in Reichert (1919: 177–207).

identifications using techniques that are neither destructive nor expensive. For the purposes of designing methodology for archaeological and paleoenvironmental research concentrating on native starch, this chapter focusses on traits that can be viewed with a binocular compound light microscope (cf. Chapter 3). More sophisticated tests that are particularly relevant for studying processed starch are discussed in Chapter 10.

RESEARCH DESIGN

The nature of the morphological description or the level of identification required for archaeological starches is partially determined by the research goals. Not every research question will demand the same amount of detail or accuracy in identification and this will affect the methodology employed. To illustrate this point, we describe four widely shared archaeological research goals and show how each one uses a slightly different approach to identification. These are arranged in order from the least to the most specific level of classification.

Level I: Presence/Absence

The most general type of subject that can be addressed by starch analysis is whether evidence for the exploitation of plants is present or not. For instance, the role of plants in early human diets is often debated, especially in comparison to the relative status of fauna. A contribution to this debate could be made by collecting and identifying starch of economically useful plants within sediments at a specific site or, more convincingly, as residues on artifacts. A related question is whether or not stone tools were used to process plants. Artifacts encrusted with plant remains will be especially important in cases where the relationship between hunting and gathering within a subsistence regime is at issue, or if one wishes to link tool use to gender roles—perhaps having made the common inference that plant exploitation in hunter-gatherer societies is more likely to have been carried out by females (e.g., Binford and Binford 1969; Gero 1991). Finally, one might want to test whether a specific feature or space within a site was used for plant processing or storage (e.g., Haslam 1999). The data necessary for addressing the research question in each of these examples is simply evidence that starch is present in soils or on the surfaces of artifacts within given contexts. All that is required to identify artifact residues as starchy, therefore, are general criteria that define starch and to distinguish starch from other microfossils.

The most commonly used chemical test to verify the presence of starch is to expose the sample to iodine-potassium-iodide (IKI), as described in Box 7.2 (see also Johansen 1940; Hall et al. 1989: 146–47; Plate 24). One needs to use extreme caution with this test because starch can stain various shades ranging from nearly black to yellow-brown (cf. Reichert 1913, 1919). In general, however, when treated with IKI, amylose becomes a deep blue color and amylopectin is blue-violet or purple (Hall et al. 1989: 146; cf. Chapter 2). Table 7.3 summarizes the different reactions between iodine and Trypan blue-stained starches as compared to other organic substances.

The iodine-potassium-iodide test has been used successfully on small amounts of starch adhering to stone tools by Briuer (1976: 482), Shafer and Holloway (1979), Loy et al. (1992), and Barton and White (1993). It should be noted that treating starch with IKI will destroy some characteristics of the granules including the extinction cross, and internal features of the granules such as fissuring will be obscured by the dark coloration. Thus, if future study of archaeological specimens might include taxonomic identifications, some samples should be reserved so that the morphological features of the granules remain intact.

For paleoenvironmental and archaeological research, optical methods are more easily applied. The most commonly employed criterion for identifying the presence of starch under a microscope is the presence of the extinction cross when viewed in polarised light (e.g., Plates 10–12, 17–20). It is important to note that the 'Maltese' cross pattern in starch granules rotates when the polariser is turned (cf. Chapters 2, 3). All unmodified starch granules will exhibit this property, although the strength of the cross can vary. There are a few other substances that have some of these same characteristics, but they can generally be discriminated from starch granules by other physical characteristics, such as lack of a rotating cross, the presence of a rainbow effect under polarised light, and the overall shape of the object. Box 7.3 describes some of the objects that might be confused with starch

Box 7.2. Iodine-Potassium-Iodide Test For Starch

Tom Loy

In conjunction with observation of its optical properties, and especially the extinction cross, the change in colour following the reaction of starch with iodine-potassium-iodide (IKI) is diagnostic of starch. Amylose combines with a weak solution of iodine-potassium-iodide to form a generally deep blue to black colour. In contrast, in the amylopectin reaction the colours range from pink to blue to light violet. The presence of waxes, or dense hydrophobic outer layers, can act to slow or reduce the intensity of the reaction. Intensity of the staining is, of course, also dependent upon the concentration of the IKI solution: generally the weaker the solution, the lighter the staining. Of diagnostic significance is the fact that iodine-stained starch will lose the stained colour (in moderate to lightly stained intensity) when heated and then regain the colour upon cooling (Johansen 1940). Although iodine reacts with many other organic compounds, the immediate staining of starch granules by IKI, especially when applied directly to a starch granule being observed on the microscope stage, is considered diagnostic of starch.

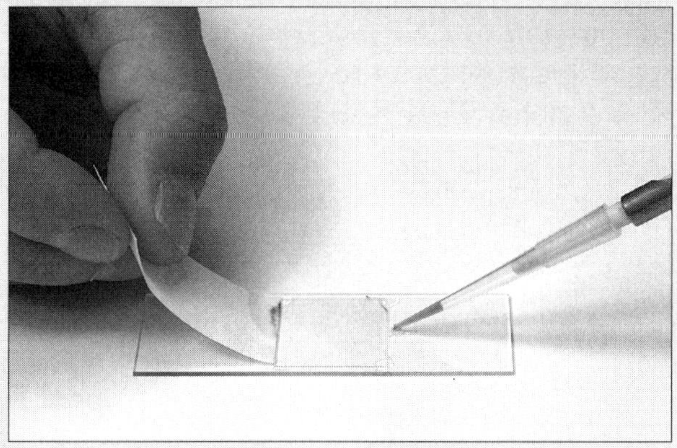

Figure 7.2. Iodine-potassium-iodide test for starch. Placement of a piece of filter paper against the cover slide causes the IKI solution to migrate across the sample. *Photo by Stuart Humphreys, Australian Museum*

Method

1. Make up a weak (1%) solution of IKI. Use 0.3 g crystalline iodine and 1.5 g potassium iodide made up to a final volume of 100 ml in ultrapure water.

   NOTE: iodine is a toxic chemical and the crystalline form sublimes to produce toxic or, in low concentration, irritant fumes. Preparation of the IKI stain must be done in a fume hood with suitable laboratory precautions. Avoid skin contact with the IKI stain.
2. Cut a small square of filter paper to the width of the coverslip. (Whatman 3MM chromatographic is sufficiently bibulous.)
3. Apply a small amount of IKI solution (~20 to 40 μl) along one edge of the coverslip of previously prepared sample (cf. Box 6.5).
4. Place the microscope slide on the viewing stage. Position and focus to observe a number of starch granules and confirm their birefringent properties.
5. Place the edge of the filter paper with the IKI solution against the opposite edge of the coverslip as shown in Figure 7.2.
6. Migration of the IKI solution across the coverslip can be monitored visually because the solution has a yellow-brown colour. The colour reaction takes place immediately when there is contact of the IKI solution with the starch.
7. The filter paper should be removed before it soaks up all the liquid under the coverslip.
8. The IKI solution can be removed with a fresh piece of filter paper by applying ultrapure water to the coverslip edge as in the original staining procedure.

   Colours ranging from pink to blue-violet and dark blue to black will be observed on true starch granules depending upon the strength of the IKI solution and the

granules and discusses how starch granules can be discriminated from them.

An important exception to the general rule are starch granules that have been gelantinised through heating (generally in the presence of water) because they do not exhibit the extinction cross (cf. discussion in Chapter 2) and may lose their morphological integrity (e.g., Plates 57, 59). As noted in Reichart (1913: 308, 334) and Gough and Pybus (1971), however, gelatinisation is a gradual process that does not affect all the granules at the same time or to the same degree. In many cases not all the granules will have undergone total gelatinisation and so it is likely that some granules with defining features will be present in a sample. It is also likely that some granules will have been protected by their position within plant tissue. It is worth noting that the IKI and Trypan blue tests are

Table 7.3. Comparisons of reactions between iodine- and Trypan blue-stained organic material (based on Flint 1994: Tables 6.2 and 6.3).

I. Aqueous iodine-stained starches, dextrins, and proteins viewed in brightfield and between crossed polars.

Material	**Reaction**
Raw, amylose-containing starches	Blue, birefringent
Raw, amylopectin starches	Reddish, birefringent
Chemically modified starches	Yellow or brown, birefringent
Pre-gelatinised, i.e., precooked starch powders	Reddish, swollen particles, which may contain individual blue-stained granules (not birefringent)
Dextrins	Blue-purple or reddish particles which quickly disperse
Proteins in cryosections and powders	Yellow

II. Trypan blue-stained starches, protein, and cellulose.

Material	**Reaction**
Damaged starch granules	Blue (no birefringence)
Starch which has lost granular structure, e.g., that present in wafers and extruded products	Pale blue
Intact starch	Unstained, birefringent
Protein	Blue
Lignified cellulose	Dark blue
Cellulose	Pale blue

Box 7.2 (continued)

properties of the starch granules (Plate 24). The IKI reaction destroys the extinction cross effect and this can be used as part of the confirmatory evidence.

To Monitor Colour Loss Following IKI Staining

The effect of loss of iodine colour upon heating is somewhat difficult to control and requires a well-organized microscopy bench and some practice.

1. Following the IKI treatment the slide is removed from the stage, placed on a hotplate, and gently, but rapidly, warmed to less than 100° C. The mount should not be allowed to dry during this step.
2. Immediately replace the slide on the stage and locate previously observed granules. (Use the stage micrometer coordinates you carefully recorded in the previous test.)
3. As the slide cools, the starch granules will regain their colour. If the granules develop a very dark, opaque colour, then this reaction should be repeated on fresh granules using a dilution of the 1 percent IKI solution.

Box 7.3. Optical Properties Of Potential Look-Alikes

Tom Loy

Starch granules can readily be identified by their reaction to polarised light: i.e., the presence of the extinction cross which rotates (cf. Plates 10–12, 17–20). Nevertheless, there are some common natural substances that have similar properties and might be mistakenly identified as starch granules, especially when observed under incident light: e.g., coprolitic spherulites; some spores; bordered pits and fragments of spiral wall thickening; some fossil pollen granules; coccoliths; and some diatoms (Plates 45–51). Many of these look-alikes are in the size range of common starches and can produce some form of the birefringence extinction cross and its rotation. Distinctive differences between these potentially misleading substances and starch are described below. The use of chemical tests, such as iodine-potassium-iodide (IKI), which is described in Box 7.2, can also be used to discriminate between starch granules and other materials.

- Faecal spherulites (Plate 45), which are composed of calcium carbonate, are derived from the dung of a number of animals and have been recovered in archaeological sites (Canti 1998, 1999). They are generally spherical, clear or white, and range in diameter from 5 to 15 microns, but may range up to 25 microns. Their mode of formation yields a highly regular quasi-crystalline structure, and because they are spherical, they can radially polarise incident light giving rise to the cross effect. The cross-polarising effect is uniaxial interference and so the cross does not rotate as the polariser is turned. When a range of accessory lenses is used in combination with cross polarisers, the normally white areas between the arms of the cross become particular colors (Canti 1998: 439) and this will discriminate them from starch granules.
- Crystalline spherulites in volcanic glass will show a weak birefringence cross. These spherulites are feldspar inclusions that form in some glasses and although fully contained in the glass, they could be mistaken for surface features along the edge of a very transparent and thin obsidian flake (Kerr 1959: 420–23). A few naturally occurring minerals can form oolites, or massed spherulites, including calcite in organic systems, dahllite, and chamosite. There are distinctive optical properties associated with each of these minerals which allow their identification (Kerr 1959). These mineral spherulites would be very uncommon within artifact or ceramic residues, but might be observed in soil samples.
- Spores have been observed which give a weak birefringence extinction cross (Plate 47). The cross is difficult to observe, however, and it does not change when the polariser or analyser is rotated. Spores usually range in colour from black through brown to shades of amber to yellow. In the lighter spores it is sometimes possible to see internal structures. Spores can vary in shape from spherical to ellipsoidal and usually lack microscopic surface features. In many cases the very smooth surface will reflect an image of the darkfield apparatus to the observer. One form of fungal spores called conidia has been dubbed 'string of pearls' (Plate 48). This fungal growth creates small (<2 microns) white spheres connected by a very thin tube (c. 0.1 micron or less) and is usually only found in the soil impinging on the surface of an artifact. Further details and excellent illustrations of fungal spores that might be confused with starch are presented in Haslam (2005).

an effective way to detect gelantinised starch (Reichart 1913: 171; Table 7.3).

A good example of the archaeological relevance of simply identifying the presence or absence of starch is provided by Williamson's analysis of stone tools from the Middle Stone Age site of Rose Cottage Cave in South Africa (Box 7.4). She found that starch residues on stone tools were more common than blood, thus suggesting that plant processing was a primary activity at the site. Her results support other evidence from the excavation that the cave was used as a home base rather than a temporary camp focusing on animal butchery. In a second case Fullagar's (1992) analysis of tool residues from the site of Bitokara Mission in Papua New Guinea demonstrates that the presence of starch is a key variable when the relative incidence of plant versus animal resources is a key issue.

Level 2: Assemblage Variation

A second category of research question benefits from the study of assemblage variation. In many cases one can identify morphological variation within the population of starch granules, but not be able to link differences to specific taxa. Although identification is always desirable, a great deal can still be achieved through the simple recognition of similarities and differences. The best examples are cases in which one is testing hypotheses that predict environmental change, differences in the use of the site through time, or alterations in the nature of artifact use. In these cases one can classify the starch into unnamed groups according to one or more criteria and then compare the assemblages on the basis of the differential representation of the groups. In most of the studies to date, the primary variables used are shape (as a

Box 7.3 (continued)

- Cellulose, due to its very regular molecular structure, will react to polarised incident light in characteristic ways (Plate 50). In incident cross-polarised light (where the field of view is dark to black) cellulose will shine brightly because its molecular structure is altering the vibration plane of the reflected light and allowing it to pass through the analyser plate. Occasionally small structures within plant cells can mimic starch by exhibiting an extinction cross: e.g., sieve plates and bordered pits in vessel walls and fragments of spiral wall thickenings (see Raven et al. 1999). In bordered pits especially, the arms of the extinction cross will move upon rotation of the polariser or analyser (Plate 49). Nevertheless, bordered pits are not solid spheres but toroid shapes with a hole or very thin membrane in the centre (like a doughnut) and therefore the cross will not be complete. Bordered pits may be mistaken for partially gelatinised granules which also show only the outer ends of the extinction cross.
- Occasionally the partially fractured exines of pollen granules can give a weak and only partial birefringence interference cross. It is relatively easy to identify this source of potential misidentification given the variety of surface features and overall shape and size of pollen granules.
- Coccoliths (Plate 51) and some diatoms might be confused with small starch granules. Coccoliths are formed on the outside of unicellular plankton (Coccolithophores) as protection from herbivorous plankton (Brasier 1980: 45–50). Their size (c. 20 microns) and spherical to elliptical shape could cause some confusion with starch: however, they are associated with deep ocean habitats and would be highly unlikely to be found in archaeological samples. Diatoms are unicellular animals that construct elaborate silica skeletal frameworks and live in both fresh- and saltwater environments. At high magnification the complex silica body is relatively unmistakable.

Box 7.4. Starch Granule Detection On Stone Tools At Rose Cottage Cave, South Africa

Bonny Williamson

Rose Cottage Cave is situated on a northeast-facing slope of the Platberg escarpment about 3 km east of Ladybrand in the eastern Free State of South Africa (29°13'S, 27°02'E, Wadley 1991, 1997). The seven-metre-deep deposit dates from almost 100,000 years BP and has yielded a complex series of Middle and Later Stone Age industries. The method of excavation used at Rose Cottage provided tight chronological control and has enabled a fine-tuned analysis of many forms of data including stone tools, faunal material, and rock art (e.g., Wadley 1996; Plug and Engela 1992; Ouzman and Wadley 1997; Ouzman 1997). A residue analysis was conducted using 336 lithic artifacts recovered from various squares and levels at Rose Cave (cf. Figure 7.7; Williamson 1997). The results are summarised in Table 7.4.

Although some studies of residues on stone tools have been able to identify the starch granules to species of origin (e.g., Loy et al. 1992; Piperno and Holst 1998; Piperno et al. 2000), this has not yet been possible at Rose Cottage Cave. The very large potential number of wild plant species that might have been present at the site over such a long period of time when there were significant environmental changes requires an extensive reference collection that was far beyond the resources of this study. An attempt was made to set up a comprehensive starch granule catalogue with modern plant species found in the area today, but difficulties have been encountered in discriminating species from each other on the grounds of shape and size. A great deal further work is required before species identification of the Rose Cottage Cave samples will be possible. For this reason, a different approach which focused simply on the presence or absence of starch granules has been used here. Despite the simplicity of this approach, it has yielded very important results for understanding the function of Rose Cottage Cave.

In this study the presence of starch granules larger than 2 microns on stone tools at Rose Cottage has been inferred to indicate the processing of plant material. This size was chosen to avoid the potential of contamination from airborne starch. Many starch granules in the order of 1 to 2 microns were found on clean microscope slides which had been left on a shelf in the microscope room over a period of two months. They are thought to have been derived from a nearby flourmill. To avoid confusion with ancient starch, granules in this size range were ignored for the purpose of analysis.

One explanation for the high frequency of starch granules on all tool types was that the

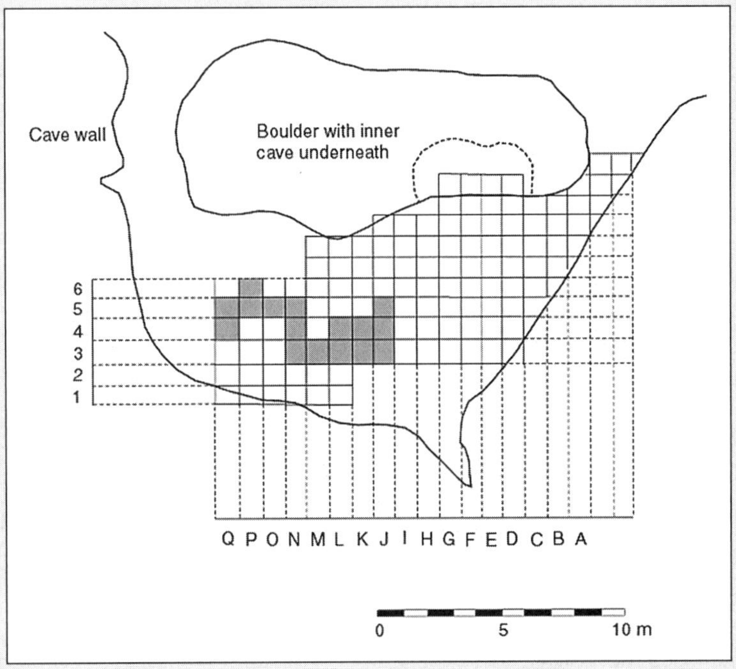

Figure 7.3. Rose Cottage Cave excavations. Samples for this study were taken from the squares with hatching. *Redrawn from Wadley 1991,* based on Tables 6.2 and 6.3, pp. 30, 32 in *Food Microscopy* by Olga Flint (Oxford: Bios Scientific Publishers for the Royal Microscopy Society, 1994).

Box 7.4 (continued)

tools might have a background level of starch granules which were derived from the surrounding deposit rather than from tool use. To test this hypothesis, soil samples were examined for the presence of starch granules and for background levels of soil organics. The results are summarised in Table 7.5. It is notable that starch is not as common in the soil as on the tools and that the size of the granules is generally smaller than those recorded on the tools (cf. Barton et al. 1998).

As shown in Table 7.5, the most notable result from the analysis of residues on stone artifacts from Rose Cottage Cave is that plant remains predominate over animal materials on all the tool types investigated. Starch in the form of a few granules or a mass of material is particularly abundant. This result supports other evidence from the site suggesting that this cave was used as a home base or camp site, rather than as a butchery site (Wadley 1997). Animal remains were processed in the cave, but the primary activities probably centred on plant food preparation, secondary processing of animal carcasses, and artifact manufacture. The tool residues show that plant foods were a significant part of the diet for the people at this site. Furthermore, if it is assumed that women are primarily responsible for the procurement and processing of plant foods in hunter-gatherer societies (e.g., Lee 1979), then it seems likely that women also made and used the stone artifacts at Rose Cottage Cave. This case study demonstrates that the simple presence of starch on stone artifacts can make an important contribution to the interpretation of diet and of site function. Although identification of the plants that people were processing on the basis of the starch granules present on the stone tools is a desirable long-term aim, at this stage residue analysis based on presence or absence is still a very valuable exercise.

Table 7.4. Rose Cottage Cave: Percentage of tools with specific types of residues.

Tool Type (Number)	Scraper (10)	Blade & Bladelet (20)	Misc. Retouch (28)	Core (47)	Chunk (60)	Side-struck flake (43)	End-struck flake (84)	End-struck cortical flake(16)	Broken flake (28)
Starch	70	50	54	64	57	49	60	81	71
White starchy	30	25	14	36	30	40	26	25	25
Plant tissue	60	45	32	57	48	56	59	69	50
Plant fibres	10	25	25	32	28	28	49	38	18
Resin	10	5		16	15		28	6	
Red blood cells				2			7		
Animal tissue				4	2		1		
Collagen			7	2		2	1		
Hair		4	4			2			

Table 7.5. Rose Cottage Cave: results of soil analysis for background levels of starch.

Square	Level	Date	Starch in soil	Starch on tool	pH
N5	G	Late Pleistocene	Few 1–2 μm, some 4 μm	2–4 μm	6.57
P6	Ja	Early Holocene	Few 1–2 μm	none	6.98
O5	H	Early Holocene	Few 1–2 μm, some 4 μm	2–4 μm	6.15
Q5	Ph	Holocene	Few 1–2 μm	2–4 μm	7.52
P5	CM	Holocene	Few 1–2 μm	none	7.56

nominal variable) and size (measured as maximum diameter).

The simplest example of this approach is by Therin et al. (1999) who examined temporal changes in the abundance of starch within soil samples taken from two archaeological sites in Papua New Guinea. In this case they divided starch granules into two different categories (2–5 microns and >5 microns). They made the assumption that the small size class represented mainly transient starch and therefore reflected the presence of general vegetation, whereas the larger category was more likely to be associated with underground storage organs. On the basis of this analysis, the authors proposed changes in the site function. At one location the major activity shifted from quarrying to residential and at both there was a shift from shorter to longer term occupation (see summary in Box 8.8).

A more sophisticated multivariate study conducted by Lentfer et al. (2002; Box 8.1) tested whether starch analysis was capable of differentiating one flora from another using starch assemblages extracted from modern soil samples derived from different forms of human land use. Granule morphology and size provided descriptors for the individual starch granules. The results show that even when measured in this rather simple way, variation in starch assemblages can differentiate among plant communities. The study gives credence to the proposal that starch can be used alongside pollen and phytoliths in paleoenvironmental analyses. Furthermore, by combining data from archaeological assemblages of starch to that gained from the analysis of other microfossils, Parr (2003a) and Lentfer (2003) have shown that one can make inferences about prehistoric land use, although none of the starch granules have yet been assigned to a particular plant taxon.

Level 3: Targeting One or More Species

Unlike Levels 1 and 2, the next two kinds of identification require that the archaeological starch granules be identified to a particular taxon. Level 3 research hypotheses target the presence of one or just a few particular species. In these cases, the hypotheses are adequately tested by the determination of the presence or absence of these key species within the archaeological contexts. Level 3 identifications can make a very important contribution to archaeological research because often only the presence or absence of key economic plants is needed to solve the research goal. Simple hypotheses related to one or two taxa are relatively easy to address because only a small reference collection and limited knowledge can suffice.

For example, one might want to test whether a particular artifact had a specialised function, i.e., was used to process a specific type of plant. One could also assess whether a tool had a single purpose or was used to process multiple food plants by determining the range of starch morphologies present and identifying plant taxa from which the residues were derived. Since starch granule morphologies vary between plant taxa, comparative data from the plants that may have been processed with the tools in question is needed. An example of this approach is provided by the analysis of bevelled pounders from Southeast Queensland in Australia (Hall et al. 1989). An understanding of the function of these tools was important for testing hypotheses about the timing of the development of specialised and intensified subsistence patterns in this region. It had been proposed that the pounders were involved in the processing of a particular species of fern root (*Blechnum indicum*) that was important to Aboriginal diets in the recent past. All that was required was the presence on prehistoric tools of starch granules derived from this one species. Microscopic study revealed the presence of *Blechnum indicum* starch granules as well as another form which could not be identified.

A somewhat similar approach is represented by Pearsall et al.'s (2004) study of starch granules and cob phytoliths from the site of Real Alto in Ecuador. The research addressed controversies about the nature of subsistence and particularly the presence and importance of maize at what is considered to be an early stage in village development. Consequently, Pearsall et al. (2004) focused solely on the question of whether maize was present at the site. By comparing the starch granules derived from stone tools with a targeted but comprehensive reference collection comprised of 9 varieties of maize, they showed that the ancient starch 'is a mixture of hard and soft kern endosperm' which was 'perhaps the result of processing a genetically plastic maize variety or varieties' (Pearsall et al. 2004: 438–39). Since these identifications matched those obtained from phytoliths extracted from the same tools, their previous argument for the use of maize at the site was substantiated.

In another key study of this type, Linda Perry's (2001, 2002a, 2004, in press) analyses of stone tools from Venezuela demonstrate that Level 3 identification can be highly significant for archaeological research. Her research tested the hypothetical function of stone tools assumed to be the microlithic teeth from manioc grater boards. The role of these is important because their presence at archaeological sites in lowland South America has been used to support models for agricultural intensification and site function. Perry was unable to detect starch granules that could be undoubtedly classified as manioc, but she did recover a wide range of starch granules with other distinctive morphologies, many of which were identified to other taxa. Her work does not support the simple classification of these tools as manioc grater teeth, suggests that the subsistence pattern was more mixed than had previously been assumed, and raises serious questions about the role of manioc in the prehistory of this region.

The Pearsall et al. (2004) study highlights a further potential of Level 3 identification: the discrimination between domesticated and wild varieties of the same plant. As described in Box 1.4, the use of starch granules to detect the presence of domestication was pioneered by Ugent. Following his lead, Piperno and Holst (1998) and Piperno et al. (2000) have identified a number of both wild and cultivated species of yam, manioc, arrowroot, and maize in residues found on Panamanian grindstones. Similarly, Perry (2001, 2002a, 2004, in press) also identified many of the same species of plants from a collection of flake and groundstone tools derived from two sites in the Middle Orinoco Basin. More recently, Piperno (in press) and Perry (2002b) have demonstrated the possibility of monitoring the history of plant domestication through the study of changing morphologies in starch granules.

Level 4: Identification of Most Taxa

Ultimately, the goal of most analyses will be to assign a significant percentage of starch granules within a sample to a particular taxon and to present quantitative data about their relative abundance in the sample. As with both pollen and phytolith analyses, species or even subspecies identifications may be possible in many cases, whereas in others, a genus or family level identification will be most appropriate. Armed with these data one could carry out a full environmental analysis of a sediment core or provide a fairly complete description of plant use as represented in a soil or residue sample from an archaeological context.

The remainder of this chapter focuses on the general principles of assigning starch granules to a taxonomic group. This process is essential for Levels 3 and 4 classification and is also relevant for the Level 2 more general approach. At this point, it is worth stressing that in order to make a taxon-specific identification in archaeological contexts, it will generally be necessary to focus on storage or reserve starch granules that commonly occur in plant organs such as seeds and roots, because transitory or transient starches are mostly small and difficult to classify (cf. Chapter 2; Haslam 2004).

The most important research tool is a good reference collection (Chapter 6) together with a set of hierarchical keys, if these are available for the relevant species. Generalised keys are discussed by Ugent in Box 7.1, and some have been developed within recent archaeological studies (Perry 2001, 2002a, 2004, in press; Piperno et al. 2000, 2004; Piperno and Holst 2004; Piperno in press). The basic procedure for identification of granules to plant taxa has been well summarised by Perry (2004: 1075–76). First one compares the basic morphology of the granules to be identified with the reference collection in terms of a set of basic characteristics discussed below. The more traits used to match unknowns with reference material, the more secure the identification. Archaeological starch granule identifications may employ characteristics of single, diagnostic granules or of starch assemblages as a whole, depending on the issues being addressed and the type of analysis needed for a positive identification. When identifying a

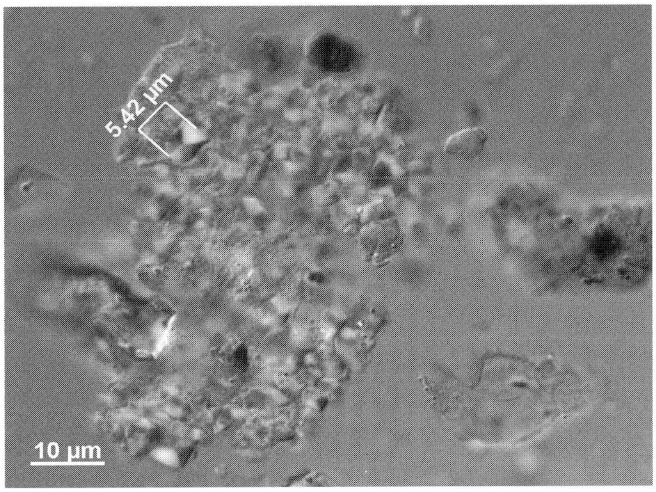

Plate 1. Starch granules identified as taro (*Colocasia esculenta*) found on stone tools dating to c. 10,000 BP at Kuk Swamp, Papua New Guinea viewed under DIC; cf. Plate 2. *Photo by Judith Field*

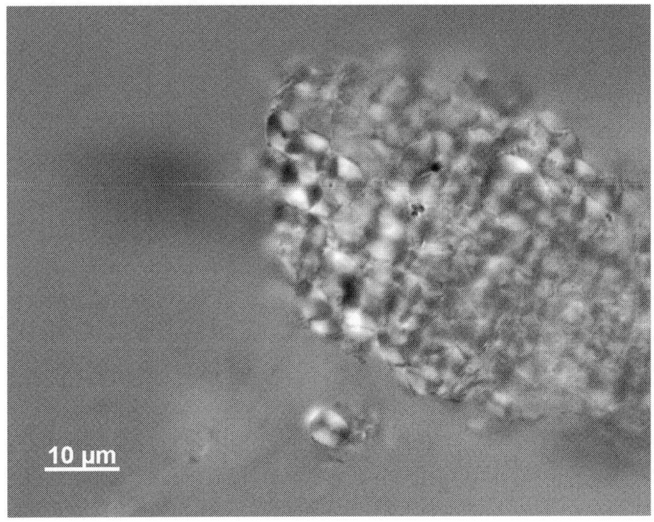

Plate 2. Starch granules identified as taro (*Colocasia esculenta*) found on stone tools dating to c. 10,000 BP at Kuk Swamp, Papua New Guinea viewed with polarising filters; cf. Plate 1. *Photo by Judith Field*

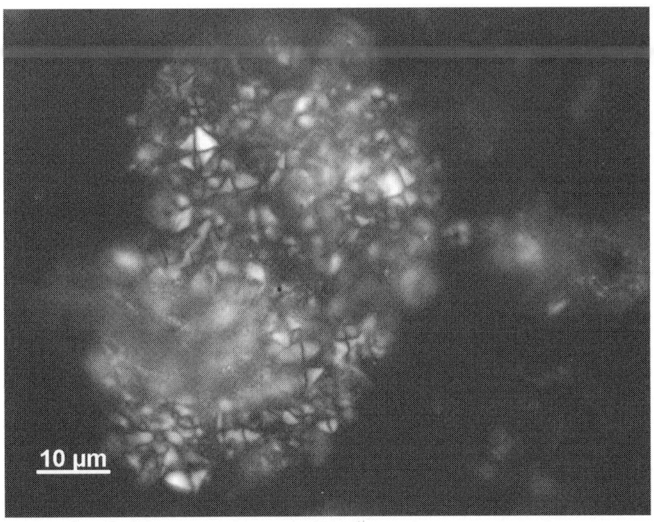

Plate 3. Starch granules of modern taro (*Colocasia esculenta*) from the Australian Museum reference collection. A diagnostic characteristic of this species is that the granules are often found in sheets as shown; viewed under DIC; cf. Plate 4. *Photo by Judith Field*

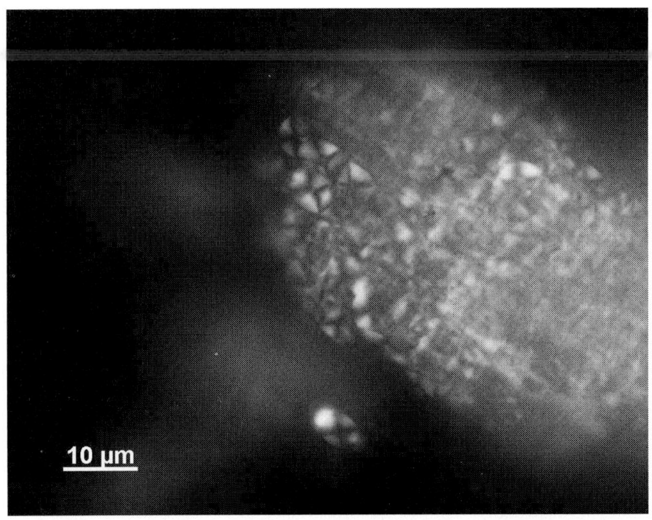

Plate 4. Starch granules of modern taro (*Colocasia esculenta*) from the Australian Museum reference collection viewed with polarising filters; cf. Plate 3. *Photo by Judith Field*

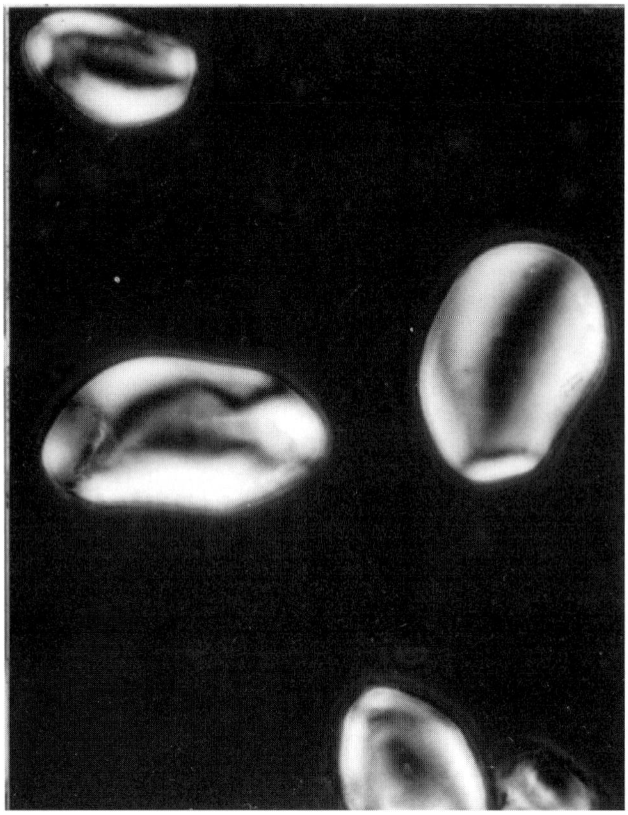

Plate 5. Starch granules of achira rhizomes from the archaeological site of San Diego (550–295 B.C.) in the Casma Valley, Peru; viewed in phase contrast and polarized light, x250. *Photo by Donald Ugent*

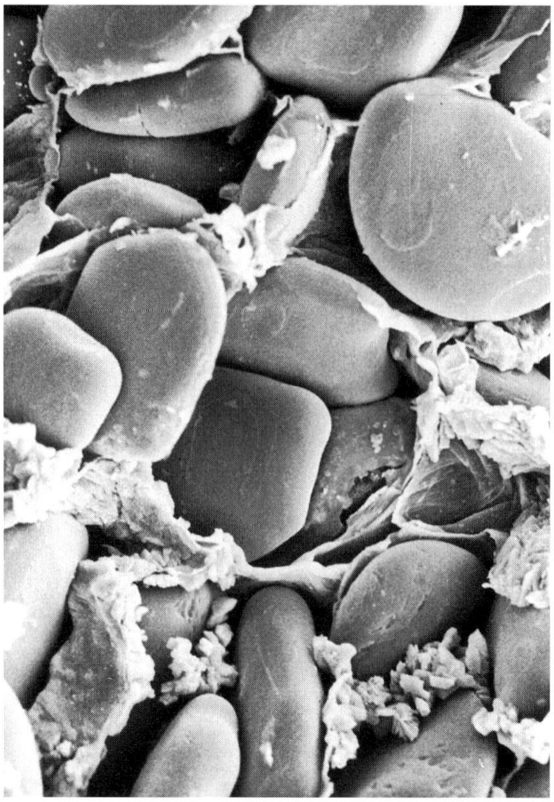

Plate 6. SEM micrograph of starch granules in an ancient rhizome of the achira plant from the site of San Diego (550–295 B.C.) in the Casma Valley, Peru. Gold-paladium coated, x1000. *Photo by Donald Ugent*

Plate 7. SEM micrograph of manioc (*Manihot esculenta*) starch granules from the Pampa Rosario archaeological site, Peru, x4000. *Photo by Donald Ugent*

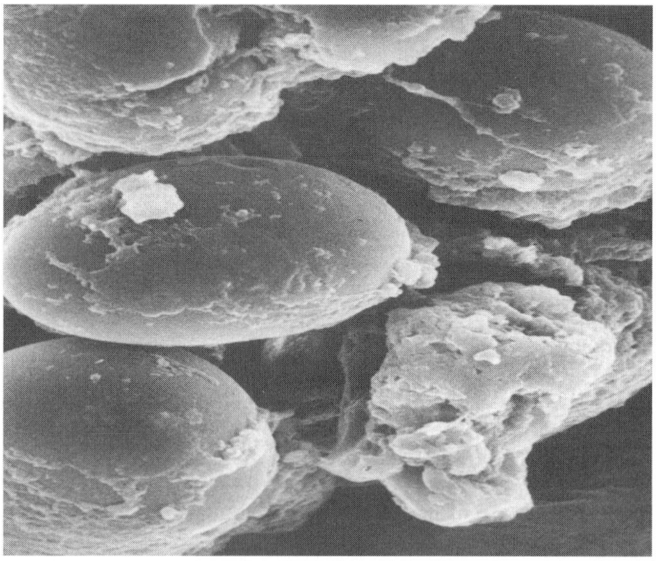

Plate 8. SEM micrograph of ancient starch granules (c. 13,000 BP) from the wild Chilean potato species (*Solanum maglia*). *Photo by Donald Ugent*

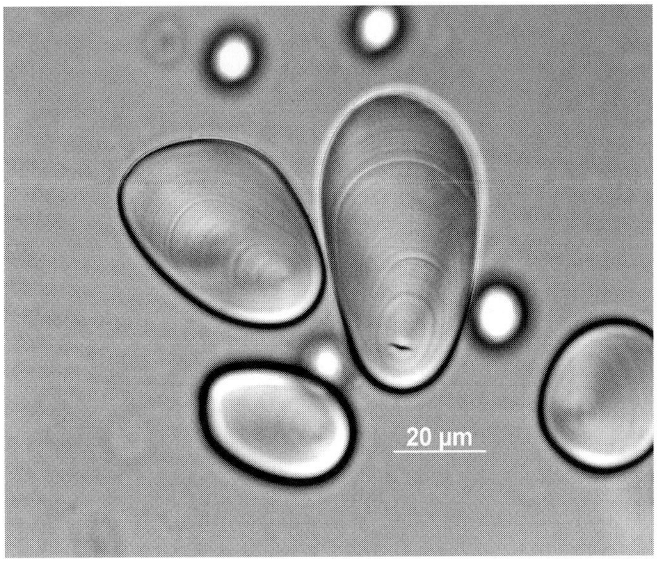

Plate 9. Starch granules of white potato (*Solanum tuberosum*, domesticated variety). The layers of starch accumulation called lamellae are visible as concentric lines on the granule; viewed under DIC. *Photo by Judith Field*

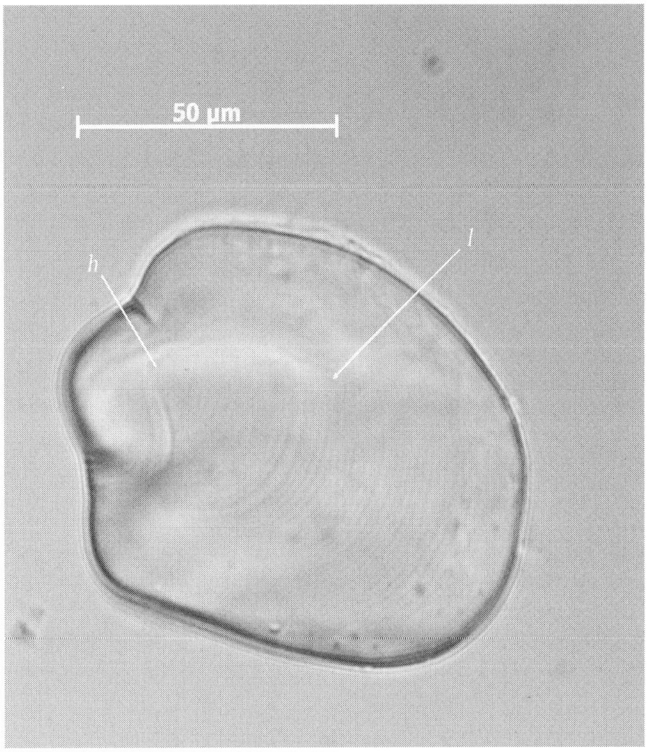

Plate 10. Starch granule of *Dioscorea pentaphylla*. The layers of starch accumulation are visible as concentric lines (l) and the hilum (h) is located towards one end of the granule (eccentric form); viewed under DIC. *Photo by Huw Barton*

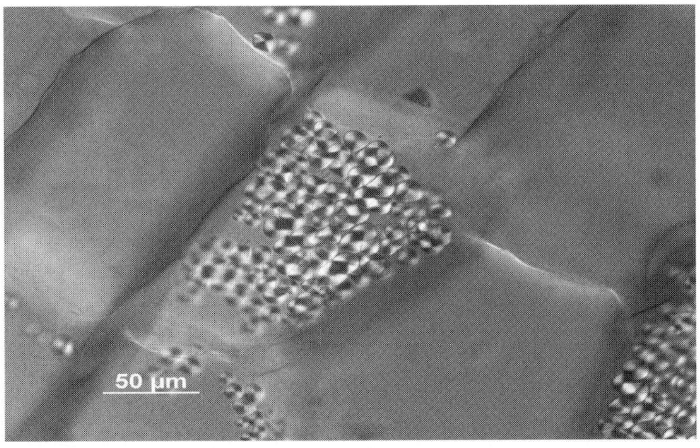

Plate 11. Hand section of a water ribbons tuber (*Triglochin procera*) shows how starch granules are packed within cells leading to flat faces or facets on the surface of the granules. The packing gives the appearance of compound granules; viewed under DIC; cf. Plate 12. *Photo by Judith Field*

Plate 12. Hand section of a water ribbons tuber (*Triglochin procera*) viewed under cross-polarised light; cf. Plate 11. *Photo by Judith Field*

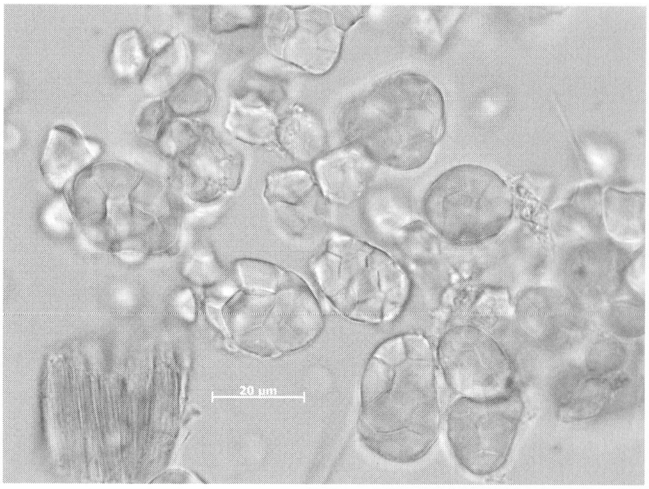

Plate 13. Compound starch granules of *Alocasia reginae* viewed under DIC. *Photo by Huw Barton*

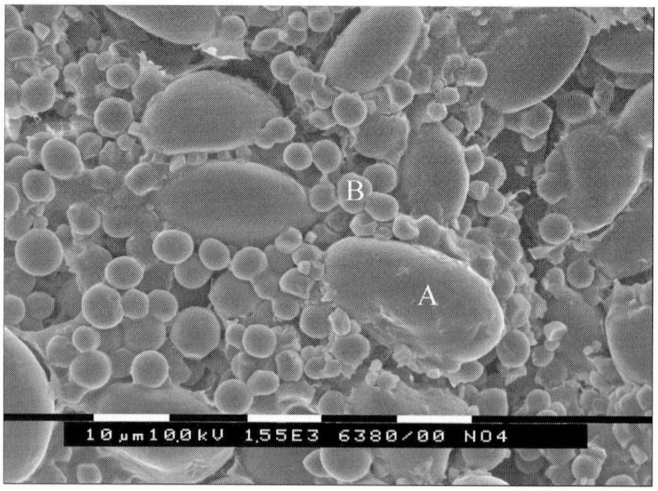

Plate 14. SEM micrograph of fractured A (large) and B (small) types of starch granules from wheat (*Triticum aestivum* cv). Magnification bar shows 10-micron intervals. *Reprinted with permission from Fred L. Stoddard, 1999, Survey of starch particle-size distribution in wheat and related species, Cereal Chemistry 76: 145–49.*

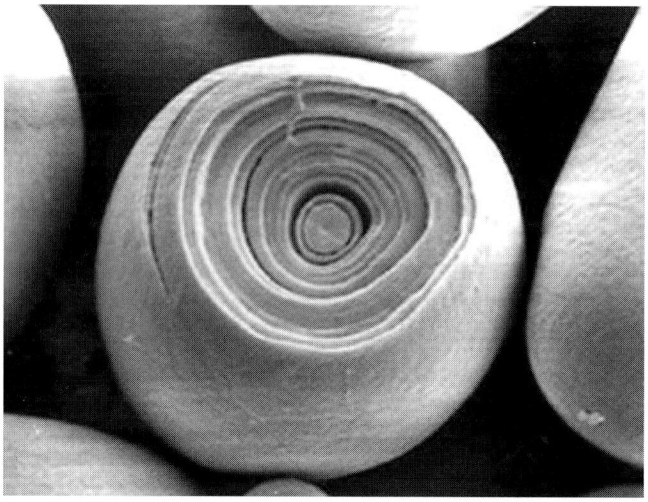

Plate 15. SEM micrograph of maize starch showing layered structure of the granule. *Image taken by Bruce L. Wagner using a JOEL JSM-35 in the Bessey Microscopy Facility, Iowa State University, Ames, Iowa, USA*

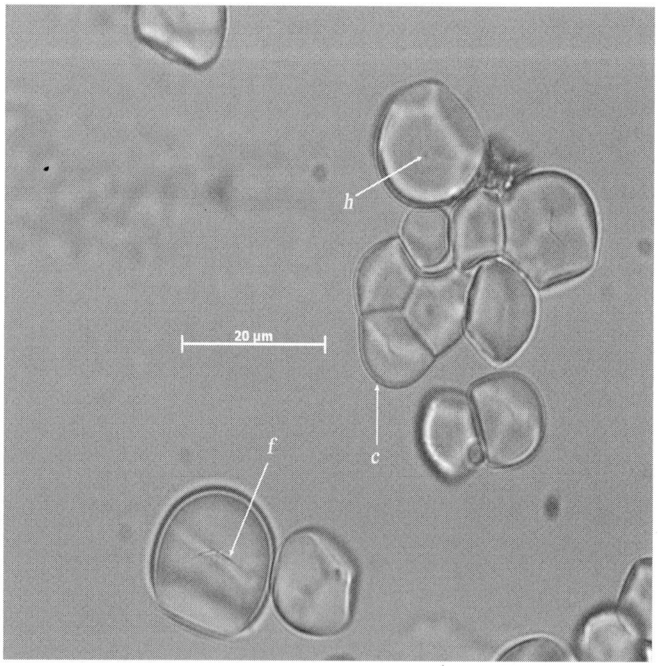

Plate 16. Granules of Araceae sp. (Hassk.) Schott with fissures in the top right and (f) fissure passing through hilum of granula, (h) hilum visible as a dark point, and (c) compound granules viewed in brightfield light. *Photo by Huw Barton*

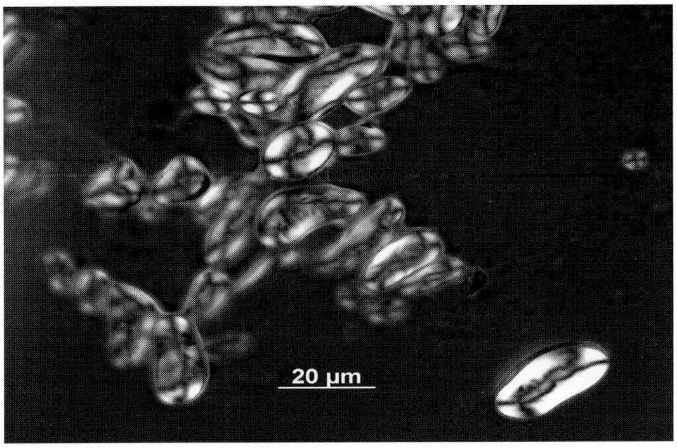

Plate 17. Starch granules from unstressed rhizomes of Australian bracken fern (*Pteridium esculentum*). *Photo by Judith Field*

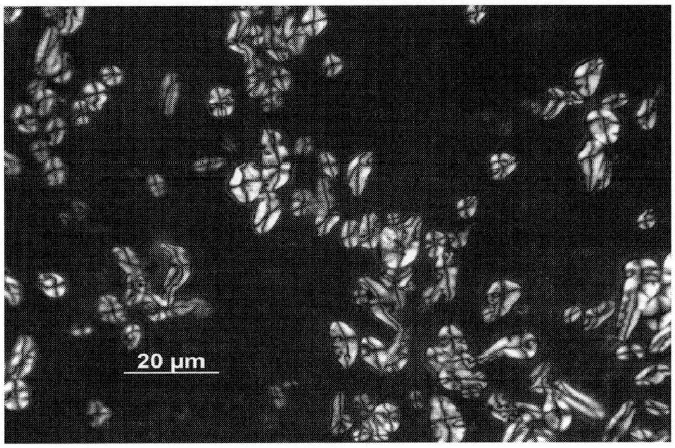

Plate 18. Starch granules from stressed rhizomes of Australian bracken fern (*Pteridium esculentum*). *Photo by Judith Field*

Plate 19. A young plant tuber *Triglochin procera* with abundant starch granules. *Photo by Judith Field*

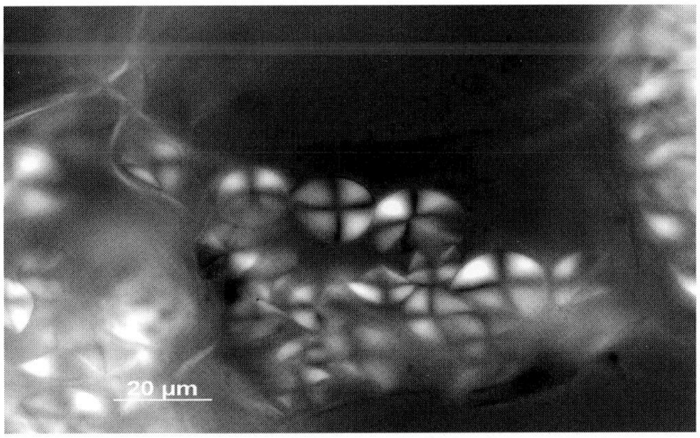

Plate 20. Mature plant tuber of *Triglochin procera* with fewer starch granules than Plate 19. *Photo by Judith Field*

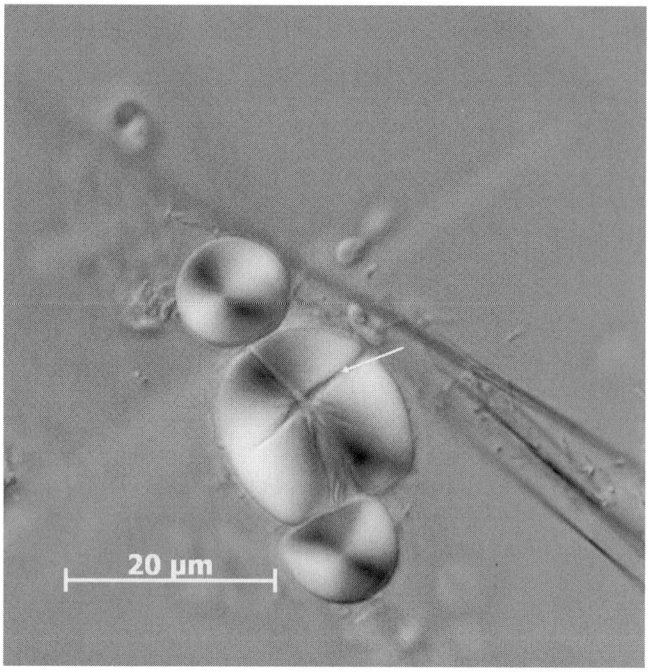

Plate 21. Starch granules of *Alocasia longiloba* viewed in DIC. Arrow points to crack across body of granule. Note how the round granules have a pyramidal appearance induced by birefringence and the polarisation cross. *Photo by Huw Barton*

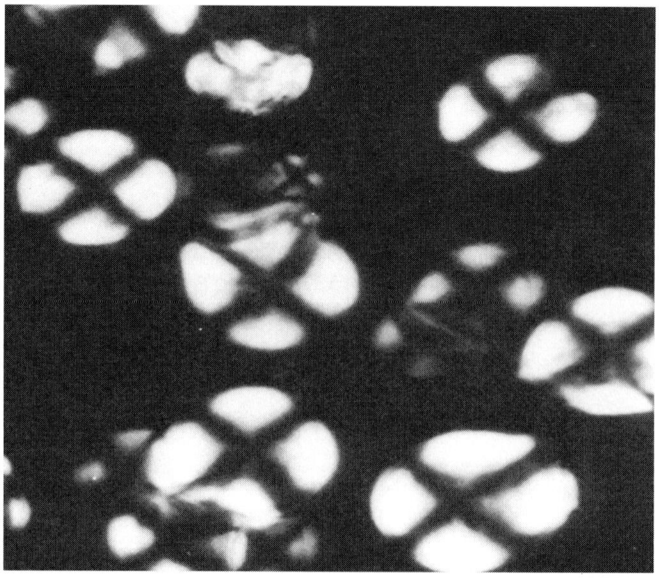

Plate 22. Starch granules of an ancient manioc root *Manihot esculenta* found at the Casma Valley site of Manchan (A.D. 1470–1532). Phase contrast micrograph with polarised light, x400. *Photo by Donald Ugent*

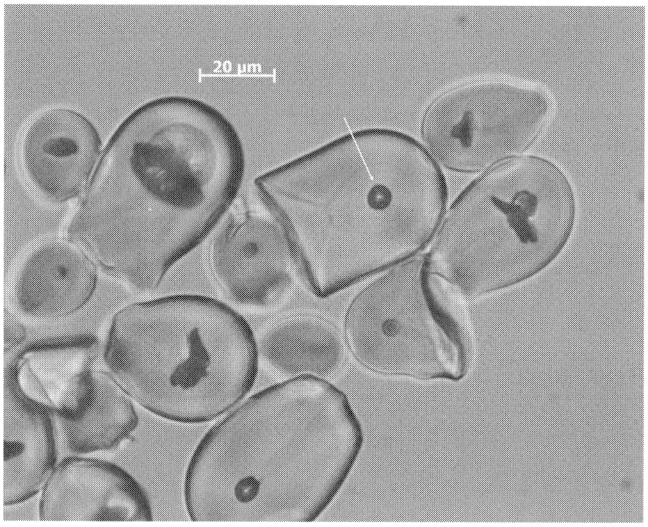

Plate 23. Examples of an open hilum (vacuoule) in starch granules from *Eugeissona utilis* viewed in brightfield light. *Photo by Huw Barton*

Plate 24. Starch granules of *D. bulbifera* var. *bulbifera* stained with iodine-potassium-iodide after leaching and cooking. *Photo by Wendy Beck*

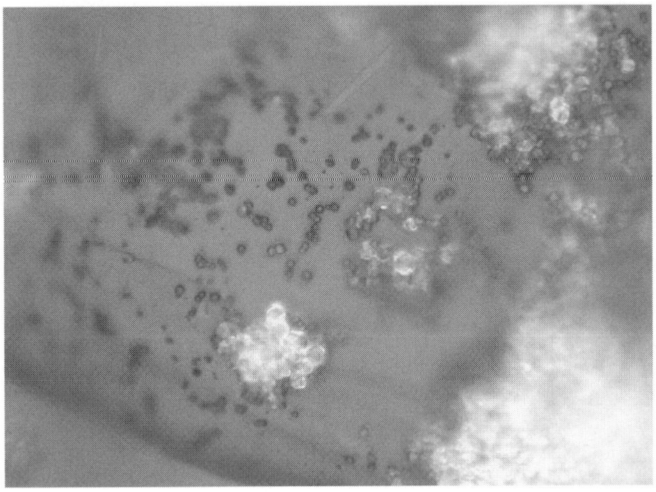

Plate 25. Reflected light brightfield image of starch granules on the surface of an obsidian flake; cf. Plate 26. *Photo by Huw Barton*

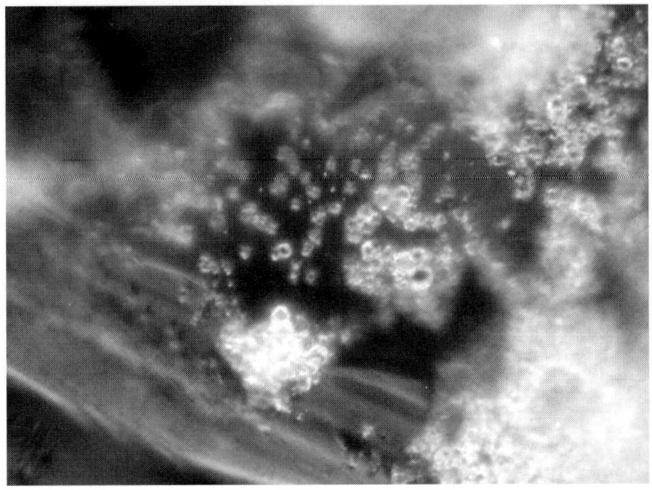

Plate 26. Starch granules on the surface of an obsidian flake viewed under darkfield light; cf. Plate 25. *Photo by Huw Barton*

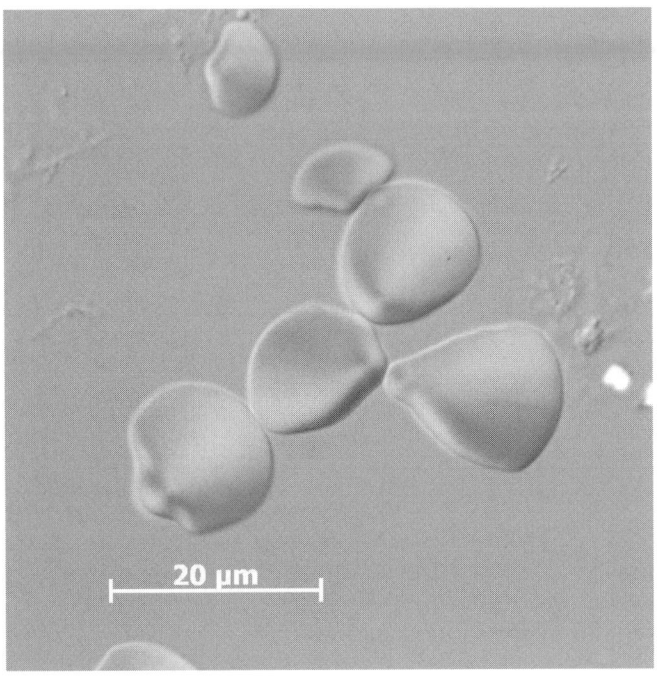

Plate 27. Starch granules of *Zingiber cassumunar* with low relief clearly resolved in DIC; cf. Plate 28. *Photo by Huw Barton*

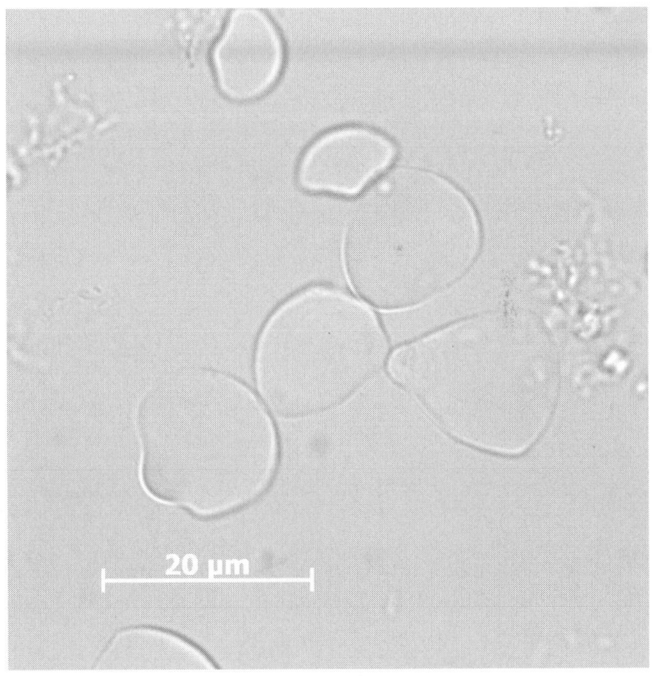

Plate 28. Starch granules of *Zingiber cassumunar* in brightfield lighting; cf. Plate 27. *Photo by Huw Barton*

Plate 29. Starch granules of cheeky yam (*D. bulbifera* var. *bulbifera*). *Photo by Wendy Beck*

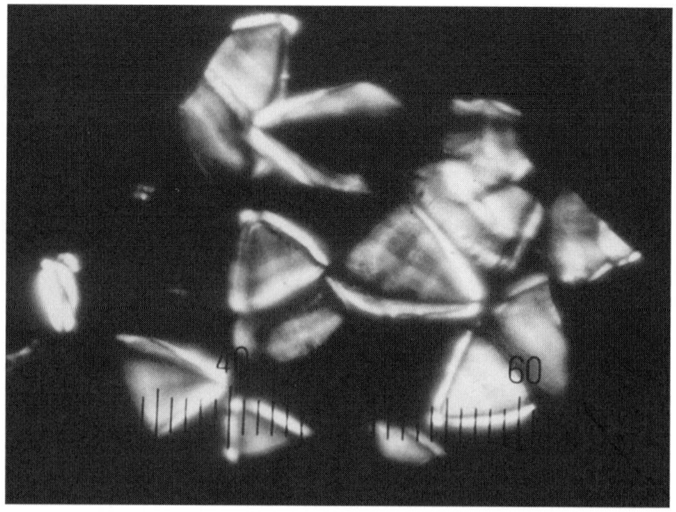

Plate 30. Starch granules of cheeky yam (*D. bulbifera* var. *bulbifera*) after leaching and cooking; cf. Plate 29. *Photo by Wendy Beck*

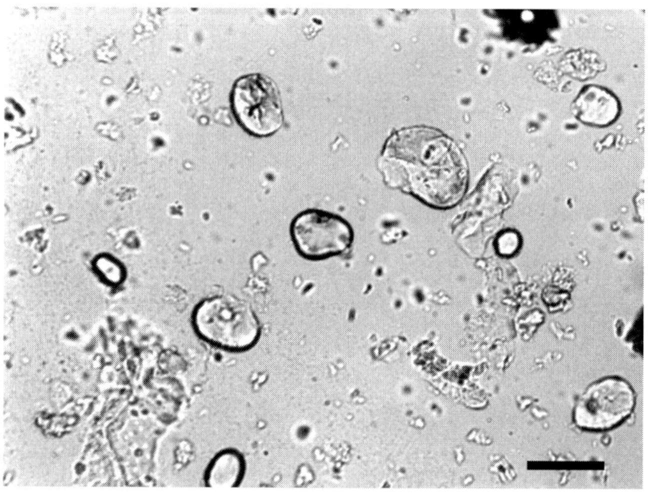

Plate 31. Damaged starch granules from common beans that have been milled. Scale is 40 microns. *Photo by Maria Babot*

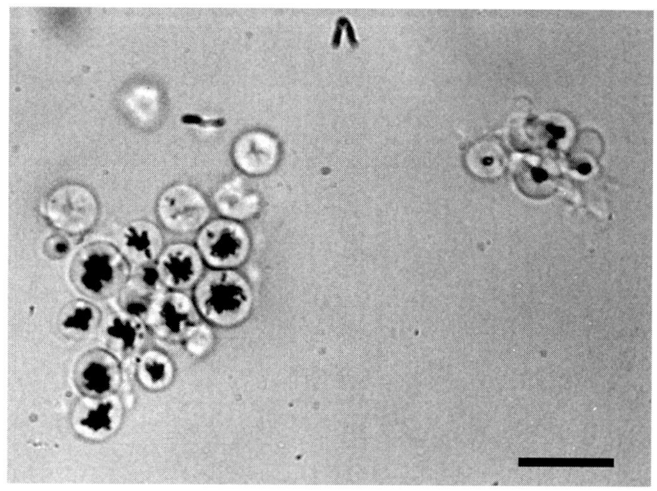

Plate 32. Damaged maize starch granules after roasting during the preparation of *tostado*. Scale is 20 microns. *Photo by Maria Babot*

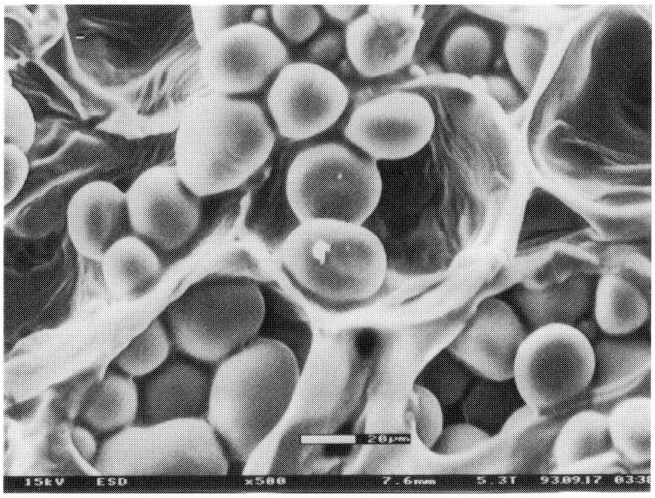

Plate 33. ESEM micrograph of starch granules in a fresh sample of *Canavalia rosea* at 500 diameters. *Photo by Gail Robertson*

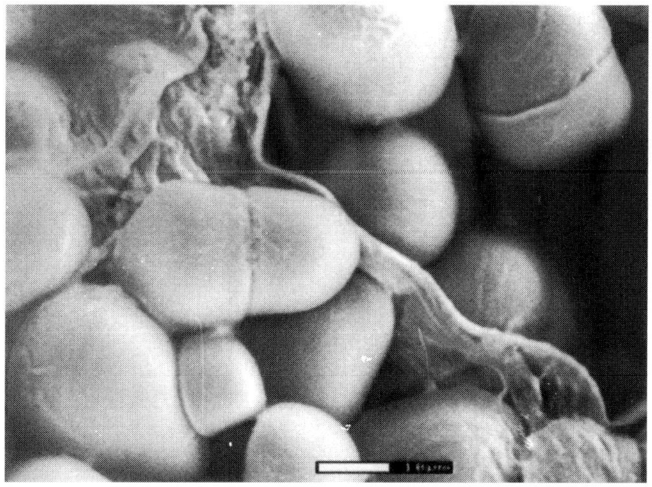

Plate 34. ESEM micrograph of starch granules in a fresh sample of *Blechnum indicum* showing compound granules at 1,250 diameters. *Photo by Gail Robertson*

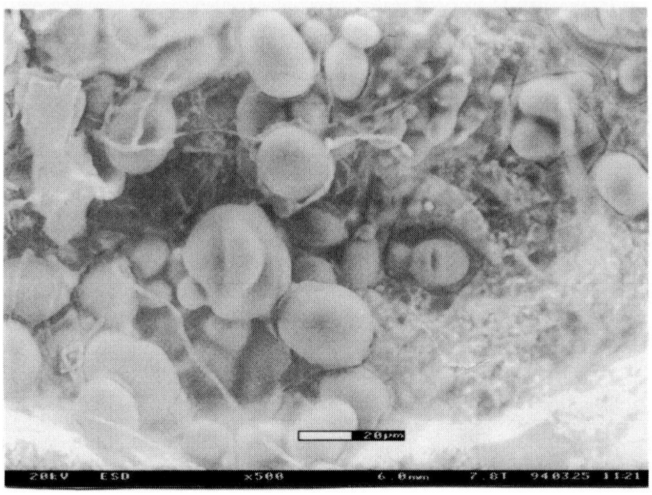

Plate 35. ESEM micrograph of starch granules of *Canavalia rosea* from residues on an experimental pounder at 500 diameters. *Photo by Gail Robertson*

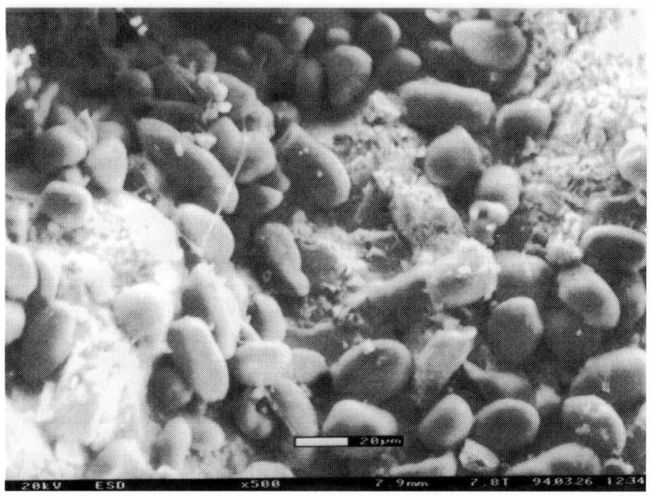

Plate 36. ESEM micrograph of starch granules of *Blechnum indicum* from residues on an experimental pounder at 500 diameters. *Photo by Gail Robertson*

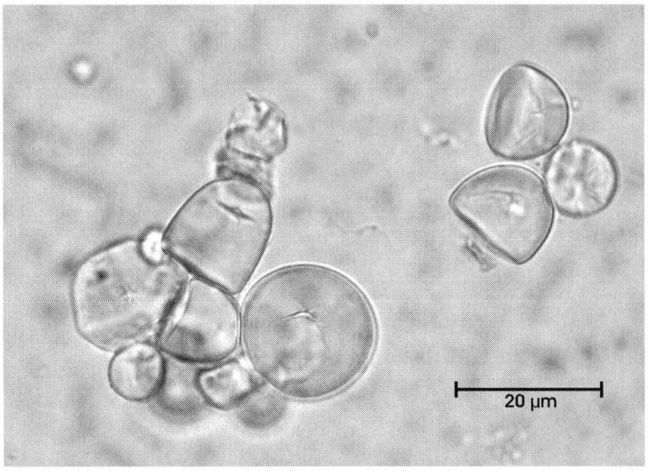

Plate 37. Starch granules of *Ipomoea batatas* extracted from a coprolite dated to younger than A.D. 1520–1690 from Triangle Flat, Great Barrier Island, New Zealand. *Photo by Mark Horrocks*

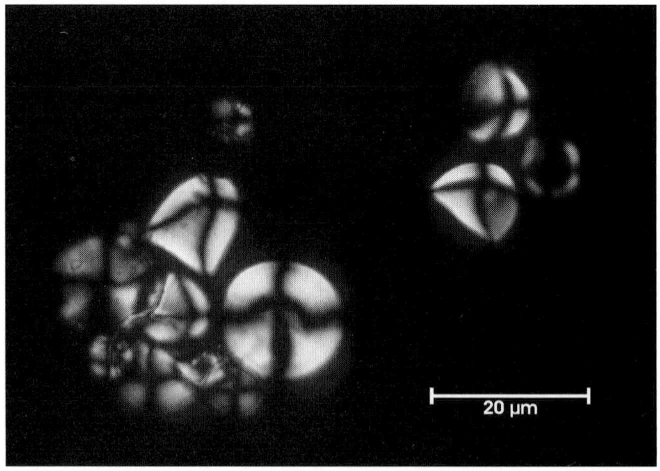

Plate 38. *Ipomoea batatas* starch granules as in Plate 37 under cross-polarised light. *Photo by Mark Horrocks*

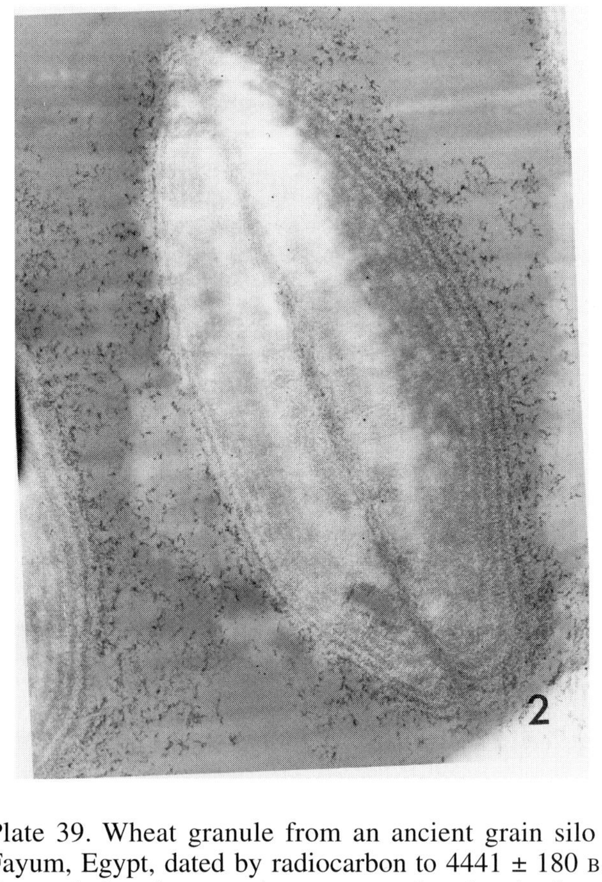

Plate 39. Wheat granule from an ancient grain silo at Fayum, Egypt, dated by radiocarbon to 4441 ± 180 B.C. Transmission electron microscope section showing lamellae. This granule gave the characteristic blue colour with iodine. *Hall and Osborne 1974: 598*

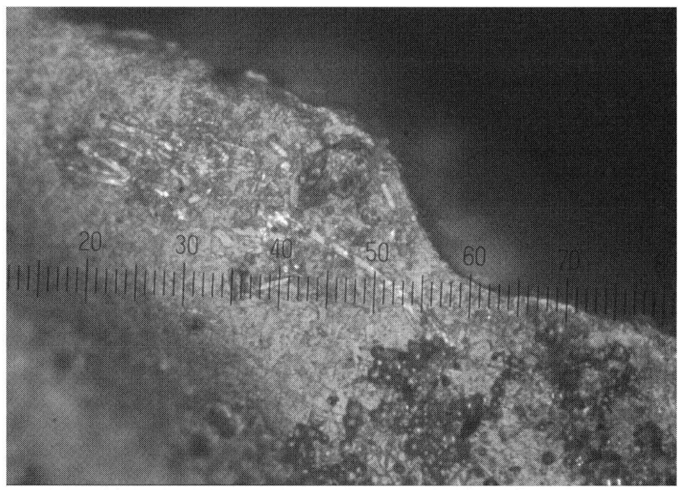

Plate 40. Starch residues of cheeky yam (*D. bulbifera* var. *bulbifera*) on the edge of a shell grater. *Photo by Richard Fullagar*

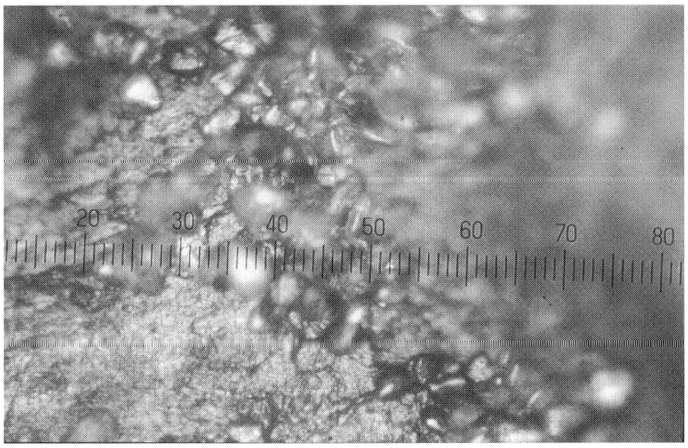

Plate 41. Starch residues of cheeky yam (*D. bulbifera* var. *bulbifera*) on the edge of a shell grater. Same as Plate 40 but a closer view. *Photo by Richard Fullagar*

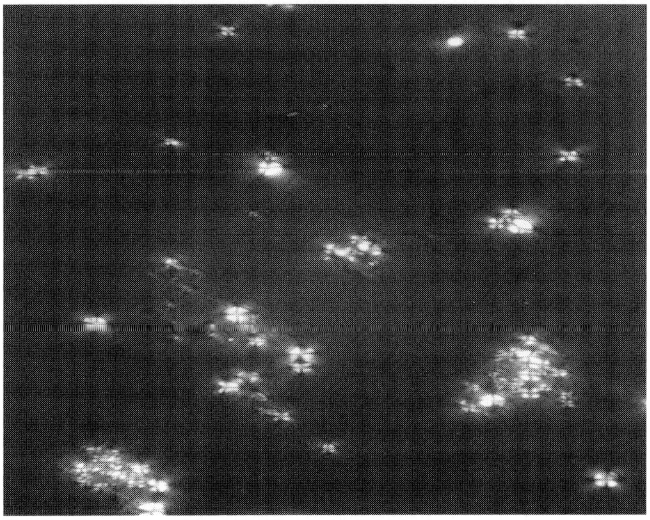

Plate 42. Starch granules from giant water lily (*Nymphaea gigantea* Hook) seeds after grinding and baking viewed under cross-polarised lighting. *Photo by Jennifer Atchison*

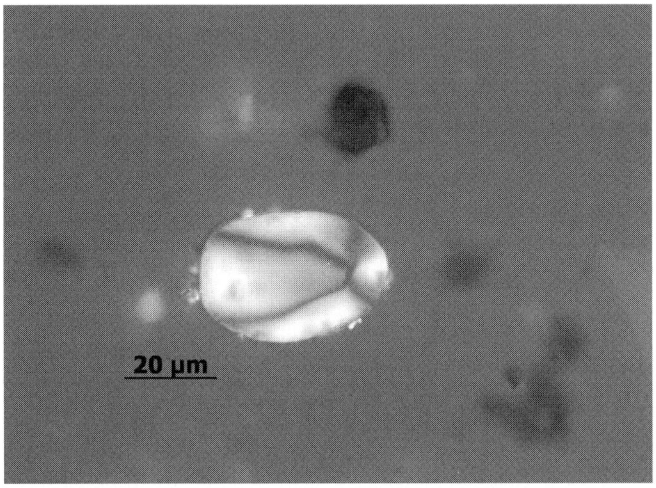

Plate 43. Type 6 starch granule from soils at Niah Cave (Hell Trench, Context 3137, Ms0059a) dated to c. 44,000 BP and identified as *Dioscorea* sp. viewed in DIC. *Photo by Huw Barton*

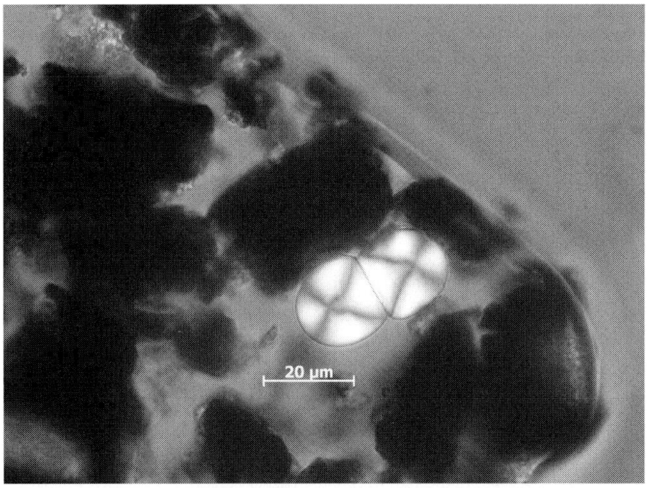

Plate 44. Type 8 semicompound starch granule from soils at Niah Cave (Hell Trench, Context 3129, Ms0055) dated to c. 40,000 BP and identified as *Alocasia longiloba* viewed in cross-polarised light. *Photo by Huw Barton*

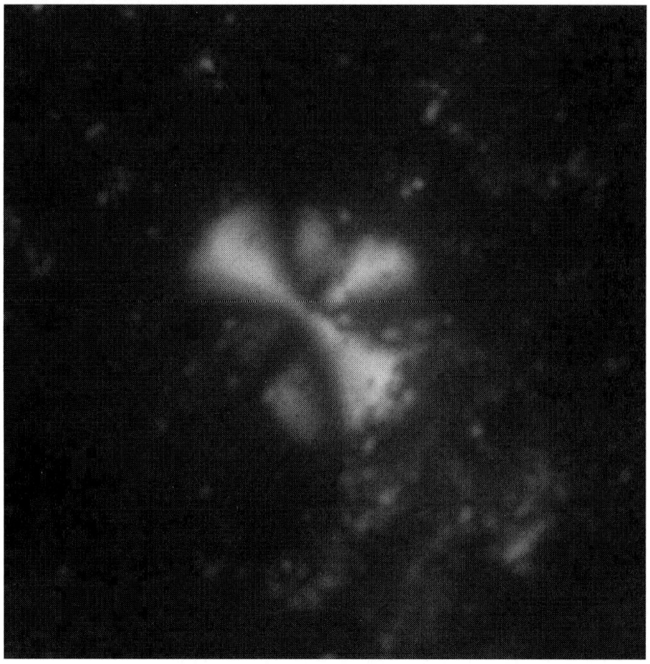

Plate 45. Gypsum spherulite (diameter c. 15 microns) derived from an Australian soil context. Brightfield, cross-polarized light at x1000. *Photo by Sue Nugent*

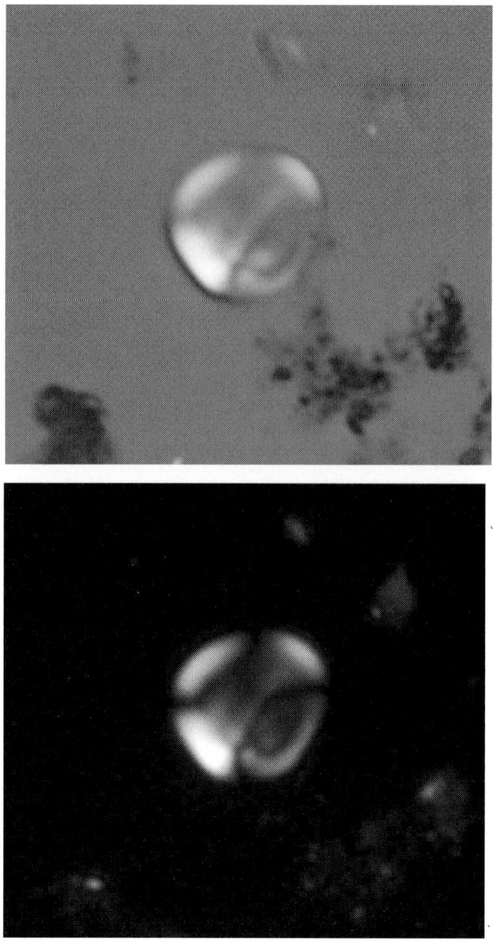

Plate 46. Avireptillian uric acid spherulite from an Australian soil (13 microns diameter). Brightfield, x1000: (top) unpolarized; (bottom) cross-polarized. *Photos by Sue Nugent*

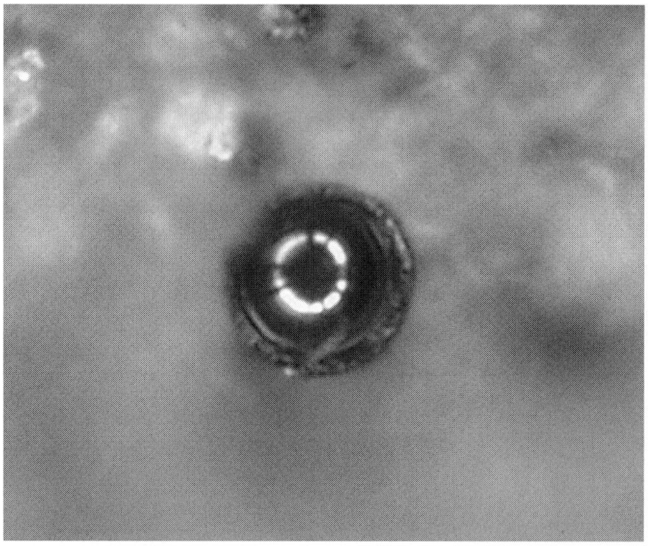

Plate 47. Spore with a reflection of the darkfield apparatus at x500. *Photo by Gail Robertson*

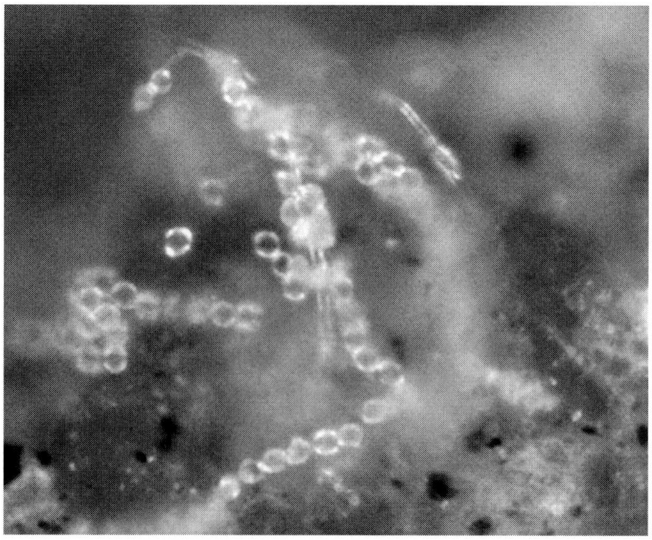

Plate 48. So-called 'string of pearls,' which are fungal hyphae with attached spores (2 microns) in reflected, cross-polarized light at x500. *Photo by Gail Robertson*

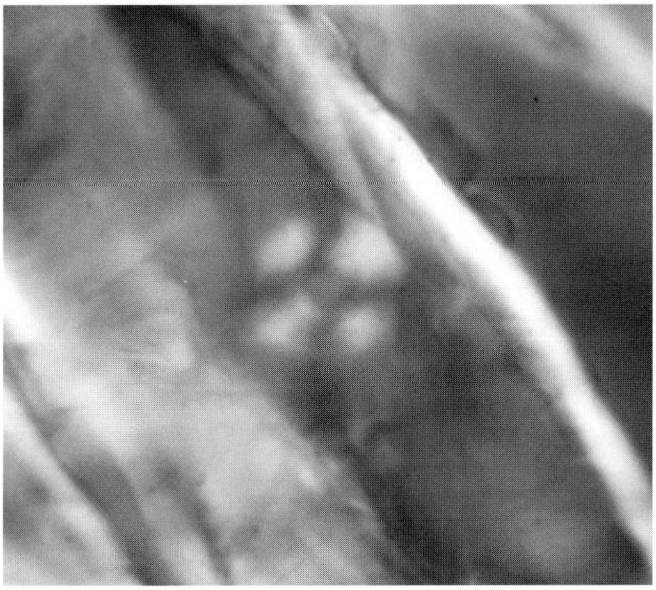

Plate 49. Bordered pit showing extinction cross caused by regular concentric buildup of cellulose (22 microns diameter). Note the central hole; brightfield; cross-polarised light; x500. *Photo by Sue Nugent*

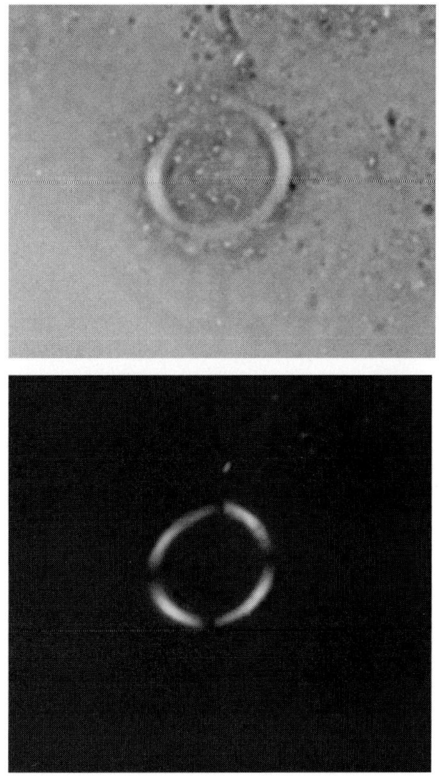

Plate 50. Annular wall thickenings of cellulose in brightfield at x1000 under (top) unpolarized and (bottom) cross-polarized light. *Photos by Sue Nugent*

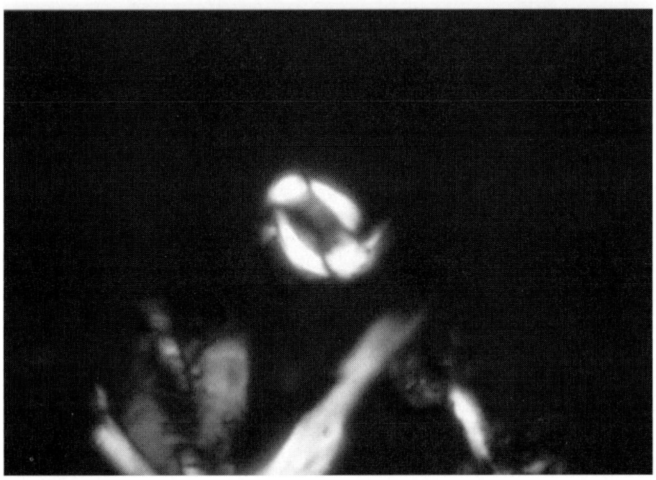

Plate 51. Coccolith from Australian soil (10 microns maximum dimension); cross-polarised light; x500. *Photo by Tom Loy*

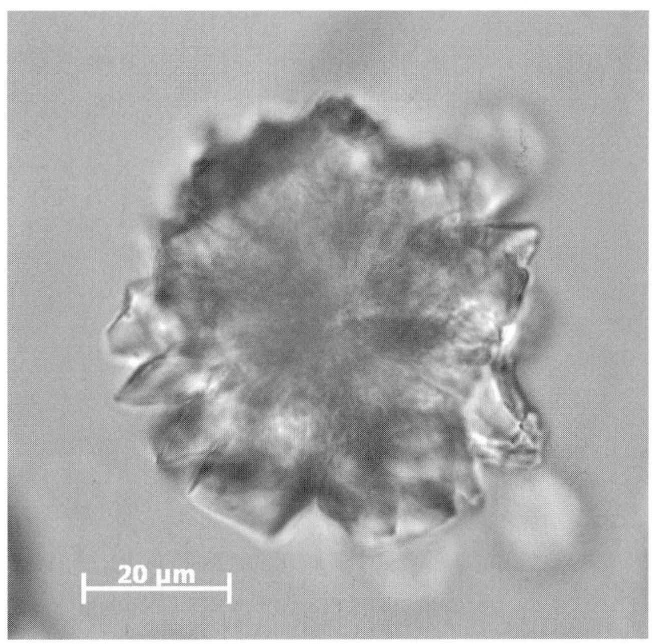

Plate 52. A crystal druse from the yam *Dioscorea pentaphylla* viewed in brightfield light. *Photo by Huw Barton*

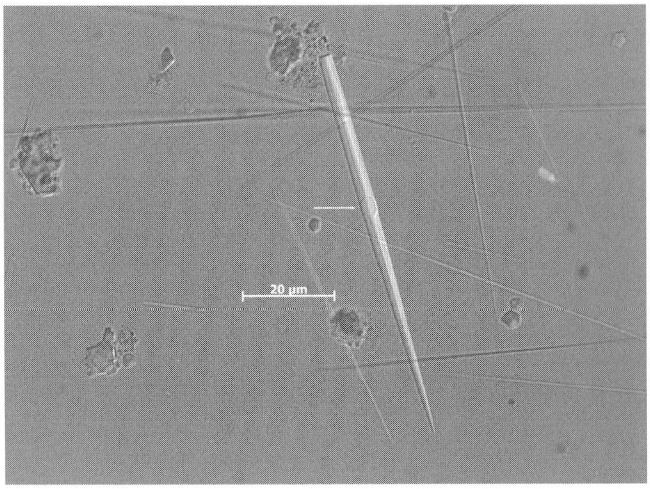

Plate 53. Large, complex raphide of *Alocasia longiloba* viewed under DIC. *Photo by Huw Barton*

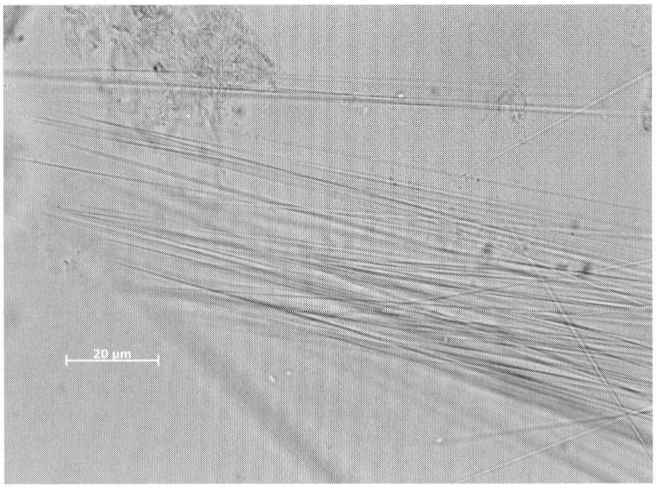

Plate 54. Whisker raphides of taro (*Colocasia* sp.) viewed in brightfield light. *Photo by Huw Barton*

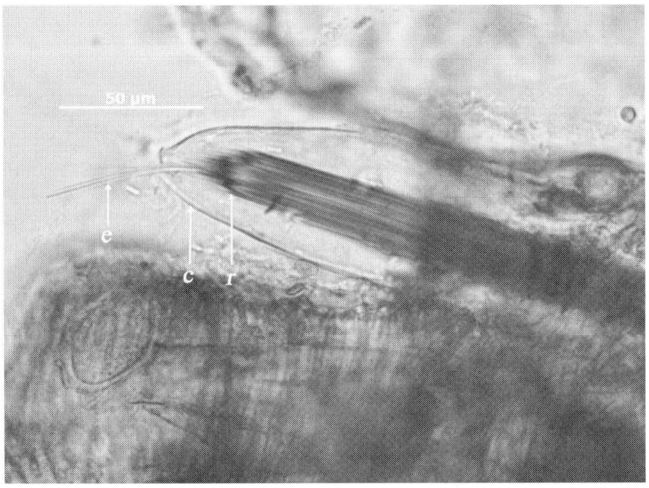

Plate 55. Ideoblast cell of taro: (e) ejected raphide crystals; (c) ideoblast cell wall; and (r) raphide bundle within cell viewed in brightfield light. *Photo by Huw Barton*

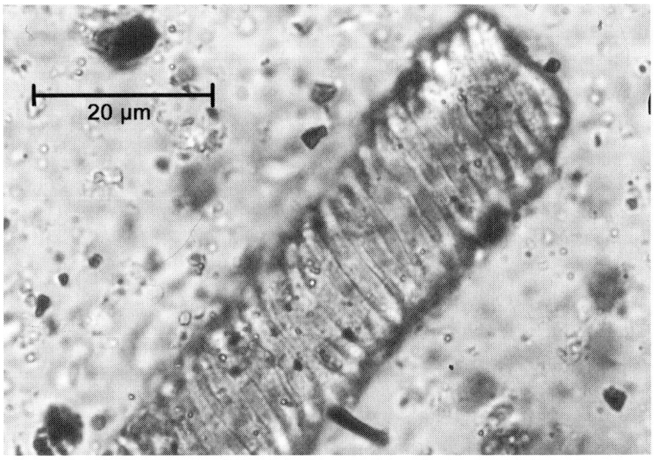

Plate 56. Fragment of xylem cell (vessel element) of taro *(Colocasia esculenta)* corm from soil profile at Rangihoua Bay, New Zealand. Taro vessels are up to c. 50 microns across with helical walls up to c. 4 microns across. *Photo by Mark Horrocks*

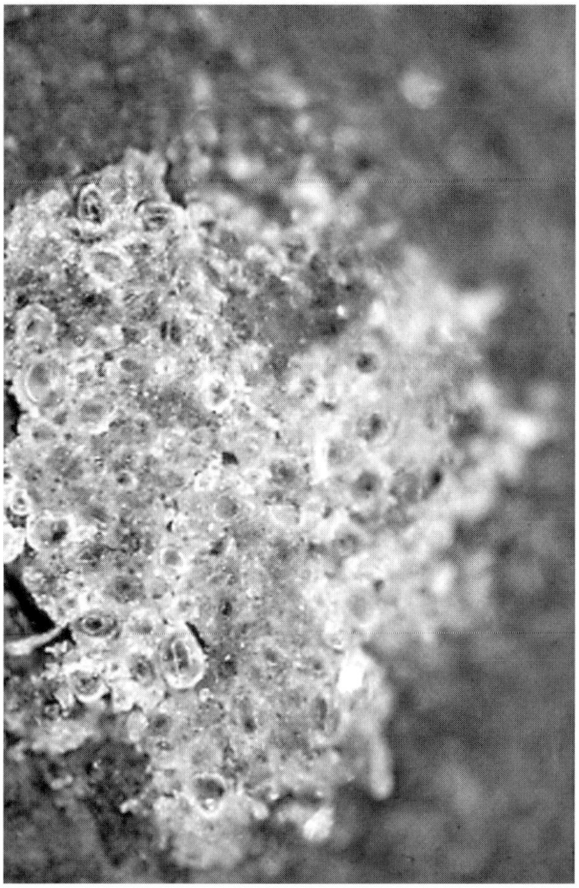

Plate 57. Starchy mass preserved on Ötzi's copper axe blade (Figure 9.1) is stained slightly green from copper oxide on the axe (magnification ca. 500 diameters). When viewed with cross-polarised light, the diagnostic interference figures were observed, but they were distorted, which is consistent with the granules having been heated to below their gelatinisation temperature. Note also the macerated cellulose and plant fibres associated with the starch grains. *Photo by Tom Loy*

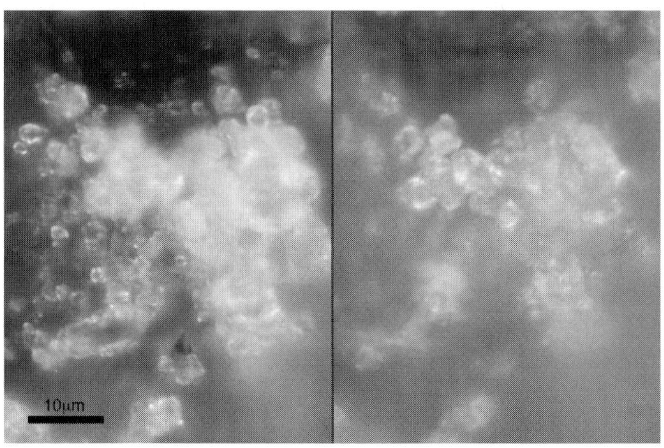

Plate 59. Experimental cooked starch granules from taro (*Colocasia esculenta*). *Photo by Alison Crowther*

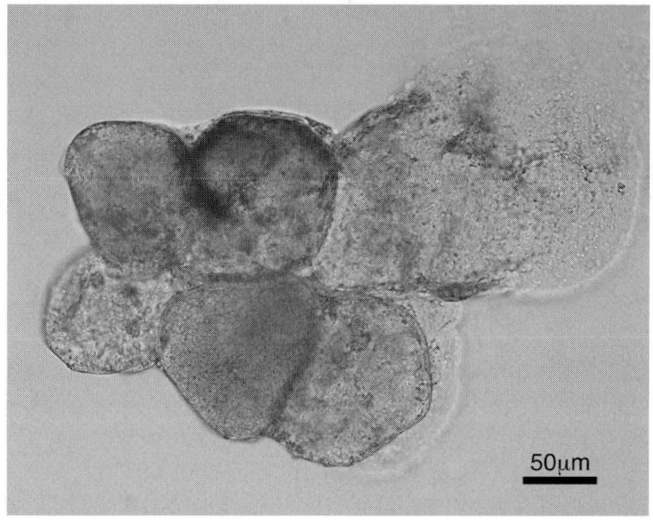

Plate 58. Partially gelantinised starch granules from taro (*Colocasia esculenta*) extracted from the surface of a Lapita potsherd (c. 3,200 BP), Papua New Guinea. *Photo by Alison Crowther*

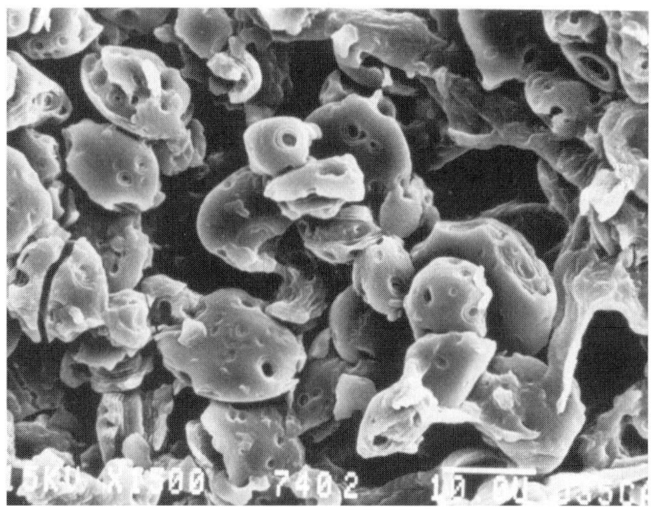

Plate 60. SEM micrograph of pits and channels created on the surface of starch granules during the production of Egyptian beer. *Photo by Delwen Samuel*

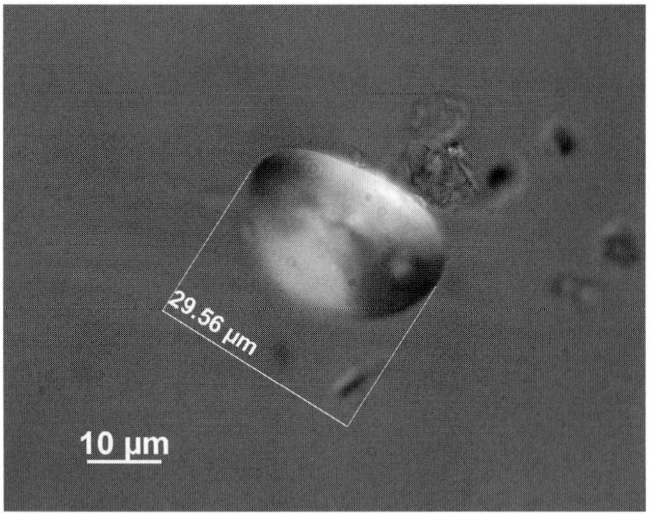

Plate 61. Starch granule recovered from the surface of grindstone CS6023 from Cuddie Springs. It is similar in shape and within the size range of Nardoo *(Marsilea drummondii)* and *Crinum flaccidum*, both plants of the arid/semi-arid zone known through ethnohistory (Gott 2002) to have been used in the recent past. Photographed using a Zeiss AXIOSKOP2 and Zeiss Axiocam HcR camera under DIC. *Photo by Judith Field*

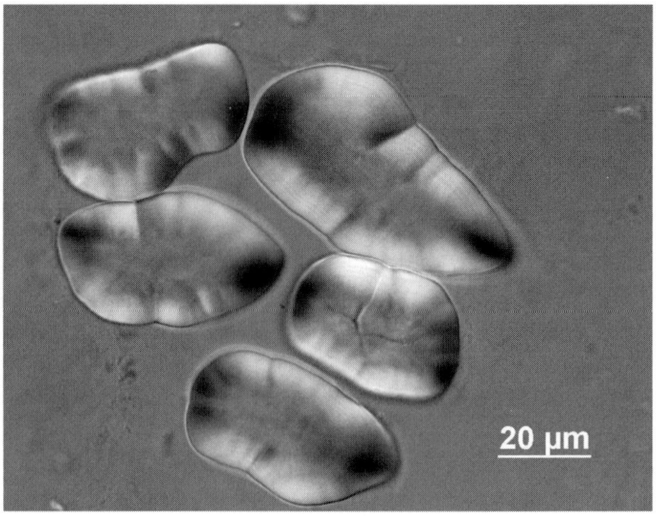

Plate 62. Starch granules of *Marsilea drummondii* (Nardoo) collected from sporocarps near the Cuddie Springs site in the semi-arid zone of southeastern Australia. Plan view showing lamellae, flattened hilum area, and folding of surface. Photographed using a Zeiss AXIOSKOP2 and Zeiss Axiocam HcR camera under DIC. *Photo by Judith Field*

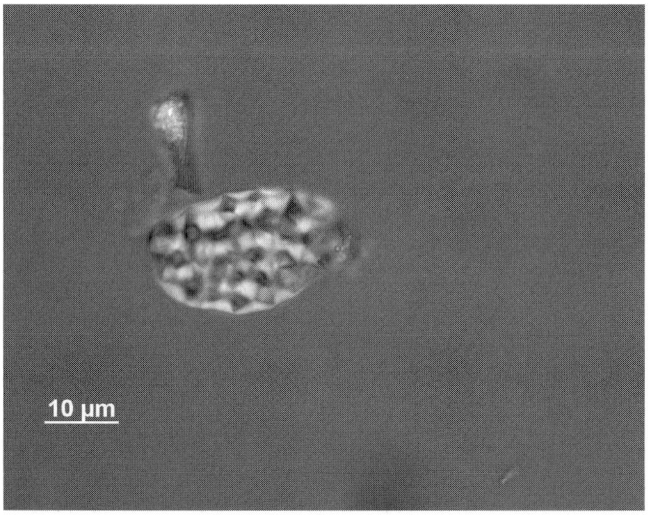

Plate 63. A compound starch granule recovered from the surface of grindstone CS6023 from Cuddie Springs found in association with grass type phytoliths is similar to the reference sample shown in Plate 64. Photographed using a Zeiss AXIOSKOP2 and Zeiss Axiocam HcR camera under DIC. *Photo by Judith Field*

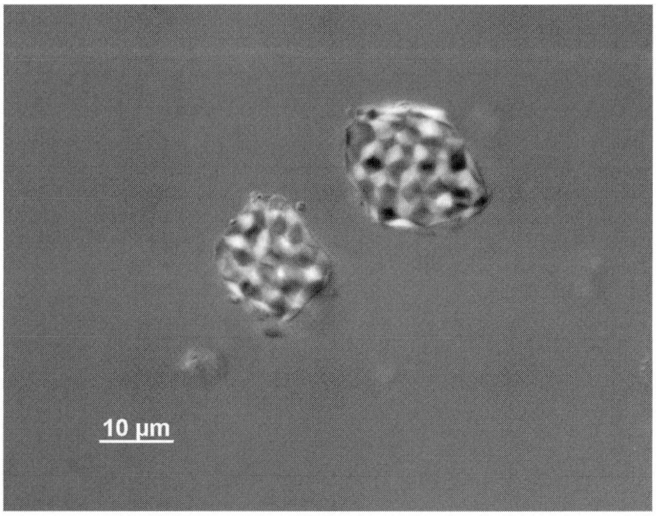

Plate 64. Starch granules from reference material of *Eragrostis parviflora* (Weeping Love Grass) collected in the same semi-arid zone as Cuddie Springs are similar to the granules recovered from prehistoric grindstones shown in Plate 63. Photographed using a Zeiss AXIOSKOP2 and Zeiss Axiocam HcR camera under DIC. *Photo by Judith Field*

single granule, as many variables as possible should be used to secure identification, but when there is a population, then all the granules should match most of variables. In order to confirm the identifications, Perry (2004: 1076) notes that one should further compare the granules to published keys, photos, and descriptions by other authors. Within her studies she adopted the very conservative approach that 'if the characteristics of the starch granule in question deviated from any source, published or otherwise, the identification was not secured.'

Piperno et al. (2004: 673) consider a population approach to identification as 'an effective and perhaps the most conservative means to distinguish individual genera and species' (cf. Piperno et al. 2000: 896–97). In this case one 'employs all the morphological characteristics that account for the population of starches in a single species as well as their quantitative frequency tendencies.' So, for example, a trait that occurs in a certain proportion of the granules in the reference collection should be present in about the same proportion in the ancient sample. Examples of case studies in which starch granules were identified are discussed later in the chapter.

SELECTION OF OPTICAL ATTRIBUTES

In choosing criteria for starch granule identification, it is important to focus on what Reichert (1913, 1919) has called the histological approach. This involves the microscopic analysis of basic attributes relevant to the shape and size of the grain. He notes that the diagnostic tests listed in Table 7.1 have serious limitations. For example, 'the polariscopic, iodine, and aniline methods are so crude that the personal equation enters largely into the determination of the values recorded' and polariscopic and aniline methods were 'by far the least useful of all of those used' (Reichert 1919: 371). Reichert favoured tests measuring the temperature of gelatinisation, but as noted above, these methods are unlikely to be useful in archaeological research because they are destructive and require relatively large samples.

In contrast, optical attributes are easy to describe and measure systematically and can be repeated on the same sample. As noted above, these have already been identified as useful by many botanists and have been used successfully in archaeological studies. The following characteristics have proved to be most important in assigning starch granules or populations to a plant taxon (cf. Pearsall et al. 2004; Perry 2001, 2002a, 2004, in press; Piperno in press; Piperno et al. 2000, 2004; Piperno and Holst 2004, Torrence et al. 2004):

1. overall granule shape taking into account two or three dimensions, contour, and surface features—e.g., spherical (Plates 1–4, 21, 33), ellipsoidal, ovoid (Plates 9, 10), or bell-shaped (Plates 7, 22); rough or smooth;
3. position and form of the hilum—e.g., centric (Plate 21) or eccentric (Plates 9, 10, 29); open (vacuoules) (Plate 23), semi-open, or closed (Plate 9);
4. fissure—e.g., present or absent; simple (straight), stellate, or y-shaped (Plate 16);
5. presence/absence of lamellae;
6. number and characteristics of pressure facets—e.g., rounded or angled edges; banded or unbanded (Plates 1–4, 11, 12, 13, 34);
7. size and morphology of the extinction cross—e.g., relative lengths and angle of the arms (Plates 10, 21).

Descriptions of these basic characteristics can be found in Chapter 2. Since starch granules are three-dimensional objects, an attempt should be made to rotate them under the coverslip. For example, a bell-shaped granule from *Ipomoea batatas* viewed from one end may appear round with a centric hilum, but on its side, it will display a very characteristic bell-shape with the hilum seen in an eccentric position (Plates 37, 38) (cf. Torrence et al. 2004: 520–24). Gently tapping the coverslip should provide enough force to mobilise most granules. Since polarised light is necessary for studying trait 6 and useful for trait 2, but traits 1, 3, 4, and 5 are most easily studied if cross-polarised light is not used, then both views should be available. Loy (Box 7.5) has also suggested that the refractive index of the granule might be a useful character, but as yet this has not been well studied.

Most established keys, such as Nägeli's (Table 7.1), begin with the distinction between simple and compound granules (Chapter 2). The problem for

Box 7.5. Measurement Of Refractive Index Using The Microscope

Tom Loy

Refractive index (ri.) of starch granules can be used to differentiate between starch granules from different plant taxa or to rule out suspected mimics. These methods are based on classical techniques used for mineralogical ri. determination and *do not* require a petrographic microscope. See Kerr (1959: 57–63) for a discussion of methods and the optical physics of refractive index determination in petrological applications. As yet, no large-scale database is available for ri. of starch granules, but preliminary evidence suggests that it may provide a useful partitioning test for starch granules (Fullagar et al. 1998: 52). There are two methods given below: one uses central illumination, and the other, oblique illumination. Oblique illumination can be achieved either below or above the stage in transmitted light. The Becke line mentioned below is a refraction effect. When the granule is observed in focus, there will be a slight 'halo' or white line following the outline of the granule. The Becke line is most clearly visible in monochromatic light. As the objective lens is raised or lowered, this halo will appear to move either into the body of the granule or out into the mounting media.

Materials

- Starch granules from control species and residues.
- Microscope slides and coverslips.
- Refractive index reference media (cf. Table 6.4). (For critical measurements a set of retraction index immersion media as used in mineralogy would be useful, although they are expensive.)
- Green filter to place between the light source and the condenser.

Method

1. For both methods, prepare a slide mount and immerse the sample in media of known refractive index and cover with a coverslip.
2. Begin with *central illumination* for the observation of *single* granules in transmitted light with a magnification of x100 or more and with the iris diaphragm partially closed (critical illumination). Illuminate with monochromatic light produced with a green filter.
3. In *central illumination* observe the faint white line near the edge of the granule (the Becke Line). Raise the microscope objective (that is, defocus by lowering the stage) and the line will move toward the medium which has the highest ri. If the image is defocused by moving the stage upward toward the objective, the line will move toward the medium that has the lower refractive index.
4. Set up *oblique illumination* (below the stage) for the observation of *many* granules. In transmitted light with a magnification of x100 or less, remove the condenser lens and insert the green filter to create monochromatic light. Then insert an opaque card below the stage, cutting off half the field of light.
5. Observe in *oblique illumination* the shadowing of individual granules. The field of view will be light on one side and dark on the other. If the shadow on a granule falls on the light side of the field of view, then the ri. is higher than the immersion liquid. If it is on the dark side, then the ri. is lower than the immersion liquid.
6. Repeat the *oblique illumination* observation in white light. If the ri. of the granule is about equal to the ri. of the immersion liquid, then one side of the granule will be blue, and the other will be red.
7. An alternative set-up for *oblique illumination* (above the objective) is also appropriate for the observation of many granules. In transmitted light with a magnification of x100 or less, with condenser lens in place, monochromatic light is produced with a green filter. Insert an opaque card above the objective lens, cutting off half the light.
8. Observe the granules as above. If the shadow is on the side toward the dark area, then the granule has a higher ri. than the immersion liquid.

identification of ancient starch is that few intact compound granules survive in the archaeological or environmental record. Generally, separate subgranules are observed. Although these are characterized by having polyhedral shapes because they were formed as part of a cluster, similar configurations can also be created as a result of crowding by small simple granules within cells (cf. Plates 11–13, 34, 36). Differentiation between simple granules with pressure facets and granula of compound granules can therefore be a complex task (Cortella and Pochettino 1994: 174). For this reason, analysts of ancient starch generally ignore the distinction between simple and compound granules and begin the process of identification by focusing on general, overall morphology. This character alone often helps narrow the range considerably, as, for example, in the case of the distinctive bell-shape of manioc starch (Plate 7), described in detail by Piperno et al. (2000; Piperno and Holst 2004). An important aspect of granule morphology is the presence or absence and differences of pressure facets which create polyhedral granules. These can also vary in terms of their number and configuration.

The position of the hilum is the next most important variable in identification. This point from which the granule grows can occur in the center of the granule, often within circular or oval granules, when it is termed centric (Plate 21). In oval granules or elongate granules, the hilum is often located at one end of the granule, when it is termed eccentric (Plates 10, 11). The hilum can also be described as slightly eccentric or very eccentric, or in rare cases a hilum may not be observed (Piperno and Holst 2004).

Lamellae and fissures are formed during the growth of the starch granule, but since they are either not present or not easily visible on the granules of all plant species, their presence can be a useful aid to identification. Lamellae are parallel bands surrounding the hilum (Plate 9). Fissures form outward from the hilum (Plate 16). In some species a number of different fissure types are present. For example, Piperno et al. (2000: 895) note that 'crossed, winged to y-shaped, or stellated fissures' are all characteristic of domesticated manioc. In contrast to fissures, which are a natural part of some starch granules, cracks are splits in the granule structure induced by an external force, such as milling, or as a consequence of dehydration (Plate 21, 31, 32; cf. Box 4.4).

The final criteria important in the identification of starches are their characteristics under cross-polarised light. Differences in the shape, size, and character of the extinction cross and its occurrence within the granule vary across plant types. Three main types represented by potato, wheat and sorghum, and cowpea (*Vigna catjang*) are illustrated by Chandrashekar et al. (1987): (1) potato has an isotropic region around the periphery of the granule (Plate 10); (2) sorghum and wheat have an isotrophic region in the center of the granule so under polarised light the granule has a central dark area and it appears as if it has a hole in the middle (known as a vacuole or closed hilum) (Plate 23); and (3) there are no isotrophic areas in the cowpea granule. Furthermore, the arms of the extinction cross can meet at various angles, which is partially dependent on granule orientation, and the extinction cross can have bent or straight arms that may appear fine to quite broad (Plates 5, 10, 12, 38). The degree (brightness) of polarisation can also vary from taxon to taxon, from fair to very high, but so far this trait has not been found to be a useful criterion for identification because polarisation is very difficult to measure accurately and can often vary between starch granules derived from within one organ of a single species (Reichert 1919: 371).

TAXONOMIC DETERMINATION

Barton's (2005) case study concerned with the identification of starch granules recovered from sediments at Niah Cave, Sarawak, summarized in Box 7.6, illustrates the procedures required when dealing with plant taxa that have not been well studied and are not included in previous starch granule keys or compendia (e.g., Reichert 1913). The first task was to group the granules into a number of putative 'types,' based on their shape, size, and any distinctive features. Next Barton compared these to reference material obtained from herbaria and supplemented with fieldwork. Five of the types were successfully matched to plant taxa. The results are important because they demonstrate that a wide range of tubers were processed at this early cave site. Improvements to the specificity of the identifications will depend on accessing a wider

Box 7.6. Starch Granules From Niah Cave Sediments

Huw Barton

Recent archaeological excavations at the site of Niah Cave in Sarawak (Malaysia) have confirmed a history of human use that dates back at least 45,000 years BP. The site contains a rich assemblage of human remains, vertebrate fauna, molluscs, pottery, stone, bone, and shell tools (Barker et al. 2002). Organic preservation within the guano-rich, but dry, limestone cave sediments is very good. Bone is well preserved and several late Holocene burials contained excellent examples of preserved basketry and woven fabric (Harrisson 1967: 189). The rationale for starch research at the site was the expectation that the processing and consumption of starchy foods might have left a 'residue' of starch which had been incorporated and preserved within defined sedimentary horizons (Barton 2005). The identification of starch granules could therefore be used to reconstruct the history of plant use at Niah.

Samples

A total of 94 soil samples were analysed from three main deposits within the cave: (1) an excavated baulk within the Hell Trench (57 samples) that dates between 45,900 ± 800 (OxA-V-2057-31) years BP and 34,880 ± 390 (though there are some age reversals in the sequence); (2) Area A (21 samples) dating from 27,960±200 (OxA-11034) near base to 915 ± 28 (OxA-14067); and (3) Block B (16 samples) from 19,650±200 (OxA-11550) to 8,630±45 (OxA-11549).

Method

Soil samples were initially sieved to remove the coarse fraction. Two 5 g samples were taken from each sedimentary unit. The lightest fraction of organics and minerals was removed using a low-density solution of sodium polytungstate. Next, the starches and any other material with a specific gravity less than 1.5 were removed in a high-density solution of sodium polytungstate. At the end of the extraction process, the final sediment was resuspended in 500 ml of ultrapure water, from which a 50 ml aqueous sample was removed by pipette. The extract was left to air-dry on a covered glass slide and mounted with glycerine.

Recording Starch

The starch granules were described using a combination of size and morphological characteristics. Maximum length and width were recorded using an eyepiece micrometer and most shape and other attributes were recorded at x200 to x1000 magnifications. Since starch granules are three-dimensional objects, it is highly desirable to try and rotate them beneath the coverslip to ensure that an accurate shape identification is made. I used glycerine as a mountant as it remains liquid, although it is prone to drying out if the coverslip is not airtight. The determination of size and/or shape alone will not be sufficient to enable accurate identification in most cases, as there is some overlap among the granules of several species. However, when combined with a more detailed assessment of the morphological characteristics of granules, such as surface features, colour, hilum position, hilum features, and the number and nature of granule facets, a higher level of taxonomic identification is possible (e.g., Loy et al. 1992; Piperno et al. 2000, 2004).

Accurate identifications of starch grains in archaeological deposits depend on comparison with modern starches. This study has focused on the identification of starchy roots and some palms known or thought likely to have grown within the vicinity of Niah Cave in the past. My reference collection of Southeast Asian species housed at the University of Leicester is not exhaustive by any means, but represents what can be put together with living and voucher specimens from various botanic gardens (such as the Royal Botanic

range of reference material for the genera already recognized at the site.

The search for agricultural origins in the Pacific region has turned to the study of starch granules as a way to monitor the presence of domesticates. Although most analyses have had difficulty in distinguishing among the numerous aroids simply on the basis of granule shape and size (e.g., Horrocks and Bedford 2005), Fullagar et al. (2005) have made progress by including a number of extra traits. In their study they found that in addition to the small size and spherical shape of *Colocasia esculenta* granules, the presence of a faceted surface together with the distribution of the granules within sheets of tissue enabled them to make confident identifications of starches found within residues on stone tools, a property also noted by Horrocks (2004: 328) (Plates 1–4).

Further examples of the procedures for classifying starch granules are provided in two classic studies in which starch granules extracted from grinding stones have provided the earliest evidence for the use of particular plants. In the first case, Piperno et al. (2000), building on initial identifications by Piperno and Holst (1998), identified granules from manioc, yams, and arrowroot on

Box 7.6 (continued)

Gardens, Kew, in London, and the Edinburgh Botanic Gardens, Scotland) and limited field collection in Sarawak. The Kew herbarium was a particularly useful source of material because small pieces of dried tubers were present on many of the herbarium sheets. Although dry and woody (mostly collected between 1916 and 1964), identifiable starch granules were still present. The oldest starches in my collection are derived from Kew's Economic Botany Collection. Preserved starches were recovered from samples of dried tissue and flour, some of which had been collected as early as 1856.

Results

A total of 201 starch granules were recovered from the initial screening of the deposits. Starch granules were recovered from around 50 percent of all slides processed. Generally only a few granules were recovered from each 50 g soil sample, though one slide yielded 100 granules (the limit of counting per slide). Recovery rates seem quite low, but it has to be emphasised that we are still at the early stages of developing this technique. Associated with the starch granules on most slides were other well-preserved organic remains such as intact cellular plant tissue, including fibres, vascular tissue, plant hairs (trichomes), phytoliths, pollen, and bird feather barbules. I was able to characterise at least fourteen distinct 'Types' of starch granules. A total of 155 granules were assigned to one of 14 distinct 'Types,' that is groups with distinctive recurrent morphologies. Five 'Types' could be confidently assigned to a specific genus and two to a species (Table 7.6). Forty-four round granules of various sizes could not be assigned because they could belong to any one of a wide range of plant taxa.

Granules from 'Type 4' could fit within either the genus *Alocasia* or *Cyrtosperma.* These could not be separated by a discriminant analysis (SPSS) comparing the archaeological sample to my reference material. My 'Type 6' granules appear very similar to the yam *Dioscorea alata* and make a good fit statistically within this group (Plate 43). However, as I am not confident that my reference material contains enough representatives of the Dioscorea family known to occur in Malaysia, its identification to species remains unsecured.

My problems with classification highlight one of the main problems faced when trying to identify single granules from an archaeological context to the level of either genus or species. Some granules may not display enough unique variables to allow confident identification or it may be that the reference collection does not contain all the relevant species. The approach that I have taken is to describe starches by as many unique variables as possible, not all of which are listed in Table 7.6. For example, coloured bands are present on *Alocasia longiloba* (Type 8) granules (Plate 44), a rather uncommon feature of this species not seen on any other examples of *Alocasia* collected thus far. Further reference material may increase the specificity of such variables.

grinding stones. In each case a combination of characters was used to match the ancient starch granules to modern reference material. For example, the diagnostic characters for manioc are given as 'rounded and unbanded pressure facet edges, centric and open hila, and no demonstrable lamellae' combined with 'crossed, winged to y-shaped, or stellated fissures.' At the same time they notc the presence of a very rare starch granule morphology but one that is a 'highly idiosyncratic form unique to manioc' (Piperno et al. 2000: 895). This study illustrates the importance of using a combination of approaches involving, first, the detection of specific characters on individual granules, and, second, applying the population approach in which a range of particular forms are used to characterise a particular plant taxon.

In a second case the task was to isolate a particular group comprised of three genera of wild grasses from others known to occur in the same region. As in the previous case, the granules had been extracted from grinding stones. Piperno et al. (2004) again used the presence of a unique combination of characters for individual granules combined with the correct frequencies of particular combinations within the population to create a

Box 7.6 (continued)

Table 7.6. Analytical type categories of starch granules defined at Niah Cave.

Type	Taxonomic Grouping	Simple Shape Category	Granule Form	Size Range (µm)	Mean Size *(l)*	Hilum Position	Count
4	*Alocasia* sp. and *Cyrtosperma merkusii.*	sub-round	single	15–20	17	centric	5
5	*Alocasia longiloba* Complex	bullet	single	23–25	24.5	eccentric	4
8	*Alocasia longiloba* Complex	round	single, semi-compound	18–35	24.8	eccentric	8
6	*Dioscorea* sp. (poss. *D. alata*)	ovate	single	20–45	34.5	highly eccentric, eccentric	4
9	*Eugessonia utilis*	ovate	single	28–60	45.2	highly eccentric	9
1†	unknown	round, sub-round	single (clustered)	1.5–75	61, 1.5	centric	7
2	unknown	polygonal	single (clustered)	15–20	16.2	centric	4
3	unknown	round, sub-round	single, semi-compound	2–10	7.5	centric	100s
10	unknown	sub-ovate	single		55	highly eccentric	1
11‡	unknown	polygonal	single (clustered)	1–2	2	not visible	100s
12	unknown	sub-ovate	single	20–30	25	centric	3
13	unknown	bean	single	18–30	24.8	centric	5
14	unknown	sub-round, disc	single	15–25	21.8	centric	7
NT	not typed	round	single				44

† Type 1 consists of a distinct cluster of granules on one slide. It had two distinct shapes: very large round granules (mean 61 µm) and very small sub-round granules (mean 1.5 µm).

‡ Not yet fully determined if this group is starch. It appears similar to *Dioscorea hispida* or *D. esculenta* but the archaeological granules are not birefringent.

diagnostic signature for each of the particular genera. The result was similar to Barton's (Box 7.6) study, in that size turned out to be a very important criterion, followed by shape. In this case, however, the proportion of particular granule types within the overall population added credibility to the identifications.

MULTIVARIATE APPROACHES

A supplementary approach to the use of taxonomic keys is to use multivariate statistical procedures to discriminate granules from various plant types on the basis of a relatively large number of variables. Preliminary differentiation of the wide range of types encountered in an archaeological deposit may be quite difficult in the initial stages of research, especially at the very early stages of an analysis where there are not very many distinctive variables, or when one is faced with a large number of species with roughly similar granules. Multivariate analysis can also make an important contribution when there appear to be only minor differences between species (cf. Barton 2005). The first use of multivariate techniques in the study of ancient starch granules was Loy et al.'s (1992; Loy 1994a) study which discriminated among a number of economically useful species found in Southwest Asia and the Pacific region.

Having found that it was difficult to distinguish a number of Aroids (taros) species which are described as 'generally subround to ovate grain shape with a variety of well-developed facets, reflecting the formation of compound grains during growth' (Loy 1994a: 104), they used multidimensional scaling to see if they could identify particular species. They recorded the following variables for 100 granules of each of 20 species of economically useful USOs, mainly taros and yams, from the Pacific and Asian region: overall shape, angularity, faceting, 'maximum and minimum dimensions, location of hilum, pattern of extinction cross and layering patterns' (Loy 1994a: 103).

Their results showed that size and shape were useful as initial discriminators, but that the presence/absence of particular raphides (calcium oxalate crystals) was also necessary for satisfactory discrimination (Plates 53–55; see Box 7.7). The presence of raphides was also used by Crowther (2001, 2005; Box 9.5) as an important trait in the identification of *Colocasia esculenta* from residues on potsherds, but Horrocks and Bedford (2005) note that whereas the commonly found raphides (whisker type) are common to several species of aroid, the species-specific types are very rare (cf. Fullagar et al. 2005) (Plate 54). Another indicator which Horrocks (2004; Horrocks et al. 2004a, 2004b; Horrocks and Bedford 2005) has found useful is the morphology of the xylem vessel elements (Plate 56). In cases where discrimination is difficult solely on the basis of granule morphology, then additional elements may prove helpful in identification of plant taxa.

In a further multivariate study of Pacific plants, Torrence et al. (2004; cf. Box 7.8) moved beyond the set of domesticates analysed by Loy et al. (1992) to include a range of 'wild' plants. As in the previous study, shape and size were monitored through a range of metrical variables, but area and a range of measurements on the extinction cross were also included. A combination of correspondence, linear discriminant, and canonical variates analyses were successful at separating the groups and classifying a significant number of the granules to the correct species. Confusion was mainly between plants of the same species, genera, or family level. The multivariate study demonstrates that starch granules from different species are truly distinguishable and reinforces the basic assumption underlying starch granule classification and identification.

Tackling the job of establishing a key for a large number of plant types for an area that is poorly researched, such as Papua New Guinea, is a daunting task. The first stage, however, can be facilitated through multivariate analyses of granule morphologies using the variables that have been shown in previous work elsewhere to be useful. Starch granules will not vary so much that one needs to start from scratch. Multivariate analyses can help establish a preliminary, working key and will demonstrate which plant types can most easily be discriminated and where further study is required. In the long run it may even be possible to use this approach to create an automated system of classification using well-established and tested keys.

IDENTIFYING DOMESTICATION

Recent work suggests that starch granule morphology might be a useful trait for discriminating between wild and domesticated varieties of the same plant species. In a pioneering article, Piperno et al. (2000) noted that the *Dioscorea* granules recovered from a grinding stone more closely resembled those of modern wild *Dioscorea* species than they did the modern domesticated yam *Dioscorea trifida*. In the same article, the authors present quantitative data to show that the starch granules in the wild relatives of domesticated manioc are generally smaller. In addition, the wild forms have 'much narrower pressure facets, simple

Box 7.7. Raphides

Tom Loy

Crystalline forms of calcium oxalate ($Ca_2(COOH_2)_2$) are found in many plant families, genera, and species. These elongated crystals are called raphides (Plates 53, 54). The crystals are formed in specialized cells and grow in their thousands as parallel aligned individual raphides (Plate 55). The raphide cells can be found in leaves, stems, petioles, and in the outer layers of tubers and corms, and to a lesser extent within the tuber/corm itself. Raphides perform two important functions within the plant: (1) as a means of calcium storage that can be mobilised by the plant when extra physiological calcium is required; and (2) as a part of the defensive system of the plant.

Related to raphides, and made of the same compound, crystalline druses are grown by some plants (e.g., *Ipomea costata, I. polpha*). Druses are roughly spherical aggregations of flat crystals that lack the regularity of properly twinned mineral crystals (Plate 52). Druses do not play a defensive role, but do store calcium.

Cells containing raphides generally have an elongated ovate shape with a circular cross-section (Plates 53, 54). The cells are turgid, with very thin walls, and are easily ruptured under light external pressure. When a plant with raphides is bitten into, the cells rupture and forcefully eject the sharp raphides into the soft tissues of the mouth. Consequently, part of the 'bite' experienced when eating a raw taro (*Colocasia esculenta*) is caused by the raphides. In the family of the Araceae (which includes taro and its less edible relatives, e.g., *Alocasia* sp.), the raphides are also apparently coated with an 'acrid principle' which has yet to be fully chemically characterised and which contributes to the unpleasantness of eating raw aroids.

Interestingly, the morphology of raphides appears to be under genetic control because raphides within any species are identical. They can therefore be used in conjunction with starch grain morphology and size analysis to increase the confidence in the identification of the plant of origin. Raphide morphology varies widely: e.g., simple elongated prisms (*Colocasia esculenta*); smooth cylindrical shapes tapering to a sharp point at each end (*Dioscorea esculenta*); and complex forms that have components of varying shape and cross-section (e.g., *Amorphophallus* sp.). In some species smaller crystals grow at the tips to form barbs. Raphide length tends to be uniform unless the crystals have been broken. Variations in the actual crystal growth axes can often result in pleiochroism (bright colours) when viewed in cross-polarised light, and a variety of bands, segments, and zones of colour can also be observed. These patterns of shape, the presence of pleiochroism, and its patterning can be used in many cases to identify at the genus/species level (e.g., the very closely related *Hypoxis* species that occur in the high veldt of north-eastern South Africa).

Further details about raphides can be found in the following sources: Bradbury and Holloway (1988); Loy (1994a); Loy et al. (1992); Sakai (1979); and Sunell and Healy (1979).

Box 7.8. Classification Using Image Analysis And Multivariate Techniques

Robin Torrence and Rebecca Conway

A multivariate approach can be very useful during the initial stages of setting up a system of classification for starch granules, particularly if the data set is largely from species not previously incorporated into earlier keys and/or nondomesticates. For example, ethnobotanical research in a New Guinea village led to the collection of 172 plant species that were regularly used (cf. Box 6.2; Therin et al. 1997). Since these could be expected to occur in the archaeological record of the region, the problem then arose as to how to create starch granule keys for so many plants. Multivariate methods were selected because they could determine whether the species could be adequately discriminated on the basis of granule characteristics in the first place and second, could identify which characteristics provided the best discrimination (cf. Torrence et al. 2004).

We began with 29 plant types which were selected because the starch granules were relatively large, had distinctive characteristics, or were very abundant. Aroids and yams from the original Loy et al. (1992; Loy 1994a) multivariate study were included for comparison. Unlike many available starch keys which have focused on cultivated or domesticated species, we included wild and feral plants. As shown in Table 7.7, the sample includes a very wide range of uses in addition to food: e.g., housing, material culture, poison, and medicine.

Recording Methods

Since the reference collection had been prepared with a permanent mounting medium, morphology could only be monitored in two dimensions. Three views (centric, eccentric, side-on) are distinguished by the position of the hilum (respectively central, off center, not visible). Centric granules consistently present an 'end-on' view from whatever angle they are observed, but eccentric granules can be observed in any view depending on the plane in which the granule is observed (cf. Chapter 2). Fifty granules of each orientation (or 'view') that occurred were recorded for each species. Fifty granules of each morphology observed were recorded since four species had two very different types of granule. A total of 848 granules observed in the eccentric view and 1,150 granules in the centric view were recorded.

Unlike previous quantitative studies, which have focused mainly on surface area and volume to discriminate among different morphologies within a single species (cf. Zayas et al. 1994; Oliveira et al. 1994), we included a wide range of variables, as listed in Table 7.8. First, the granules were viewed under normal incident light to assess their shape and the presence/absence of striae, lamellae, vacuoles, a particle at the hilum, and raphides. Second, each granule was photographed under cross-polarised light and measurements were taken using UHSCS Imagetool software.

Multivariate Techniques

The eccentric (15 species) and centric (20 species) were treated separately for the purposes of analysis. To begin with, the nonmetric variables were converted into presence/absence, if that was not already the case. Of the three forms of discriminant analysis, the nearest neighbour technique applied to scores derived from a canonical variates analysis of the combined metric and nonmetric data produced the best results (cf. discussion in Torrence et al. 2004: 522–24).

Results

The so-called 'confusion' matrices shown in Figures 7.4 and 7.5 show how the granules are classified using the rules generated by the discriminant analysis. One can see which granules are most often correctly attributed. Where there is an error, one can find out to which incorrect species the granules were classified. For

Box 7.8 (continued)

Table 7.7. Species used in a multivariate classification of starch granules.[1]

Species	Family	Common name	Use	Local?
Alocasia magnifica	Araceae	wild taro	forest herb; not used	Yes
Alocasia sp.	Araceae	taro	cultivated herb; corm and leaves edible	Yes
Amorphophallus campanulatus Bl.	Araceae	elephant foot yam	wild herb; also cultivated; corm and leaves edible	Yes
Colocasia affinis	Araceae	taro	wild herb	No, SE Asia
Colocasia esculenta (L.) Schott	Araceae	taro	cultivated herb; tubers eaten	Yes
Colocasia fallax	Araceae	taro	wild herb	No, SE Asia
Cyrtosperma chamissonis (Schott) Merr.	Araceae	giant swamp taro	wild herb; also cultivated; corms, leaves and flowers edible	Yes
Xanthosoma sagittifolium (L.) Schott	Araceae	singapore taro	cultivated herb; corms eaten and young leaves edible	Yes, recent
Caryota rumphiana Mart.	Arecaceae	fish tail, black palm	forest palm; used for building and crafts; starch sometimes extracted from trunks; pith eaten	Yes
Metroxylon sagu Rottb.	Arecaceae	sago	wild or cultivated palm; starch extracted from trunks; young leaves eaten	Yes
Barringtonia asiatica (L.) Kurz	Barringtoniaceae		coastal strand tree; seed parings used for fish poison	Yes
Canarium indicum L.	Burseraceae	Pacific almond, galip	forest tree tended for nuts; sometimes cultivated	Yes
Ipomoea batatas (L.) Lam.	Convolvulaceae	sweet potato	cultivated vine; tubers and leaves eaten	Yes
Dioscorea alata L.	Dioscoreacae	greater yam	cultivated vine; tubers eaten	Yes, recent
Dioscorea bulbifera L.	Dioscoreacae	yam	cultivated vine; bulbils and tubers eaten	Yes
Dioscorea esculenta (Lour.) Burk.	Dioscoreacae	yam	cultivated vine; tubers eaten	Yes
Dioscorea nummularia Lam.	Dioscoreacae	yam	cultivated vine; tubers eaten	Yes
Dioscorea pentaphylla (L.)	Dioscoreacae	yam	cultivated vine; tubers eaten	Yes
Dioscorea rotunda	Dioscoreacae	yam	cultivated vine; tubers eaten	Yes
Manihot esculenta Crantz	Euphorbiaceae	cassava, tapioc	cultivated shrub; tubers, leaves and seeds eaten	Yes, recent
Hernandia nymphaeifolia (Presl.) Kub.	Hernandiaceae		coastal strand and forest tree; trunk used for canoes	Yes
Hibiscus manihot L.	Malvaceae	aibika	cultivated shrub; leaves and young stems eaten	Yes
Artocarpus altilis (Park.) Fosb.	Moraceae	breadfruit	cultivated tree; also in forest; leaves, seeds, flowers, and fruit eaten	Yes
Musa sp.	Musaceae	banana	tall cultivated herb; fruit, flower buds, flowers, pith, and stem bases eaten; leaves used for wrapping and cooking	Yes
Horsfieldia laevigata (Bl.) Warb.	Myristicaceae		forest tree; timber used for house construction	Yes
Syzygium malaccense (L.) Merr. & Perry	Myrtaceae	jambu	forest and garden fruit tree	Yes
Lygodium circinatum (Burm. f.) Sw.	Schizaeaceae	climbing fern	forest fern; stem used for bags, baskets, armbands, and binding	Yes
Solanum torvum Sw.	Solanaceae		weedy shrub in gardens and regrowth; juice from leaves used to treat malaria and colds	Yes
Zingiber officinale Roxb.	Zingiberaceae	ginger	cultivated herb; used for magic, medicine, and spice	Yes

1. Information from Peekel 1984; French 1978; C. Lentfer fieldnotes.

Box 7.8 (continued)

Table 7.8. Variables used in a multivariate classification of starch granules (after Torrence et al. 2004).

Variable	Type of variable
1. Orientation of starch grain (view)	3 Nominal categories
2. Area of whole view	Numeric
3. Area of polarization cross	Numeric
4. Length (maximum dimension)	Numeric
5. Breadth	Numeric
6. Maximum dimension of the polarization cross	Numeric
7. Maximum distance between long arms of the cross	Numeric
8. Distance of hilum from end of grain	Numeric
9. Maximum angle of the cross	Numeric
10. General shape	6 Nominal categories
11. Refined shape	17 Nominal categories
12. Striae	Presence/Absence
13. Lamellae	Presence/Absence
14. Margins	2 Nominal categories
15. Vacuole	Presence/Absence
16. Particle at hilum	Presence/Absence
17. Flat facet	Presence/Absence
18. Style of cross	3 Nominal categories
19. Raphides	Presence/Absence

example, looking at the rows in Figure 7.4, 39 of the 50 *Dioscorea nummularia* granules have been correctly allocated to that species and 11 have been assigned to other species, although these are all members of the same genus. Alternately, columns show that 39 granules were correctly classified to *D. nummularia,* whereas 18 granules from three other species were incorrectly assigned to this species, but only one of these was a granule from a different genus. One can therefore be highly confident that granules of *D. nummularia* would be classified correctly to species levels using the variables we recorded and that an assignment to genus would be even more accurate.

The confusion matrices can also be examined to see which species are easiest to classify and where the least successful classifications take place. It is heartening to know that 100 percent success rates were achieved for the eccentric orientation of *Dioscorea alata*, *Musa* sp., and *Metroxylon sagu* and for centric views of *Dioscorea esculenta* and *Horsfieldia laevigata.* Overall the results for the eccentric view were much better than the centric orientation.

In a pilot study to this one, Jackes (1997) was unable to discriminate among the *Dioscorea* species of *nummularia, bulbifera*, and *alata*, but our study has had much more success. The least successful classification involved *Dioscorea nummularia* and *Dioscorea pentaphylla*, but these were mainly confused with other species of the same genus. *Hibiscus manihot*, a common garden plant rather like spinach in which the leaves are eaten, was misclassified on both views with several 'wild' species, all of which tended to get confused among each other. These results confirm the expectation that starch from USOs is likely to be more distinctive than that from fruits and leaves which contain transient granules.

The worst overlap for eccentric grains was between one morphology of *Caryota rumphiana* and all of *Syzygium malaccense.* On closer observation we noted that one of the morphologies of *Caryota rumphiana* is quite distinctive, whereas the other is not. This also

Box 7.8 (continued)

	No. of cases	Barringtonia asiatica	Caryota rumphiana	Colocasia fallax	Dioscorea alata	Dioscorea bulbifera	Dioscorea nummularia	Dioscorea pentaphylla	Dioscorea rotunda	Hernandia nymphaeifolia	Hibiscus manihot	Ipomoea batatas	Metroxylon sagu	Musa sp.	Syzygium malaccense	Zingiber officinale
Barringtonia asiatica	50	39	0	0	0	0	0	2	0	0	1	8	0	0	0	0
Caryota rumphiana	100	0	65	8	0	0	0	0	1	4	8	0	0	0	14	0
Colocasia fallax	48	0	11	24	0	2	1	2	1	3	4	0	0	0	0	0
Dioscorea alata	50	0	0	0	46	0	0	0	4	0	0	0	0	0	0	0
Dioscorea bulbifera	50	0	1	3	1	36	6	0	0	1	0	0	0	0	0	2
Dioscorea nummularia	50	0	0	0	1	4	39	0	6	0	0	0	0	0	0	0
Dioscorea pentaphylla	50	1	0	3	0	1	0	33	0	1	5	6	0	0	0	0
Dioscorea rotunda	50	1	1	2	1	2	11	2	26	0	2	0	2	0	0	0
Hernandia nymphaeifolia	50	0	11	2	0	0	0	0	0	33	2	0	0	0	2	0
Hibiscus manihot	50	1	10	2	0	0	0	4	0	1	27	1	0	0	4	0
Ipomoea batatas	50	9	0	0	0	0	0	5	0	0	0	36	0	0	0	0
Metroxylon sagu	100	0	0	0	0	0	0	2	0	0	0	0	98	0	0	0
Musa sp.	50	1	0	0	0	0	0	0	0	0	0	0	0	49	0	0
Syzygium malaccense	50	0	13	0	0	0	0	0	0	1	1	0	0	0	35	0
Zingiber officinale	50	0	0	0	0	0	0	0	0	0	0	0	0	0	0	50

Figure 7.4. Multivariate classification of starch granules eccentric view: confusion matrices based on the discriminant analyses. *Reproduced with permission from Torrence et al. 2004: figure 7*

	No. of cases	Alocasia magnifica	Alocasia sp.	Amorphophallus campanulatus	Artocarpus altilis	Barringtonia asiatica	Canarium indicum	Colocasia affinis	Colocasia esculenta	Cyrtosperma chamissonis	Dioscorea esculenta	Dioscorea pentaphylla	Hernandia nymphaeifolia	Hibiscus manihot	Horsfieldia laevigata	Ipomoea batatas	Lygodium circinatum	Manihot esculenta	Solanum torvum	Syzygium malaccense	Xanthosoma sagittifolium
Alocasia magnifica	50	39	0	0	4	0	0	0	0	0	0	0	1	0	0	0	0	0	0	6	0
Alocasia sp.	50	0	44	6	0	0	0	0	0	0	0	0	0	0	0	0	0	0	0	0	0
Amorphophallus campanulatus	50	0	33	16	0	0	0	0	0	0	0	1	0	0	0	0	0	0	0	0	0
Artocarpus altilis	50	7	0	0	32	0	0	0	0	0	0	0	0	0	0	0	0	0	0	11	0
Barringtonia asiatica	50	0	0	0	0	20	1	0	1	0	1	1	0	0	0	9	0	16	0	0	1
Canarium indicum	50	0	0	0	0	0	28	0	4	0	0	6	0	6	0	0	3	0	3	0	0
Colocasia affinis	50	9	0	0	10	0	0	7	8	0	0	1	1	0	0	0	2	0	1	11	0
Colocasia esculenta	50	0	0	0	0	2	1	0	33	6	0	1	0	0	0	1	5	0	0	0	1
Cyrtosperma chamissonis	100	0	0	0	0	3	1	0	2	49	0	2	1	0	0	1	1	10	1	0	29
Dioscorea esculenta	50	0	0	0	0	0	0	0	0	0	50	0	0	0	0	0	0	0	0	0	0
Dioscorea pentaphylla	50	0	0	0	3	0	3	3	1	5	0	23	0	3	0	0	1	2	1	3	2
Hernandia nymphaeifolia	50	1	0	0	0	1	0	0	0	5	0	0	40	0	0	0	0	1	0	0	2
Hibiscus manihot	50	0	0	0	3	0	11	2	1	0	0	5	0	16	0	0	1	0	5	6	0
Horsfieldia laevigata	50	0	0	0	0	0	0	0	0	0	0	0	0	0	50	0	0	0	0	0	0
Ipomoea batatas	50	0	0	0	0	9	0	0	1	0	0	1	0	0	0	20	0	14	0	0	5
Lygodium circinatum	50	0	0	0	0	0	5	0	9	9	0	3	0	1	0	0	17	0	4	0	2
Manihot esculenta	100	0	0	0	0	8	0	0	0	0	0	3	1	0	0	6	0	73	0	0	9
Solanum torvum	50	0	0	0	2	2	7	1	3	3	0	2	3	1	0	0	2	0	12	11	1
Syzygium malaccense	50	2	0	0	9	0	0	2	0	0	0	0	1	0	0	0	0	0	3	33	0
Xanthosoma sagittifolium	100	0	0	0	0	10	0	0	1	8	0	2	0	0	0	7	1	17	0	0	54

Figure 7.5. Multivariate classification of starch granules centric view: confusion matrices based on the discriminant analyses. *Reproduced with permission from Torrence et al. 2004: figure 8*

Box 7.8 (continued)

appears to be true of the centric views of *Cyrtosperma chamissonis* and *Manihot esculenta*. Piperno et al. (2000) have also noted that not all morphologies of *Manihot esculenta* are equally useful in identification. In these sorts of cases a population approach taking into account all the potential morphologies for a species is the most appropriate strategy.

Another way to evaluate the data is to plot the centroids of each species based on the first two canonical variate axes. Figures 7.6 and 7.7 show a clear degree of separation in most cases, although again the discrimination is better for the eccentric rather than the centric view and there are some overlaps between the groups, although these are normally separated when the full set of axes are used. Finally, the structure coefficients ('scores') from the canonical variates analyses indicate that the best characteristics are the dichotomous variables which focus on the 'ornamentation' of the granule followed by size, with those relating to shape as the least useful criteria. A combination of all the variables, however, provided much better discrimination than any smaller subset of them (Torrence et al. 2004: 522–23, 528).

Significance

The multivariate analyses confirm the assumption implicit in previously proposed starch granule keys that a number of metric and nonmetric variables can successfully discriminate among plant types with a high degree of consistency. The results are also quite important because the statistical procedures used indicate that the probability of failing to distinguish between two of the plant types used in this study is reasonably low for almost all cases and decreases markedly if one targets families or genera rather than species.

The results also show that additional focus on the surface features of the granules

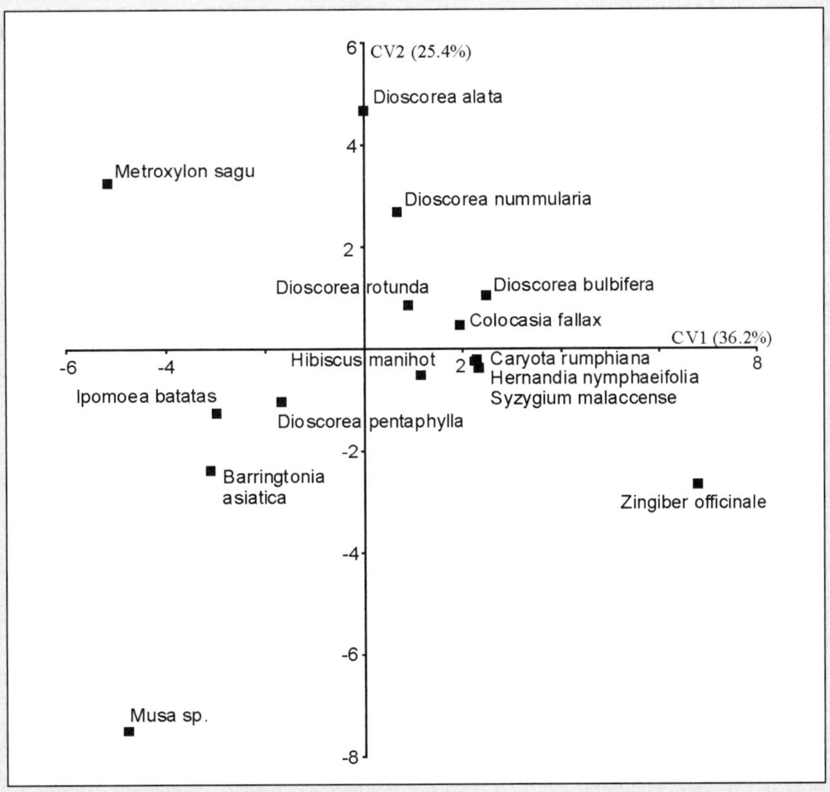

Figure 7.6. Multivariate classification of starch granules eccentric view: centroids of each species plotted on the first two canonical variates. *Reproduced with permission from Torrence et al. 2004: figure 9*

Box 7.8 (continued)

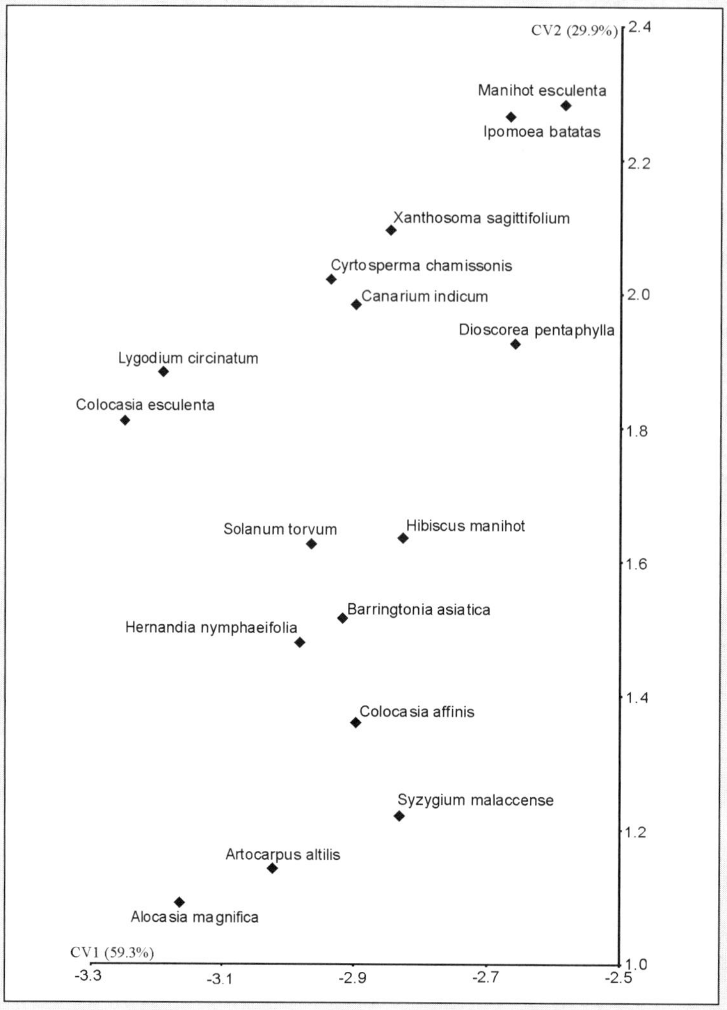

Figure 7.7. Multivariate classification of starch granules centric view: centroids of the majority of species plotted on the first two canonical variates. *Reproduced with permission from Torrence et al. 2004: figure 11*

(ornamentation) as opposed to the characteristics of the extinction cross could improve the discrimination between plant taxa. The study highlights the importance of including 'wild' species at an early stage when establishing systems for classification. Although this appears to be a difficult task in a region such as Papua New Guinea where there are numerous plant species, not all of these have significant quantities of storage starch and most will not make an important contribution to the archaeological or paleoenvironmental record. For example, this study began with a reference collection of 172 species but was able to eliminate the vast majority of these because they were characterized by very small, undiagnostic transient starch or granules (Therin et al. 1997).

At an early stage of setting up a classification, multivariate analysis can be extremely useful because it helps identify the relative contribution of the variables that could potentially be important in discriminating among groups, and it points out which groups are most similar or different and where additional variables may be required. A final important point is that the multivariate classification has a very broad utility because it can be applied by researchers who are unfamiliar with the specific morphologies of particular species.

fissures and mostly eccentric hila' that differ from the range of fissure types and centric hila of the domesticated form (Piperno et al. 2000: 895).

Perry (2002b) has confirmed and extended Piperno et al.'s (2000) study through a comparison of modern wild and domesticated forms of *Dioscorea esculenta* as well as a comparison of modern and archaeological samples of *Ipomoea batatas*. One important finding was that the overall shape of the granules of both species was stable through time. Although her results confirm the size difference between modern varieties of manioc, she found no simple direct correlation between age of sample and size or uniformity of granules. Perry (2002b: 344) notes that since a wide range of factors can contribute to granule size, a larger study is necessary to factor out the relationship between granule size and growing conditions of the crops. Following on from this observation, her work did, however, identify differences in starch granules between lowland and highland forms of manioc that suggest different routes of domestication. These preliminary findings indicate that a study of granule populations in terms of size and especially uniformity have potential for monitoring the history of human use and manipulation of plants.

CONCLUSIONS

Identification of ancient starch on any of the first three levels described here can be achieved with minimal effort and can make important contributions to archaeological research. Continued improvements in the description, classification, and identification of starch granules will result in further advances in starch research, particularly for level 4 identification, but the large corpus of information on starch granule morphology already compiled by investigators from around the world offers a valuable guide to the interpretation of archaeological starch granule assemblages. Whenever possible, the results of ancient starch analyses should be evaluated in tandem with phytolith, pollen, and macrofossil studies (e.g., Pearsall et al. 2004; Denham et al. 2003; Lentfer 2003; Parr 2003a; Kealhofer et al. 1999). In the humid tropics where starch analysis will provide virtually the only means for studying the use and cultivation of the underground storage organs that dominate the diets in these areas, the construction of rigorous keys for root crop identification is a very high priority for paleobotanical research. In these and other regions, large regional reference collections still need to be constructed. In addition, detailed comparisons of starches from modern cultivated and domesticated plants with those from close wild relatives could benefit studies aimed at reconstructing the timing and processes of plant domestication.

The case studies reviewed here demonstrate that starch granules are being routinely identified to various levels within archaeological research. Furthermore, the next three chapters which deal with archaeological and environmental case studies show that a significantly high number of plant taxa are already identifiable through the morphological characteristics of their starch granules.

CHAPTER 8

STARCH IN SEDIMENTS

Robin Torrence

ANALYSIS

Within archaeological research, starch has been retrieved from five main contexts: (1) desiccated plants; (2) sediments; (3) residues adhering to artifacts; (4) macroscopic food remains and material cultural items; and (5) coprolites. Methods for their study together with case studies are described within the next three chapters, beginning here with starch extracted from sediments and including coprolites which use similar methods.

The wide range of questions that can potentially be answered by studying starch preserved in sediments can be usefully divided into two categories based on the scale of the interpretation: (1) landscapes, and (2) specific contexts such as settlements or activity areas. The first group focuses on the reconstruction of past plant communities at the broadest scale of the environment or landscape. Like pollen or phytoliths, the analysis of starch has the potential to monitor environmental changes based on samples taken from sediment cores, although this has not yet been attempted. Starch granules extracted from sediments have been used to reconstruct specific habitats such as those associated with patterns of human land use (e.g., gardens, secondary forest).

In the early stages of a new method in archaeology, it is important to demonstrate that the types of inference used to reconstruct past human behaviour from material remains are justified. Along these lines a recent study focused on modern habitats in Papua New Guinea provides an important confirmation that variations in starch assemblages do reflect differences in land use patterns. As described in Box 8.1, Lentfer et al. (2002) found that it was possible to separate different categories of land use (garden, plantation, secondary forest) from a range of habitats on the basis of multivariate analyses of starch granule morphology and size from samples of modern topsoils. Furthermore, the results were improved when combined with a parallel study of phytoliths. This case study demonstrates that starch extracted from sediments does have a direct relationship to past history of human land use within a region or at a particular locality, especially if starch analysis is incorporated into a broader study with several kinds of plant macrofossil.

A second category of research on starch extracted from sediments focuses on the smaller scale of individual archaeological sites or separate contexts within these. For example, starch can be used to help identify specific human activities at a particular site or location (e.g., plant processing, storage, midden), the function of a specific feature, or to monitor spatial patterning in human behaviour. The aim of this chapter is to describe the procedures used for studying starch extracted from sediments and then to illustrate how the results of these analyses can be used to interpret changes in subsistence practices and land use.

Sampling

The basic stages in the analysis of starch from sediments involve the following procedures listed in the order in which they take place: (1) sampling; (2) extraction of starch; (3) slide mounting and viewing; and (4) interpretation. To begin with, sediment samples are collected from the environmental and/or archaeological record. The choice of a procedure for taking samples from these contexts will depend on the nature of the archaeological problem and the specific questions addressed. Sampling a sediment core or stratigraphic profile to obtain information about environment or habitat change requires a good understanding of the way the sediments were formed. In addition, the number of samples required will be determined by the desired temporal resolution. Pearsall (2000: 399) provides an excellent introduction to field sampling methods for microfossil research.

Box 8.1. Starch Assemblages Track Modern Environmental Variation

Carol Lentfer and Michael Therin

A recent study has demonstrated that phytoliths extracted from soil samples collected from a range of modern environmental settings in Papua New Guinea was able to monitor modern environmental variation, although some habitats were discriminated better than others (Boyd et al. 1998). Could these results be replicated using starch granules?

Sediment samples were collected from a variety of habitats in West New Britain, Papua New Guinea (Table 8.1), representing foreshore, closed forest, regrowth forest, new gardens, old gardens, coconut plantations, and villages (e.g., Figure 8.1). Plant species and relative abundance were recorded from within a 10 x 10 metre quadrant at each location (Figure 8.2). Pinch samples of soil from the uppermost 4 centimetres, excluding the litter, were collected haphazardly across the quadrants (Figure 8.3). In the absence of an adequate starch granule key for this region, an assemblage-based approach using multivariate techniques was employed (cf. Powers et al. 1989).

Assessing Morphological Variation

Starch was extracted from 15 sediment samples using the heavy liquid separation protocol developed by Therin (Box 8.4). Using an Olympus BX60 light microscope, slides were initially scanned under polarised light at a magnification of x200 within transects the width of the field of view in order to locate the starch granules. Once located on the slide, 100 to 150 granules were assessed at a magnification of x500. The shape and distinctive features of each granule were noted under nonpolarised light. Next, the shape and location of the extinction cross were recorded under polarised light. Maximum length and width were measured with the width perpendicular to the length. Each starch granule was drawn and assigned a morphological type using an expandable key based upon previous work by Therin (1994, 1998b; Therin et al. 1999) (Figure 8.4).

The first stage of analysis was to reduce the range of variability in the data to a

Table 8.1. Summary of localities used to study modern environmental variation in Papua New Guinea.

Site	Location	Habitat description
Gl2	Garala Island	Sandy beach with *Ipomoea pescaprae* Roth
Gl3	Garala Island	Closed forest with understorey of gingers and palms
K2	Kaula Island	Closed forest with palm understorey
GMH3	Garua Island	Closed regrowth forest with understorey of ginger
IG	Garua Island	New garden with taro
GFEK	Garua Island	Coconut plantation with grass and weedy understorey
GBL	Garua Island	Littoral with grasses, *Ipomoea pescaprae* Roth and legumes
Gu3	Garu Village	Overgrown garden with bananas, cassava, and weeds
Gu5	Garu Village	New garden with sweet potato, singapore taro, and sugar cane
Gu7a	Garu Village	Newly prepared garden with taro
Gu7b	Garu Village	Newly prepared garden with taro
Gu7c	Garu Village	New garden planted with taro
Gu8	Garu Village	Freshly cleared forest after 10 years regrowth
Gu11	Garu Village	Village area with grasses under palms
Nv	Nave River	Logged closed forest with ginger understorey

Box 8.1 (continued)

Figure 8.1. Habitat zones in the Papua New Guinea study area range from a taro garden in the foreground to abandoned gardens at various stages of regrowth visible in the background. *Photo by Robin Torrence*

Figure 8.2. The relative abundance of all plant species were recorded at each soil sampling site. *Photo by Robin Torrence*

Figure 8.3. Pinch samples of soil from the surface were collected across 10 x 10 metre quadrants in the sampled localities. *Photo by Robin Torrence*

manageable level. To begin with, scatter plots and linear regressions depicting length/width relationships were used to created subtypes based on size differences. Outliers resulting in low regression coefficient scores were considered to be atypical of their specific shape category and were deleted from the analysis along with all subtypes represented by less than 1 percent in any sample.

Based on this revised data set, principal components analysis was conducted using Euclidean distance measures and a correlation matrix. This initial analysis produced 14 principal components with eigenvalue scores greater than 1. Sample and vector plots were examined for subtype/site association. Subtypes that were always associated with the same sites were assumed to have been derived from the same plant species and were therefore recombined. This was necessary because many plant species, e.g., the common potato, have populations with large size variation (e.g., Reichert 1913). The final principal component analysis was conducted with the reduced set of 64 subtypes (e.g., Figure 8.5).

Results

The first three principal components explained 45 percent of the variation in the starch assemblages. These enabled discrimination of the starch samples into different environmental zones. For example, comparison of the

Box 8.1 (continued)

information in Table 8.1 with the plots in Figure 8.5 shows separation of four distinct habitat groups: (1) the logged forest site Nv on the upper left of the plots; (2) the small island sites K2, G13, and G12 on the right of the plots; (3) the coconut plantation sites GFEK and GBL and one of the garden sites Gu5 on the middle left and lower left; and finally (4) garden sites IG, Gu3, Gu7a, Gu7c, Gu7b, village site Gu11, and regrowth forest sites GMH3 and Gu8 which are clustered in the middle of the plots. It is notable that the coconut plantation sites GFEK and GBL were situated in the strand plain and could also be regarded as foreshore

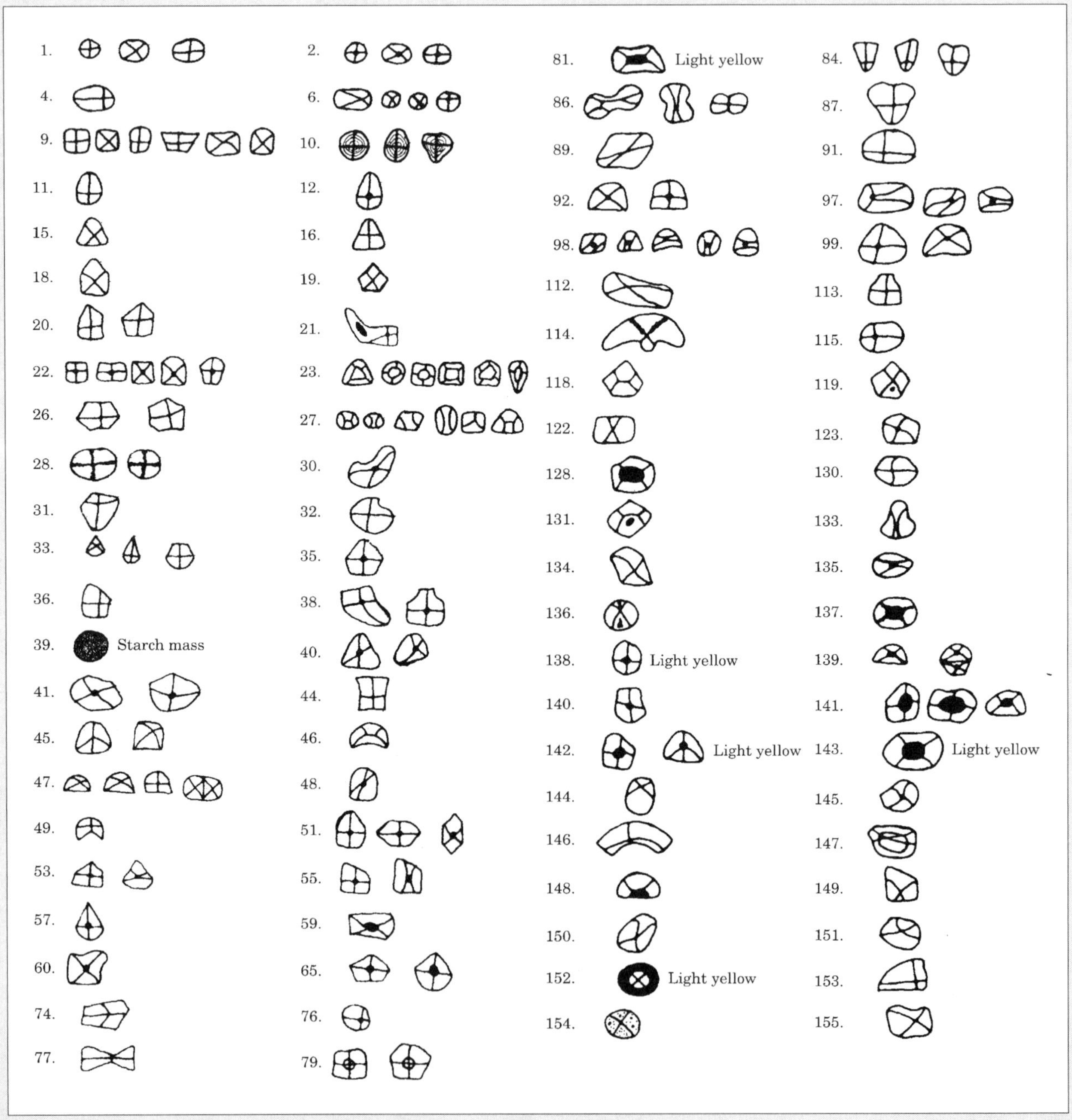

Figure 8.4. Expandable key of starch granule morphologies (Therin 1998). Not drawn to scale. *Reproduced with permission from Lentfer et al. 2002: figure 2*

Box 8.1 (continued)

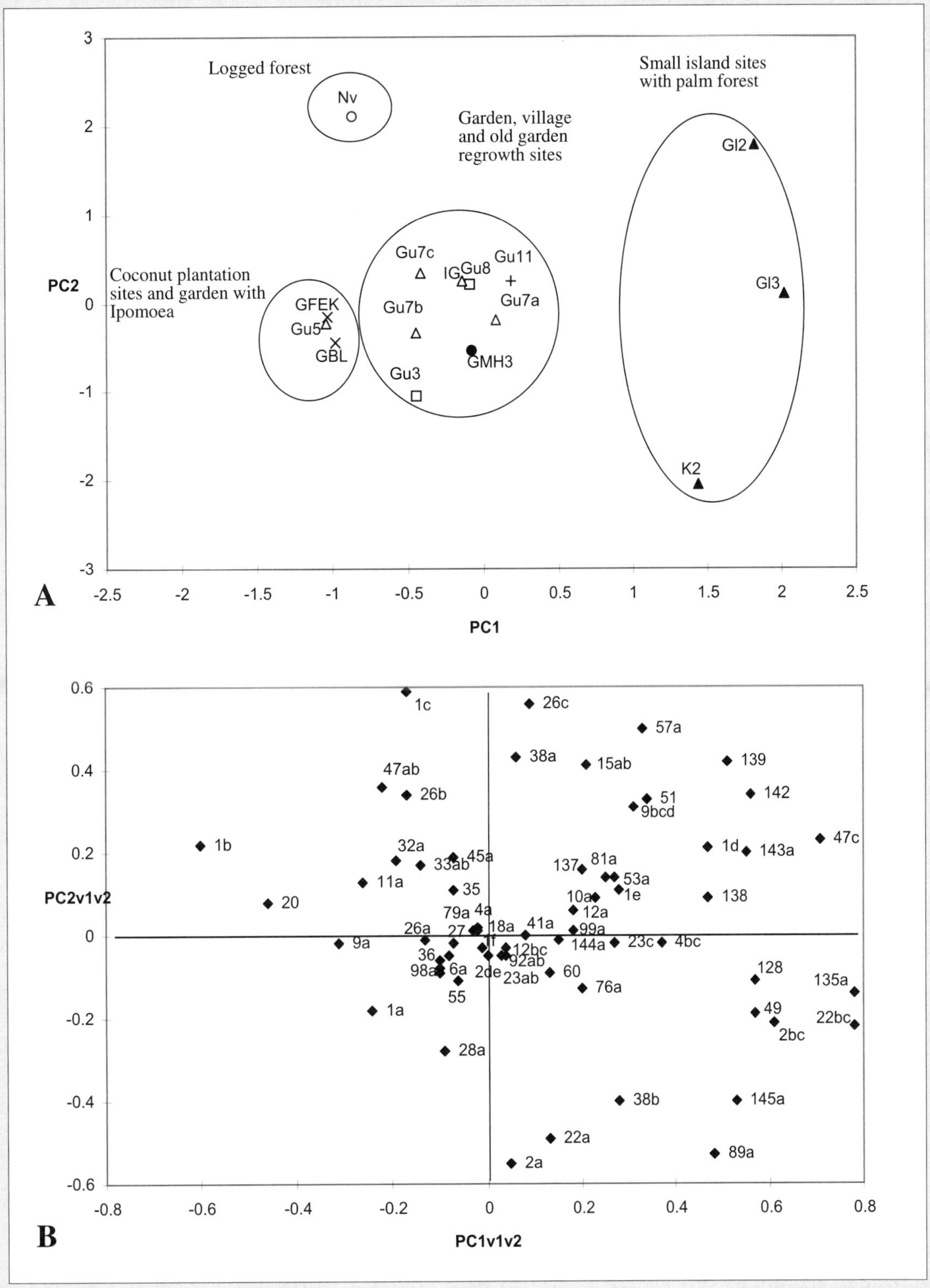

Figure 8.5. Biplot of the first two principal components which explain 33 percent of variation in the assemblages. A is the sample plot and B is the vector plot. ▲ palm forest; • old garden; △ new garden; x coconut plantation; □ regrowth forest; + village; o logged forest. *Reproduced with permission from Lentfer et al. 2002: figure 4*

Box 8.1 (continued)

sites. These locations are typically dominated by a dense ground cover of *Ipomoea prescarpe* L. (a close relative of the sweet potato). This may explain the association of these sites with the garden site Gu5 which was planted under sweet potato (cf. Table 8.1).

Discussion

The starch analysis clearly differentiated between three environmental zones: (1) small island sites; (2) a logged forest site; and (3) cultivated sites. The position of the cluster in the middle of the starch sample plot (Figure 8.5) is notable. These sites, mainly representing gardens, alternate between garden and regrowth during the current swidden system. They have diverse plant compositions derived from different environments, i.e., cultivated, herbaceous regrowth and tropical forest. It is expected that starch granule assemblages from these sites would also be variable unless crops producing clearly diagnostic starch in abundance dominated the assemblages. If not, high variation within starch assemblages—reflecting the nature of a diverse range of environments—would cause these particular sites to fall in the middle of the plots. This is clearly the case here. It also accounts for the relatively low level of variation in the assemblages explained by the first three principal components. It is notable that although no clear associations of sites with types are indicated, site distribution patterns within the sample plots are nevertheless very important for interpretation.

Currently, the division of shapes into subtypes based on size distribution is not definitive (cf. discussion in Lentfer et al. 2002). At this stage what is important is that the classifications adequately demonstrate the potential of assemblage-based methodology for determining site function (especially garden versus nongarden). The analysis of starch granules extracted from sediments clearly has the potential to discriminate among different environments and kinds of land use and is therefore appropriate for archaeological and paleoenvironmental interpretation.

The analytical procedure could be improved if the various starch categories (types and/or subtypes) were assigned to plants using an adequate reference collection. Species/site associations should be used to prioritise species for further detailed investigation. For example, a detailed study of the type 1 complex of starch granules, which is important for differentiating habitats, could be quite beneficial. In this case, taro (*Colocasia esculenta* L.), the staple crop cultivated in West New Britain gardens, has small, round starch granules characterised by the subtype 1a. The distribution of subtype 1a in the vector plots correlates with the distribution of garden sites in sample plots (Figure 8.5). This type may be derived from taro (cf. Loy et al. 1992). Other subtypes within the Type 1 complex are also shown by the analysis to be important for habitat differentiation and are possibly derived from wild taros and related species. It is known that many wild and semi-wild aroids are common in the forest understorey in the region. In the initial stages of further investigations, this group of plants should therefore be a major focus of investigation.

Conclusions

An assemblage approach of starch extracted from a range of documented modern environments in the humid tropical region of Papua New Guinea using a combination shape/size distribution analysis and multivariate analysis clearly demonstrates the potential of starch granules for differentiation between habitats. Variation in starch assemblages was particularly good at discriminating between forests and gardens, which indicates that this approach will be beneficial for studying the history of human land use. Although it would be preferable to identify the granules to plant taxa, and research should be directed to achieve this aim, a great deal can be gained by using size and morphological data alone.

To date, detailed published starch analyses applied to environmental or landscape reconstruction have been based on sampling archaeological sites either by taking one sample per stratigraphic unit taken as a bulk sample (Fullagar et al. 1998; Therin et al. 1999) or as evenly spaced or continuous samples from the exposed section of an excavation (Therin 1994; Therin et al. 1999). In Box 8.2 Lentfer describes how stratigraphic profiles were sampled from archaeological excavations in Papua New Guinea. It is worth noting that in this case multiple samples were taken from the same profile because the extraction procedure for starch differs from other microfossils, such as phytoliths, which provide complementary evidence. In addition, relatively large sediment samples were required so that a number of additional techniques such as particle size and chemical analyses could also be applied to the analysis of the profile.

The number and placement of samples obtained from archaeological sites will depend on the nature of the questions asked. The sampling may focus on particular types of contexts, such as hypothetical storage pits or garden features (Horrocks et al. 2004a, 2004b; Horrocks 2004), or range across contexts (Balme and Beck 2002; Lentfer et al. 2002). It is generally a good idea to take at least one sample from each excavated context. This is particularly important if one is planning to study the residues on stone tools. In order to assess if starch residues adhering to tools reflect their past uses, it is necessary to compare them with the starch from surrounding sediments (cf. Chapter 9; Box 9.7; Kealhofer et al. 1999).

It is difficult to be precise about sample sizes because the density of starch preserved in the sediment will obviously vary from case to case. Ideally, one would run a series of pilot samples from a particular setting to gauge the general range of starch preservation and abundance and then establish reasonable sample sizes for collection. Microwave extraction of starch could provide a useful way to screen samples and make decisions about further sampling (Box 8.6). In the studies reviewed later in the chapter, residue samples ranging from less than 1 g to up to 5 g of sediment have been processed, although when the quantity of starch recovered has been very low, it has been necessary to run additional samples.

Many environmental and archaeological studies are based on more than one type of microfossil since each one has the potential to complement and enhance the others by targeting different plant groups. In these cases it is sensible to take either very large or multiple sediment samples since it is difficult to find a single extraction technique that is appropriate for a range of different types of microfossil. For example, starches are particularly susceptible to many chemicals, as discussed in Chapter 5 and Box 8.3 (Coil et al. 2003: 1002; Pearsall et al. 2004: 428). To answer a number of archaeological questions, one may only be interested in the presence/absence of starch or in the types of plants represented. However, if one wishes to measure the relative density of starch in the samples in order to make comparisons, for example across the site or among different contexts, then a standard quantity of sediment should be processed for all cases.

Extraction

Having acquired the samples, the next step is to extract the starch granules from the sediment sample. As described by Coil et al. (2003: 996–1001), most microfossil extraction procedures use the same basic methodology with the following general steps: (1) sample preparation (sieving, drying, or soaking); (2) disaggregation and deflocculation to break up the constituents of the sample into single particles; (3) removal of undesired particles (sands, silts, minerals, organics); (4) slide mounting; and (5) viewing. The various protocols that have been used for the extraction of ancient starch are summarised in Table 8.2. The major differences between them are the various emphases on disaggregation and deflocculation and the chemicals used to remove other particles that make slide viewing difficult.

The earliest version of a protocol for extracting ancient starch was developed by

Table 8.2. Summary of methods used to extract starch from sediments with heavy liquids.

Reference	Samples	Sample Size	Drying and Disaggregation	Deflocculation	Removal of Organics	Heavy-liquid flotation
Fullagar et al. 1998; Therin et al. 1999: 451; Atchinson and Fullagar 1998: 115	sediment	1 g	oven dry only	none	none	caesium chloride
Barton et al. 1998: 1234	residues and sediment	0.01–0.5 g	oven dry only	none	none	caesium chloride
Therin 1994 in Therin et al. 1999: 451	sediment	2 g	oven dry only	none	none	caesium chloride
Lentfer et al. 2002: 690 Box 8.4	sediment	5 g	oven dry, pulverise	Calgon, soak, shake, stir, centrifuge	hydrogen peroxide, sieving	sodium polytungsgate
Pearsall et al. 2004: 428	residues	various	oven dry, sieve, mix	EDTA, shake, centrifuge	hydrogen peroxide	caesium chloride
Korstanje 2003: 107 Box 8.5	sediment	not given	oven dry, distilled water	sodium hexametaphosphate, gravity separation, sieving	not stated	zinc iodide
Horrocks et al. 2004a: 254	sediment	2–5 cm^3	not stated	Calgon, soak, shake, stir, centrifuge	none	sodium polytungsgate

Box 8.2. Sampling Sediments For Starch Analysis

Carol Lentfer

As with other plant microfossils, such as phytoliths, pollen, and spores, methods adopted for sampling sediments for ancient starch analysis should follow a systematic procedure best suited to the stratigraphy and the research question. Although methods will vary according to site characteristics, it is better to oversample than undersample. This will facilitate flexibility in the research design and avoid unforeseen analytical and interpretive problems. It is therefore recommended that (1) sampling use smaller rather than larger intervals, and (2) enough sediment/sample is collected to allow for additional extraction and/or other analyses, if required. Sampling generally proceeds from the top of the section or core (most recent) to the base (oldest).

General Guidelines

A number of general guidelines provide the basis for ensuring the samples are adequate for starch analysis.

- *Avoid contamination.* Clean back faces of section wall at each sampling point prior to taking samples. Clean all traces of sediment from sampling tools (e.g., trowels) before the collection of each sample. Use clean bags or containers and seal them securely.
- *Take accurate measurements of sample locations.* Secure a measuring tape against the section wall. It is helpful to score horizontal lines demarcating sample intervals across the wall (e.g., every 5 cm).
- *Label sample bags.* Location (e.g., GPS reading, site name, sampling or excavation unit, etc.), date of collection, sample number, sample depth (e.g., 20–25 cm), and name of collector should be clearly noted on the samples. Use permanent ink marker, but be aware that permanent ink can wear off plain plastic surfaces. To avoid losing labels, use sample bags provided with a special surface for writing on, or write on masking tape.
- *Take accurate notes.* Sample depth and number should be noted. A description of the sediment (e.g., colour, texture, inclusions, etc.) and any additional information helpful to analytical interpretation (e.g., presence of stone tools, pottery, bone, hearths, plant macroremains) should be included.
- *Sketch the section profile and label appropriately.* A section drawing showing layers, sample locations, and sample numbers is important.
- *Photograph the section wall and sample locations after sampling.* This is optional but can be quite helpful if questions about the stratigraphy arise later.

Case Studies

Various strategies can be used for sampling sediments depending on the particular circumstances. For instance, one may sample walls of archaeological excavation trenches, roadside or other cuttings, or sediment cores. The two examples given below serve to illustrate how the sampling strategy might vary according to site characteristics and the nature of the research problem.

Continuous Column

Sampling was conducted within the walls of a 3.1-metre-deep test pit which was excavated at the archaeological site of FAO, located on Garua Island, Papua New Guinea (Figure 8.6). The stratigraphy revealed in the pit reflects the location of this site in a region prone to frequent volcanic eruptions. The upper section (0–1.1 m) comprises reasonably well-defined layers of slightly weathered tephra (volcanic ash resulting from the catastrophic plinian eruptions of nearby volcanoes) alternating

Box 8.2 (continued)

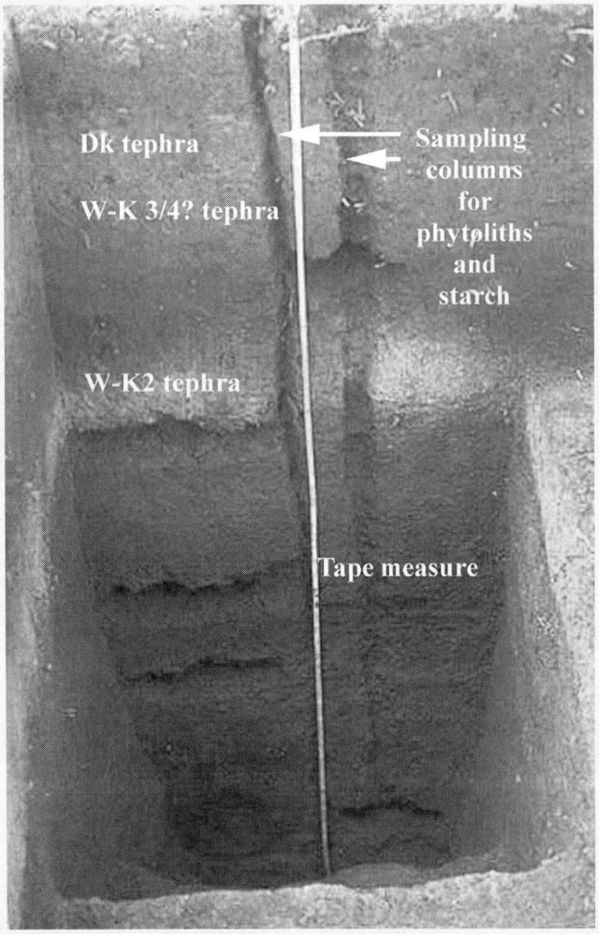

Figure 8.6. Stratigraphic diagram showing continuous sample column in test pit 1000/1000 at site FAO, Garua Island, Papua New Guinea.

between each of their respective palaeosols. The lower section (1.1–3.1 m) comprises layers of deeply weathered clays, differentiated by variations in texture and color. Prior to sampling it was known that the upper tephra, derived from the Dakataua volcano, located about 25 km to the north and the lowermost obvious tephra belongs to the W-K2 eruption of the Witori volcano, which is 55 km to the southeast. The source of the thinner tephra layer between the Dakataua (Dk) and W-K2 tephra layers is unknown. Likewise, the source of the clay deposits in the lower section has not been determined, although from the distribution of artifacts, they were thought to be derived from additional but highly weathered tephras and palaeosols. Analysis of the various tephra and palaeosol layers in the upper section presented an ideal opportunity to investigate impacts of volcanic eruptions on vegetation and people, patterns of vegetation recovery following eruptions, and changing patterns of land use that might have been associated with the introduction of Lapita-style pottery after the W-K2 eruption.

For this reason, the north wall of the test pit was cleaned back and, to allow for fine resolution determination of change, sediment samples were taken at 5 cm intervals down a continuous column, to 15 cm below the base of the W-K2 tephra layer. Below this, in the less well-defined layers of clay, samples were taken at 10 cm intervals. Two columns of samples were collected: one for starch and the other for phytoliths (Figure 8.6). At least 100 g of sediment was taken for each sample allowing for starch/phytolith analyses as well as particle size, x-ray diffraction, metal, and total carbon content analyses. Trowels were scraped clean and the wall was cleaned back before every sample was collected. Clip-lock sample bags were labelled with location and baulk details, date, collector's name, sample number, and depth interval. Additional notes with sediment descriptions were taken and the section profile was drawn and labelled. Finally, a photograph was taken to document the position of sample columns and sediment layers. Subsequently, analyses of the starch and phytoliths have provided interpretations of changing landscapes and human land use at this location (Box 8.8; Therin et al. 1999; Parr et al. 2001a; Lentfer 2003).

Targeting selected layers

Liang Bua is a limestone cave on Flores Island in Indonesia with a complex stratigraphy characterised by interbedded layers of flowstone, in-washed sediments including clays and tephras, and rockfall. With a long sequence of hominid occupation from at least 91,000 years ago up to the late Holocene and the

Fullagar, Loy, and Cox who chose caesium chloride because it was the only heavy-density liquid with the appropriate specific gravity available in the laboratory at the time. The first protocol was very simple because the object was simply to see if starch had been preserved and could be studied. Case studies using this procedure have included Therin (1994), Therin et al. (1999), Fullagar et al. (1998), Atchison and Fullagar (1998), Vinton (1995), and Balme and Beck (2002). A slightly modified version was adopted by Barton et al. (1998) in an attempt to prevent the detrimental effects of caesium chloride on starch preservation.

The improvement of protocols is part of the necessary research involved in the study of plant microfossils and this is also the case with starch. Experiments that compare the efficiency of various techniques can be very useful (e.g., Lentfer and Boyd 1998). As described in Box 8.3, Michael Therin has conducted a number of experiments to test various aspects of the Fullagar et al. (1998) protocol. He noted that a significant proportion of starch granules were destroyed by submersion in caesium chloride (CsCl) and so sodium polytungstate ($Na_6(H_2W_{12}O_{40})$) was substituted. Therin also added a number of steps to cope with the clay-rich soils of Papua New Guinea and to ensure proper mixing before a final sample was extracted for analysis. His most recent protocol, which is presented in Box 8.4 (cf. Lentfer et al. 2002: 690), contains all the basic elements for extraction techniques. Horrocks et al. (2004a: 254) have also used a modified version of the Therin protocol.

In another set of experiments reported in Box 8.5, Alejandra Korstanje tested a protocol designed to extract a wide range of microfossils simultaneously (cf. Korstanje 2003). She found that starch survived the full protocol which involved separation using deflocculation with sodium hexametaphosphate ($(NaPO_3)_6$), gravity separation in distilled water, and flotation in Zinc iodide (ZnI_2). Korstanje noted a problem previously reported by Therin (1994). Once starch had been mounted onto slides and analysed, it did not

Box 8.2 (continued)

findspot of the *Homo floresiensis* fossils, it is among the most significant archaeological sites in Southeast Asia (Morwood et al. 2004). Stone tools have been collected from various layers in the cave in association with premodern and modern humans together with shells and bones from a wide array of extinct animals including *Stegodon*. Fossil starch, phytoliths, and pollen microfossils, if present in the cave sediments, have the potential to investigate patterns of paleoenvironmental change and availability of plant resources. Analyses of plant microfossils in sediments, combined with residue and use-wear analysis of stone tools, can also provide information about the use of plant resources. With these research questions in mind and given the complexity of sedimentary processes involved in the deposition of cave sediments, a targeted sampling strategy was adopted. Rather than sampling in a continuous column down the walls of the excavations, only selected sediment samples were taken from the layers. Occupation horizons and sedimentary layers likely to be derived from palaeosols were selected. Replicate samples were taken for starch, phytoliths, and pollen extractions. The collection procedure followed the guidelines listed above.

Conclusions

As with sampling for any plant microfossil, the collection of sediment samples for starch analysis should follow a procedure that takes adequate care with recording sample details, labelling samples, and avoiding contamination. Since starch granules gelatinise at temperatures over 50° C, precautions may need to be taken. Storing sediments in alcohol or methylated spirits may be helpful for researchers faced with quarantine restrictions when transporting samples from one region into another. In these cases, it is preferable to dilute solutions to 70 percent and then make sure to rinse samples thoroughly with pure water prior to commencement of starch extraction.

Box 8.3. Experiments For Testing Starch Extraction Techniques

Robin Torrence and Michael Therin

The ideal method for extracting starch from sediments should yield a very high proportion of the original starch from all types of sediments. In addition, a good method will recover starch assemblages that are representative of the original starch population in the sediments. The extraction of starch from sediments is still a relatively new technique and consequently not enough is yet known about the effectiveness of various procedures. For this reason Michael Therin conducted several simple experiments designed to compare alternative techniques. The first group was devised to investigate to what extent the chemicals used in the procedures had a deleterious effect on starch abundance. A second set focused on potential problems caused by mechanical factors, such as the dispersal of starch within a solution. The experiments reported here represent only a few of the tests required to evaluate current methods thoroughly and to identify where improvements may be required.

Effects of Chemical Solutions

The first series of experiments assessed the effects of various chemicals on the preservation of starch. Starch is known to gelatinise in the presence of acids and salts (Chapter 5), and so it seemed likely that some solutions commonly used in the extraction of other microfossils might reduce the quantity of starch that was extracted. In addition to the heavy-density liquids caesium chloride (CsCl) and sodium polytungstate ($Na_6(H_2W_{12}O_{40})$). The commonly used deflocculant Calgon was evaluated. If there is a large amount of clay in some sediment samples, it may be necessary to deflocculate the sediment as part of the extraction protocol. As the deflocculation process takes 24 hours, it is important to evaluate the corrosive effects on starch of the chemical used. Similarly, starch can be exposed to the heavy-density liquid for significant amounts of time during the extraction procedure.

Horrocks et al. (2004a: 254) report that they soaked modern starch overnight in Calgon with 'no apparent change in the appearance of granules.' The experiment reported below focuses on the effects of the chemicals on the abundance of the starch rather than their morphology.

Methodology

One ml of ultrapure H_2O was added to a small amount of industrially processed maize starch in an Eppendorf™ tube. To distribute the starch equally throughout the sample, the Eppendorf™ tube was vortexed prior to each sample being taken. A 10 μl sample was removed by pipette from the surface of the suspension and placed on a microscope slide. Five transects were made across the slide at equal intervals, starting at the edge of the coverslip, and the starch granules were counted. Sampling and analysis was repeated at six time intervals: 0 hours, 1 hour, 5 hours, 24 hours, 48 hours, and 1 week.

The procedure was repeated for solutions of sodium polytungstate and caesium chloride made up to a specific density of 1.79 and for a 5 percent solution of Calgon in water. The tests involving water and sodium polytungstate were repeated three times, whereas those involving caesium chloride and Calgon were conducted twice.

Results and conclusions

Both water and sodium polytungsgate do not appear to have a corrosive effect on starch at normal room temperature. Over a one-week period there was no significant decrease in the number of granules observed. In contrast, caesium choride was found to have a deleterious effect. After one hour in caesium choride, the

Box 8.3 (continued)

decline in the samples was only 10 percent and 0 percent, but only 57 percent and 68 percent remained after 5 hours. The counts appeared to stabilise at about those levels and only decreased gradually after that point.

From this experiment it is clear that sodium polytungsgate has the lesser corrosive effect of the two heavy-density liquids examined. The results corroborate the experience of a number of researchers in the Australian Museum laboratory where gelatinisation of starch granules processed using caesium choloride (cf. Fullagar et al. 1998 extraction procedure) has been observed during the counting procedure. It seems likely that heat from the reflected light used in the microscope accentuated the process of decay. If caesium choride is used, the extracted samples should be washed thoroughly before mounting on slides and not stored in the solution. Another point in favour of sodium polytungsgate is that when it dries, it does not form as many crystals as caesium chloride, thereby facilitating examination of the slides.

The 5 percent Calgon suspension was shown to be more corrosive than the heavy-density liquids. After one hour 82 percent and 85 percent of the original granules were present, but the numbers declined drastically to only 35 percent and 59 percent remaining after 5 hours, although the further drop within the next 48 hours was minimal. Given the results of this experiment, the potential destructive effects of Calgon will need to be balanced against the benefits of deflocculation using this chemical.

Starch Distribution within a Suspension

During the course of the previous experiments, a further potential problem with extraction protocols was identified. If care is not taken to vigorously shake as well as vortex the samples, starch granules are not equally distributed within a suspension. This means that samples drawn from it are not equivalent. A second set of experiments was therefore conducted to examine the nature of starch distribution within a suspension. The first aim was to see if starch is distributed evenly throughout a suspension when treated in different ways. The second was to determine if the order in which samples are extracted from the same point in a snap-cap microcentrifuge tube (e.g., Eppendorf™) alters the amount of starch present in each sample.

Methodology

A 10 ml suspension of industrially processed maize starch and ultrapure H_2O was prepared in five test tubes. These were then treated in two ways. In the first experiment test tubes were only vortexed, whereas in the second the samples were both shaken and then vortexed. Following the appropriate treatment, three 1 ml amounts of the suspension were extracted into snap-cap microcentrifuge tubes as follows: (1) top, (2) middle, and (3) base of the suspension.

Each snap-cap microcentrifuge tube was then vortexed or shaken and vortexed as appropriate. Next, five 10 μl samples were pipetted from each tube and mounted on microscope slides. The microscope slides were labelled '1' to '5' in the order in which they were extracted from the solution. In order to prevent the microscope slides drying out, only one set of slides was prepared and analysed at a time. Five transects across each microscope slide were made and the total number of starch granules counted.

Results and conclusions

The results indicate, first, that vortexing without shaking leads to large variation in starch counts among samples from the same solution. Second, if the suspension has been shaken as well as vortexed, the location within the tube from which the sample is taken has no impact on the number of starch granules in the sample. Third, the order in which the samples were taken also did not affect the starch counts, when the sample was mixed correctly. These experiments show that vigorous shaking as well as

preserve beyond a day or so. Jenny Atchison (personal communication, 1994) reports that while scanning slides she observed starch granules exploding. The problem in these cases appears to be the detrimental effects of the heat generated by the lighting used with the microscope. Clearly, further experimentation would be helpful in quantifying the effects of different extraction protocols on starch preservation.

Therin (Box 8.3) notes that there is a conflict between the need for Calgon as a deflocculant and its detrimental effects on the preservation of starch. His protocol therefore includes a number of steps to rinse out this chemical. In contrast, Pearsall et al. (2004: 428–30) employed caesium chloride in a protocol rather like Therin's modified version, but they reduced the quantities of hydrogen peroxide (used to eliminate organic material) to protect the starch. They also found that many sediment samples taken from grinding stones were so clean that the flotation step was not necessary. In these cases they were able to avoid using caesium chloride.

Variations between the various protocols used so far have been found to be successful because sediments differ in the degree to which they contain organics and clay. Consequently, protocols will require adjustment to fit specific cases. Therin's procedures were developed for very clay-rich soils that require a good deal of deflocculation and settling out of the clay particles, but in other situations when the organic content of the soil was high, the results were less satisfactory (Michael Therin, personal communication, 1999). Organic-rich sediments pose a problem to starch extraction because in most instances the chemicals usually used to remove this component are likely to be very aggressive to starch. It is a good idea to run a number of pilot studies and to modify the steps according to the nature of the specific samples. It is also wise to investigate whether and how the chemicals used affect starch and to test the efficiency of the protocol for extracting starch. This can be achieved with modern starch as reported by Therin (Box 8.3), Korstanje (Box 8.4; 2003), and Horrocks et al. (2004a: 254–55).

A very different approach to starch extraction using microwave digestion has been piloted by Parr (2002; Parr et al. 2001b; Parr and Farrugia 2003). As described in Box 8.6, microwave digestion is very fast because it does not involve disaggregation, deflocculation, or separation. The method was devised to extract phytoliths, but starch granules have also been recovered. However, Parr (2003a) only included the presence/absence of starch in his analysis of landscape change in Papua New Guinea. Jeff Parr (personal communication, 2003) thinks that the microwave technique could best be used to test sediments for starch presence and relative density. Following the microwave test, standard extraction techniques may be preferred because the final sample is often cleaner and, therefore, much easier to study with normal optical methods.

Potential contamination of samples during the fieldwork and in the laboratory is an important aspect of methodology. Starch and other materials are often present in the modern environment and can therefore be unintentionally incorporated into archaeological samples at all stages of data collection and analysis. As discussed at more length in Box 8.7, care needs to be taken to monitor the occurrence of modern starch and prevent it from

Box 8.3 (continued)

vortexing is required in order to distribute starch granules equally within a solution prior to sampling for analysis.

Postscript

One potential problem with the experiments reported here is that it is assumed that starch granules are consistently distributed across the slides, i.e., that one count is adequate. Carol Lenfter (personal communication, 1998) has pointed out that one should count granules within specific grids using a cytometer and then repeat this procedure several times. The resulting count reported is the mean and standard deviation from multiple counts, rather than a single count.

Box 8.4. A Protocol For Extraction Of Starch From Sediments

Michael Therin and Carol Lentfer

The protocol described below was devised to extract starch granules from heavy clay soils with small amounts of organic material. It can therefore be altered to account for different soil types. Several points need to be stressed before beginning. First, the laboratory needs to be kept free of potential sources of contamination (e.g., food, dust, powdered gloves; see Box 8.7). Second, only use filtered water (distilled water or tap water filtered through 1 μm in-line filters) for final rinsing of glassware and mixing chemical solutions.

I. Deflocculation and Removal of Clay

1. Dry sediment[1] and sieve through a 2 mm sieve. Weigh out 5 g of sieved sediment.[2]
2. Crush[3] the sediment with a mortar and pestle until it is a fine powder. Transfer sediment to a 200 ml beaker.
3. Place 20 ml of 6 percent hydrogen peroxide (H_2O_2) in the beaker and stir.[4] Leave for 30 minutes. Add another 20 ml of 6 percent H_2O_2. Leave for another 30 minutes.
4. Put sediment suspension through a 250 μm sieve[5] using water to wash the fine sediment fraction through the sieve. Discard the >250 μm fraction or store in container for subsequent examination.[6]
5. Pour contents of the collecting pan into (a) a 600 ml beaker, or (b) 50 ml centrifuge tubes. For beakers: to allow all residue with starch granules to settle, still-settle in beakers for 1 hour.[7] For centrifuge tubes: centrifuge suspension in tubes @ 2500 rpm for 3 minutes.
6. Beakers: Remove supernatant[8] with an aspirator; discard. Transfer remaining suspension to a 200 ml beaker and allow it to sit for 40 minutes, allowing all residue with starch granules to settle, and proceed with step 7. Centrifuge tubes: Decant supernatant and proceed to step 8.
7. Beakers: Remove supernatant from beaker with an aspirator and discard. Transfer remaining suspension to a 50 ml centrifuge tube.
8. Fill tube to 50 ml with 5 percent Calgon[9] (@ 35° C), shake or vortex, and centrifuge @ 2500 rpm for 1 minute.
9. Decant supernatant containing suspended clays and discard.
10. Repeat steps 8 and 9 until the supernatant is clear. (Note that heavy clays will require several treatments to remove all clay particles).
11. Rinse by filling tube to 50 ml with H_2O and centrifuge @ 2500 rpm for 2 minutes.
12. Decant supernatant and discard.
13. Repeat steps 11 and 12 twice.
14. Dry sediment @ 40° C in oven.

II. Extraction of starch granules using heavy liquid flotation[10]

15. To remove particles lighter than starch, add 8 ml of sodium polytungstate ($Na_6(H_2W_{12}O_{40})$ heavy liquid solution (@ specific gravity 1.3) to 50 ml centrifuge tube containing dried sediment.
16. Centrifuge @ 2500 rpm for 12 minutes.
17. Discard supernatant.
18. To separate starch from the heavy fraction, add 5 ml of sodium polytungstate heavy liquid solution (@ specific gravity 2.0).
19. Centrifuge @ 2000 rpm for 10 minutes.
20. Decant supernatant into a new 50 ml graduated centrifuge tube. (Note: this supernatant contains the light residue fraction with the starch granules. Do not discard.)
21. Repeat steps 18 and 19. Decant into the same 50 ml tube used in step 20.
22. Allow starch granules to partially settle by diluting the heavy liquid solution to

Box 8.4 (continued)

lower its specific gravity. Fill tube with starch residue and supernatant to 50 ml with ultrapure H_2O. (Note: the starch granules will begin to sink to a lower level in the tube.) Centrifuge @ 3000 rpm for 10 minutes.

23. The starch granules are still floating in suspension at this stage. Pour off and discard roughly two-thirds of the supernatant[11] without the starch granules, leaving roughly 15 ml of suspension in the centrifuge tube. (Note: examine the suspension before decanting to make sure that no starch granules are lost.)
24. To lower the specific gravity of the supernatant and allow all starch granules to fully settle in the tube, repeat steps 22 and 23 three times. Decant most of the supernatant in the final procedure.
25. Rinse by filling tube to 50 ml with water and centrifuge @ 2500 rpm for 2 minutes.
26. Decant supernatant and discard. Repeat rinsing process to remove all traces of sodium polytungsgate.
27. Dry remaining suspension @ 40° C in oven.

III. Mounting

28. Add 500 μl of ultrapure water to the dried suspension.
29. Shake and vortex. Using a pipette, place 60 μl on a microscope slide and dry.
30. Mount in a suitable medium.

Explanatory Notes

1. Drying sediments

Sediments should be dried prior to extraction. This allows correct dry weights to be measured. Also, excessive moisture can change recommended concentrations of chemicals and reduce extraction efficiency. Note, however, that excessive heating can cause gelatinisation of starch granules, a process characterised by loss of birefringence, starch granule degradation, and solubilisation of starch into solution. It is therefore recommended that oven-drying temperatures for wet sediments should not exceed 50° C. To be safe, it is preferable to dry sediments for longer periods at temperatures below 40° C.

2. Subsampling

Subsamples should be truly representative of sediment samples. An efficient method for ensuring homogeneity of the parent sediment sample is the cone/quarter subsampling procedure. Stir the dried sediment thoroughly and form it into a mound. Divide the mound into four equal parts and take small amounts of sediment from each quarter. Re-mound the remaining sediment sample making sure it is well mixed again. Repeat the procedure until the required amount of sediment has been sampled.

3. Crushing

Sediments can also be crushed with a pestle and mortar prior to the initial sieving procedure. This is necessary if dried sediments comprise clods or large soil aggregates.

4. Oxidation with hydrogen peroxide (H_2O_2)

When hydrogen peroxide is added to highly organic sediment, a vigorous exothermic reaction can result. If this is not closely monitored, beakers can overflow. Stirring can help prevent overflow, but if reactions are too strong, they can be slowed by simply immersing beakers into iced water. To assess the potential strength of reactions and avoid overflow, a small amount (e.g., 5 ml) of hydrogen peroxide can be added to the sediment initially. Stir and add additional amounts as required.

5. Sieving

Nylon or wire mesh sieves can be used. Wash sediments through with distilled water. To avoid contamination, clean mesh thoroughly before sieving each sample.

contaminating the sample. For example, it is important to make regular tests for the presence of starch granules and other microscopic material by putting out traps, usually in the form of clean microscope slides. If one is informed of the likelihood of modern material getting into archaeological samples, then sensible adjustments can be made to methodology and to interpretations.

Slide Mounting and Viewing

In the last step of starch analysis from sediments, the extracted material is mounted onto a slide and viewed as appropriate for identification and counting. Therin dried the material before viewing to prevent further destruction of the starch by leaving it in a liquid (Box 8.4), whereas Horrocks et al. (2004a: 205) transferred the residue directly onto the mounting media (glycerol jelly or oil, in

Box 8.4 (continued)

6. The >250 μm sediment fraction

Rather than discard the >250 μm sediment fraction, it is often wise to store this fraction for further examination (e.g., charcoal content).

7. Still-settling of starch granules

Allow time for all starch granules to settle out of suspension. The still-settling time required varies according to depth of settling column, specific gravity of particles, and liquid viscosity (Lentfer et al. 2003). If excess water is used to wash sediment through the sieve, extra time will be needed for all starch granules to settle (e.g., in a water column 10 cm high, starch granules with diameters of 2 μm, 3.9 μm, and 5 μm will require 25.5 hours, 6.7 hours, and 4.1 hours to settle, respectively).

8. Removal of supernatant

Supernatant is the liquid left after sediment residue has settled. It can be removed by simply tipping it out carefully in a single pouring motion or using a pipette or an aspirator (i.e., a simple suction device consisting of a plastic or rubber tube with a t-junction, attached to a tap).

9. Calgon

Calgon is a commercial cleaning product made from sodium hexametaphosphate ($(NaPO_3)_6$). It is a deflocculant which results in dispersal of clay particles in solution, allowing their removal by decantation.

10. Heavy liquid flotation

Sodium polytungstate ($Na_6(H_2W_{12}O_{40})$ is a heavy salt used to make heavy liquid for separating the light fractions of the sediment residue (e.g., starch granules with specific gravity of c. 1.5 g/cm^3) from heavier fractions (e.g., biogenic silica with specific gravity of c. 2.3 g/cm^3, and other siliceous sand, silt, and clay particles with specific gravity of c. 2.6 g/cm^3, and heavier minerals). The starch protocol uses a lighter heavy liquid solution initially to clean light, nonstarch particles from the residue and then a heavier solution to separate the starch from particles comprising biogenic silica and other minerals. Make up the solution by dissolving the sodium polytungstate powder in filtered water until 1 ml of the solution weighs 1.3 g/ml or 2 g/ml. Heating may be required but the solution should always be weighed at room temperature. Otherwise use a pyrometer. (Note: Sodium polytungstate is considered to be a relatively safe and inexpensive chemical and is recommended over and above other chemicals often used for heavy liquid solutions, e.g., caesium chloride, zinc iodide, potassium iodide, and cadmium iodide.)

11. Recycling sodium polytungstate

Sodium polytungstate can be recycled. Rather than discarding the heavy liquid supernatant, pour it into a storage container. Simply filter the diluted solution through fine filters (2 to 3 μm filters should suffice) and dry it in an oven to restore it to its solid state or a solution with the desired specific gravity.

Box 8.5. Designing A Protocol For The Simultaneous Recovery Of Multiple Microfossils

M. Alejandra Korstanje

A range of different macrofossils is required to study the complex mix of crops, irrigation systems, and animal husbandry that characterized prehispanic agricultural practices in the high semi-arid valleys of the Andes in northwestern Argentina. For example, it is likely that the crops included both high-altitude taxa (tubers and pseudocereals) and middle-altitude taxa (maize, beans, chili pepper, squashes). In this case starch and phytoliths are both required since neither type of microfossil is present in sufficient quantities in all the likely plants. For the Andes the identification of starch granules is especially important due to the variety and importance of tuber crops, most of which do not produce silica phytoliths (Korstanje 2003): (*Solanum tuberosum, Oxalis tuberosa, Tropaeolum tuberosum, Ullucus tuberosus, Lepidium meyenii, Pachyrhizus ahipa, Canna edulis, Smallanthus sonchifolius*). It is also probable there were complex relationships between husbandry and agricultural practices (fodder, fertilizers, alternating the use of corrals as agricultural fields) and irrigated and nonirrigated systems may have co-occurred.

Assuming poor organic preservation in open agricultural sites, phytoliths should be an important alternative source of data (Korstanje 1996). Information from diatoms, faunal spherulites, starch granules and other microfossils could also be very useful, but the practicalities of various sophisticated laboratory conditions and the large number of extra hours needed to run samples for each specific type—for processes already known to be time consuming—create a serious obstacle (Coil et al. 2003). Ideally, one would obtain the desired range of microfossils from the same extraction protocol. In addition, it would be useful to have most types together in the same slide, because this could give the researcher an overall view of the multivariate patterns that differentiate various forms of land use.

The ideal microfossil extraction procedure would follow the basic steps described by Coil et al. (2003: 996), but involve few chemicals to avoid damaging susceptible microfossils. After trial and error the following procedure was developed. Samples are dried in the oven at low temperatures until their weight is stabilized. They are then deflocculated overnight with sodium hexametaphosphate ($(NaPO_3)_6$). Clays are separated and eliminated by gravity using only distilled H_2O. Decantation times are checked according to Stoke's law. Next, samples are sieved through 50 to 55 µm mesh to remove the sand-size fraction (which is kept apart, but stored for study of large phytoliths), and silts are optionally fractionated in three classes (fine, medium, and coarse) with the same gravity method as for clays. Finally, the remains are floated using the heavy liquid zinc iodide (ZnI_2) prepared at the density of 2.3. If the solution appears to show a 'snow dispersion' at any point of the procedure, add warm distilled water (less than 50° C) or a few drops of glacial acetic acid. Since problems had been encountered with sodium polytungstate (Coil et al. 2003), we chose the recyclable salt zinc iodide, which was adequate for our soils.

While this simple, low-chemical protocol proved to be successful for most microfossils, we were not confident about starch recovery. Clearly damaged granules (those that maintain birefringence and the Maltese cross but have lost all other characteristics) as well as dubious starch (lacking all diagnostic traits) were recovered along with well-preserved granules. The possibility was then raised that the damage might have been caused by the extraction protocol. In addition, when examined later, starch granules seemed to disappear from slides that had been scanned. It seemed likely that as with CsCl, the salt ZnI_2 might be detrimental to starch preservation (Therin 1994; cf. Box 8.3).

this case). Different mounting materials are described in Chapter 6 (cf. Table 6.4).

LANDSCAPE STUDIES

The most successful archaeological studies using starch extracted from sediments have focused either on the presence or absence of starch or on specific contexts where the diversity of starch has been low and specific identification to taxa has a good potential for success. As Therin et al. (1999), Atchison and Fullagar (1998), and Lentfer et al. (2002) have discovered, the very wide range of starch granule forms that can be recovered from sediments is a major challenge given the limited number of keys available for identification, especially for tropical nondomesticates. As with other more developed fields of microfossil study, the problem of identifying granules to plant taxa will be surmounted as the discipline develops.

All the studies so far are pioneering, but the results achieved demonstrate that the study of starch in sediments has a great deal of potential. As discussed in Box 8.1, Lentfer et al. (2002) discriminated among a number of modern habitats in Papua New Guinea by using multivariate analysis on starch granules described merely by their morphology and size. If one assumes that the same general categories of land use existed in the past and were comprised of roughly the same starch-bearing plant communities, then it would be possible to compare the results of multivariate analysis of starch extracted from ancient sediments with those collected from known modern settings. The proposed use of modern control samples to assist archaeological studies could be usefully expanded to other parts of the world.

Using samples of sediment from different layers at the Jinmium rock shelter in northern Australia, Atchison and Fullagar (1998) demonstrated that the density of starch granules per unit volume changed through time. They proposed that the differences might be linked to changes in site usage, but did not take the analysis further. In contrast, by comparing and contrasting variations in the density and composition of starch at two sites in the province of West New Britain in Papua New Guinea, Therin et al. (1999) proposed a reduction in settlement mobility during the past c. 10,000 years. Their study, described in more detail in Box 8.8, focused on differences in the relative quantity of two size classes of starch as a first-order estimation of the relative contribution of general vegetation (<5 microns) and starchy tubers and roots (>5 microns).

The results showed that the relative proportion of the two sizes of starch granule varied through time. At one site a switch from general vegetation (small size class) to starchy plants (large size class) was detected, whereas not only did the relative occurrence of starchy plants (large size class) in the other site begin much higher, but it also increased in the most recent periods. The key observation was that through time the two sites become increasingly

Box 8.5 (continued)

Experimental tests with modern and ancient starch were therefore run to test the effects of ZnI_2. Different combinations of solution densities, mounting media, time, and temperature conditions for drying were examined (Korstanje 2003). The results demonstrate clearly that it is the combination of humidity and heat that damages or destroys starch granules and not the ZnI_2 by itself. It can be concluded that this heavy liquid may be used safely for starch extraction if the slides are dried at c. 18° C. As a corollary, the progressive disappearance of starch from the slides is likely to be the effect of heating by the microscope light over a relatively long period (c. 6 hours and above) (cf. Therin 1994). Finally, an equivalent level of damage on starch granules produced by food processing, as described in Boxes 4.4 through 4.6, was present before and after the extraction protocol. This result shows that the presence of these altered granules in archaeological samples is meaningful and not an artifact of the heavy liquid procedure.

similar in the density and composition of starch. A similar convergence through time was observed in the density of stone artifacts. Furthermore, Fullagar's (1992) use-wear and residue study of artifacts from sites located in the same region found high levels of variation in the types of tasks undertaken at particular sites in the earliest period and a shift to homogeneity in tool functions through time.

When combined, these various data sets suggest that in the early period when mobility was high and sites were only occupied for short periods, they were used for a limited number of activities. Through time, as people occupied the sites for a longer time, a wider range of activities was represented. In addition, the presence of starchy plants reached relatively high levels at all places, which may indicate a shift towards an emphasis on gardening of starch tubers. This case study demonstrates that the analysis of starch from sediments has an important role to play in the larger study of land use, particularly when it is integrated into a wider analysis involving other types of data. More recent studies of landscape in West New Britain have tried to integrate counts of starch granules into a broader phytolith analysis with some success, but the starch has generally only been used as a surrogate measure for the possible presence of gardens (Lentfer 2003; Parr 2003a).

The incorporation of data on starch granules within a wider landscape study of Dauar Island, Torres Strait, Australia, was used to identify the earliest horticulture on the island (Parr and Carter 2003). Although conclusions about human land use

Box 8.6. Microwave Extraction Of Starch

Jeff Parr

Microwave sample preparation systems were first used for the extraction of pollen from recalcitrant sediments. More recently, several researchers have used microwave systems to extract phytoliths from herbarium specimens (Parr et al. 2001b) and sediments, diatoms from estuarine sediments (Acker and Russell 1999; Parr et al. 2004), and starch from archaeological sediments (Parr 2002). There are two main types of microwave sample preparation systems: (1) the focused microwave systems using open vessels, and (2) the more commonly used rapid pressurised type, which uses sealed vessels. In the pressurized method samples are placed in the vessels with various acids, sealed, and heated by the microwave under pressure to digest organic material and remove unwanted materials adhered to the outside of the microfossils. After digestion the samples are poured through a 150 to 250 μm mesh and the microfossils captured with a 5 μm filter and mounted onto a microscope slide.

Parr (2002) found significant numbers of starch granules during extraction of phytoliths from archaeological sediments with a pressurised microwave sample preparation system. In this case he used 0.25 g of sample and 3 ml nitric acid (HNO_3) and 3 ml hydrochloric acid (HCl). The maximum temperature reached was 120° C at a pressure not exceeding 150 psi for 15 minutes. Since the pressurised microwave preparation systems provide fast results (i.e., less than one hour), they are ideal for preliminary inspections to gauge the preservation and quantity of starch granules in sediments and/or for simultaneous extraction of phytoliths and starch. For example, decisions can be made about which samples should be processed further and what sample sizes of sediment would be needed to obtain an adequate number of granules for analysis. However, to isolate starch granules from other microfossils, it is necessary to use one of the established methods of starch extraction employing heavy liquid separation (cf. Box 8.3).

Box 8.7. Post-Excavation Contamination And Measures For Prevention

Tom Loy and Huw Barton

Post-excavation contaminants should be a serious concern to every researcher of ancient starch because sample sizes are often quite small and can therefore be seriously skewed by even small amounts of contaminants. Four major sources of post-excavation contamination within the research environment should be carefully controlled for: (1) air conditioning; (2) water; (3) transfer of modern food to artifact surfaces; and (4) gloves worn to protect artifacts and soil samples from touch-transfer of contaminants.

Airborne Contaminants

Air conditioning in a lab is a dubious blessing. Unless the air is well filtered near the outlets into the lab, an amazing collection of airborne contaminants may be observed. Starch, cellulose and synthetic fibres, hair, bird feather fragments, clay or silt soil particles, pollen granules, and phytoliths are all potential contaminants through air conditioning systems. In some extreme cases it is useful to place filters over the air supply outlets. In labs that are not air conditioned, open windows provide access for a similar suite of contaminants. In Williamson's analysis of artifacts from South African sites (cf. Box 7.3) starch granules were observed overlying numbers on artifacts covered by clear nail polish. It was found that the prevailing winds blew the contaminants from a flour mill a few blocks away directly into the laboratory where artifacts were laid out for study.

It is vital that the airborne loading of contaminants be continually monitored by placing microscope slides at various locations in the laboratory or even in the field setting. Periodically the slides should be coverslipped in a permanent mounting medium and examined to see what the nature of contaminants has been. The slides should be archived for reference. The procedure should be conducted about once a month.

Water

Depending upon the country, tap water contains a multitude of potential contaminants and should *never* be used in ancient starch analyses. The minimum requirement for water used for making wet mounts or removing samples from artifact surfaces is that it be distilled and filtered through 0.22 micron filters. Ultrapure water, which has been distilled or processed through reverse osmosis and then been passed through charcoal and ion exchange resin columns (e.g., Milli-Q, Elga systems), is preferable.

Food

Although it should be obvious to a starch researcher and is standard biochemical practice, food should be banned from the laboratory. It is important, however, to be aware that in some field settings excavators may contaminate artifacts with food. Recently, patches of corn (maize) starch were observed on the surface of tools from the Sterkfontein site in South Africa. Maize is a staple of the modern African diet, but the artifacts are about 2 million years old, so the New World maize was clearly not an authentic residue. The situation might have been more problematic if, for example, a modern worker in Honduras touch-transferred some maize from lunch onto artifacts dating from the period following its prehistoric introduction to the region. The obvious solution is to have workers wash their hands after eating. Requiring workers to wear starch-free gloves in the field setting to forestall such contamination is overkill.

Gloves

Protective gloves have been used for some time to avoid the transfer of modern materials (including DNA) to the surfaces of artifacts through handling. In the very early days of

Box 8.7 (continued)

ancient starch research, powdered gloves (usually vinyl or latex) were often used heedless of the fact that the powder was, more often than not, starch. Removing gloves often created a shower of airborne starch particles. Labs that had previously been used for biochemical or other biological research where powdered gloves had been routinely used also represent an environment full of starch granules waiting (and willing) to become airborne. Attempts to elicit information about the types of starch used by various manufacturers failed largely because the source of the starch varied according to the price and availability of starches. Using reference collections gathered for ancient starch research, studies of powdered gloves revealed that corn starch (*Zea maize*), rice starch (*Oryza sativa*), and to a lesser extent taro (*Colocasia esculenta*), arrowroot (*Maranta* sp.), and manioc (*Manihot* sp.) have all been used by different manufacturers (more or less in this order) as well as in different batches from the same manufacturer.

A recent study illustrates how serious the potential for contamination by powdered gloves could be. In response to warnings about increased health risks from latex allergies caused by starch-powdered latex gloves (Baur and Jager 1990; Sussman and Breezhold 1995), a survey of starch particle counts in hospital air was conducted (Newsom and Shaw 1997). Their study found that starch particle counts could vary from 0 to >110 granules (maximum countable) for each 30 liters of air sampled (i.e., from 0 to >3,667 starch particles per m^3). Maximum starch counts were recorded on five occasions when a pair of powdered gloves was unpacked in the room containing the sampling equipment (Newsom and Shaw 1997: 156).

Starch particles were collected from the air through the use of a 'personal air sampler' (Casella, Bedford, UK) fitted with a cellulose acetate filter. The sampler was set up as a static, free-standing unit at bed height, adjusted to sample 2 liters of air per minute and operated for 15 minutes (30 liters air). The sampler head was fitted with a 0.5 μm Cellulose acetate filter (Millipore). Each of the rooms was sampled on nine occasions at hourly intervals between the hours of 9 A.M. and 5 P.M. on five consecutive days (45 samples). Only one room was sampled per week. Three additional rooms were sampled for one day only. Total starch particle counts were determined by placing the acetate filter from the air sampler in a Petri dish containing four drops of Lugol's Iodine solution (iodine 50 g, potassium iodide 100 g, purified water 1 liter) for a total of 3 minutes. Starch particles stained black and were counted using a x10 hand lens (Newsom and Shaw 1997: 155–56).

The summary results of the study conducted by Newsom and Shaw (1997) are presented in Table 8.3. Starch particles were found to be readily detectable in hospital air in rooms in which starch-powdered gloves were worn, and some starch could be detected in the laboratory office and psychiatric ward where powdered gloves were not used. Rooms lacking mechanical ventilation with moderate glove usage had relatively high starch particle counts, whereas the ventilated rooms maintained moderate starch counts even when glove usage was very high, e.g., Theatres 1 and 2.

Following the introduction of powder-free gloves to the Accident and Emergency unit, airborne starch particle counts fell considerably, but some starch was detected and may have derived from occasional uncontrolled use of powdered gloves or from elsewhere or in pre-prepared procedure packs (Newsom and Shaw 1997: 157). The low counts of starch particles in the laboratory office, which was not mechanically ventilated, are thought to result from occasional visits by technicians wearing gloves. The very low counts of starch particles in the psychiatric ward is not explained by Newsom and Shaw, but could be considered as the level of background noise in an environment in which powdered gloves were routinely worn elsewhere in the building.

The overall conclusion of the study is that the use of starch-powdered gloves, particularly

Box 8.7 (continued)

Table 8.3. Counts of starch granules in hospital air (after Newsom and Shaw 1997).

Site	Number of tests	Mean count	SD	Pairs of gloves worn
Intensive care 1**	45	7.60	0.70	9
Intensive care 2**	45	21.25	0.70	3
Intensive care 3	44	42.14	0.71	6
Theatre 1**	41	10.07	0.86	25
Theatre 2**	45	12.91	1.04	58
Scrub room 1**	45	17.73	1.04	18
Scrub room 2**	44	16.58	1.07	17
Accident & Emergency	45	17.78	1.07	4
Accident & Emergency*	40	4.74	0.44	2
Laboratory office	9	1.56	0.23	0
Psychiatric ward	9	0.86	0.18	0

Mean count = mean of stated number of tests expressed as cubed root of count per 30 litres of air. The overall mean count was 13.8 granules/30 l of air.
* powder-free gloves trial
** mechanically ventilated rooms

opening packs of gloves, can contribute significant amounts of airborne starch to rooms in which they are used and contribute starch particles to rooms in which use is either infrequent or incidental. In short, do not allow the use of starch-powdered gloves in any rooms or adjoining rooms in which you intend to conduct residue studies of ancient starch.

Finding unpowdered examination gloves can be very difficult and their cost can also be prohibitive (in some countries this remains so). In the absence of unpowdered gloves, an alternate approach is to don the gloves and wash the hands in soapy water, rinsing in clear water and drying thoroughly. Removal of the gloves should be done either well away from the lab environment or underwater in a sink or bucket containing a small amount of soap or detergent. Tests using this method resulted in only the very occasional transfer of one or two granules to clean microscope slides used as controls. Some researchers have used cotton gloves as an alternative to powdered gloves, but this is a dangerous ploy because the copiously shed cotton fibres then become contaminant fibres on artifacts. Others have used common dishwashing gloves with some success, although dexterity suffers from the glove's ill fit and thickness and allergic reactions are common. Unfortunately, some people are quite allergic to the latex in the unpowdered gloves and can develop an ideopathic skin rash. Barrier cream or 2 percent cortisone creams may help reduce the reaction, but a doctor's advice should be sought. In the absence of suitable gloves for manipulation of artifacts, large forceps which have had their ends wrapped in cling wrap work well. If caught without any forceps or gloves, another trick is to unroll some food-quality cling wrap, or parafilm, or even plastic sandwich bags (new, of course) and stretch the film well, then poke the thumb and first two fingers serially into the plastic. The cling wrap coats the finger and thumb with a tight layer that can easily be discarded when finished.

Box 8.8. Landscape Change As Detected By Starch Granules

Robin Torrence

In the first attempt to extract starch granules from archaeological sediments, Fullagar and his colleagues used bulk soil samples taken from the layers of a deeply stratified site at Bitokara Mission (designated FRL in the Papua New Guinea national register). Their pilot study, conducted in 1993 (Fullagar et al. 1998), concluded that starch granules were present and their abundance had changed through time. Following on from this successful test, Therin (1994) replicated the FRL study using samples collected from the nearby site FAO (Box 8.2). Therin improved the heavy liquid extraction method and doubled the sediment sample size (Therin et al. 1999: 451). As with the FRL pilot project, he found chronological changes in the quantities of starch present in the different layers at FAO.

Although both these studies demonstrated that starch granules were preserved within archaeological contexts, the authors felt hampered in their ability to interpret their data further. In the absence of an appropriate reference collection, they could not confidently identify starch granules to plant taxa. Despite this limitation, Therin et al. (1999) made a significant contribution to the interpretation of human behaviour at FRL and FAO by taking a comparative approach and by integrating the data on starch with other classes of information.

Inter-site Comparisons

Before comparisons could be made, it was necessary to standardize the separate results from FRL and FAO. First, the findings were grouped into comparable chronological units. Second, the counts were transformed into mean number of starch granules per 2 g of sediment. Not surprisingly, the improved Therin method had consistently retrieved a larger number of granules per gram of sediment processed. Third, to counter the effects of the varying methods, the analysis focused on the relative differences among the layers, rather than the absolute amounts of starch recovered.

Having standardized the data, changes in the density of starch at the two sites can be compared (Figure 8.7). In interpreting these data, it is important to note that layers 3, 5, and 8 represent thick deposits of volcanic tephra (ash) in which the starch is mainly derived from taphonomic factors rather than human action. At FRL the relative density of starch granules increased slightly over time. In contrast, at FAO it decreased radically from the highest point at the beginning of the sequence and then stabilized later on. By about layers 4 or 5, the frequencies of starch at the two sites followed roughly the same patterns.

Next, the starch granules were divided into two size classes. Based on a preliminary assessment of reference material from the region (Therin et al. 1997), the first group, granules >5 microns, should mainly represent the starch-bearing, edible part of the plant, such as the roots and tubers that were likely to have been preferred food sources. The second group, granules <5 microns, should on the whole be derived from nonedible parts of the plant such as leaves and stems and represent general vegetation. Although there are notable exceptions to this dichotomy, namely the important food plant taro whose starch granules are in the second group, the classification provides a rough and ready means for comparison of changes through time. Finally, because small granules are more likely to be mobile through the sediments (Therin 1998a), only very gross changes in their numbers should be significant.

As shown in Figure 8.8, when the starch granules are sorted into size classes, the pattern of change at FRL and FAO again shows that beginning with major differences between the sites, they gradually become more similar through time. It is important to note that at FRL the frequency of small granules is always less than for the large granules, which suggests that

Box 8.8 (continued)

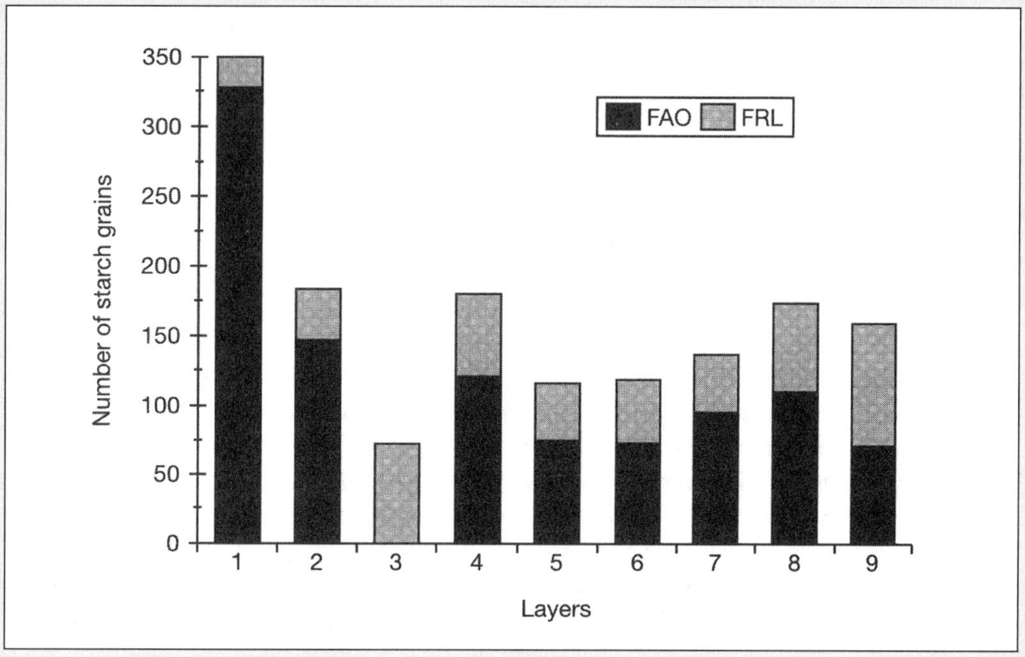

Figure 8.7. Temporal changes in the density of starch granules in sediments at the FAO and FRL sites. *Reproduced with permission from Therin et al. 1999: figure 23.7*

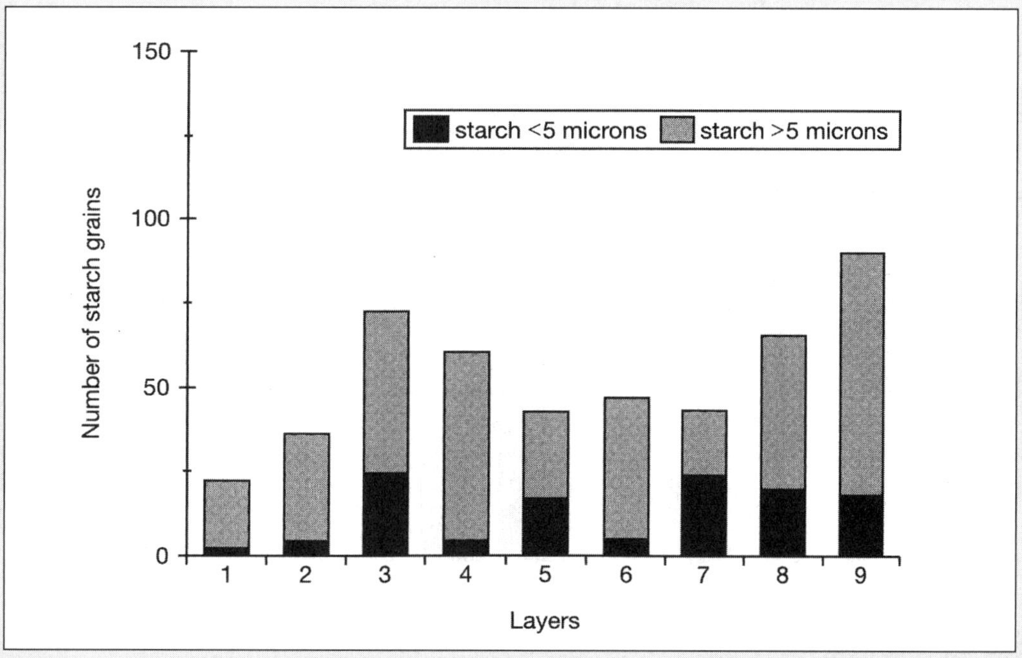

Figure 8.8. Changes through time in the density of two size classes of starch granules at the FAO and FRL sites. *Reproduced with permission from Therin et al. 1999: figure 23.8*

and anthropogenic changes were largely based on results of a phytolith study, the authors claim the presence of yams and a native form of *Ipomea* which was known to have been used in the recent past.

Activity Areas

Moving from the scale of landscape to that of a single site, Balme and Beck (2002) made a pilot study to test whether starch could play a role in reconstructing the patterning of human activities. Their major concern was whether the original distribution of starch at a rock shelter site had been significantly modified by post-depositional processes or physical attributes of the site. They compared the density of extracted starch granules within 86 sediment samples collected from across the floor of Petzkes Cave with spatial patterning in pH, moisture content, and sediment compaction as well as shelter height (related to human trampling) and the location of the drip line. As demonstrated by the data presented in Box 8.9, the patterning of starch has its own integrity. Beck and Balme attribute the pattern of starch distribution to human activities, such as food preparation in the warm, sunny part of the shelter and sweeping.

The presence of starch granules from particular plants can also be instrumental in the identification of the function of particular archaeological features of contexts. For example, Haslam's study of maize starch preserved in the soil within a rubbish pit at the site of Copán in Guatemala suggests that maize was ground by individual households (Box 8.10). Similarly, a study by Horrocks et al. (2004b; cf. Horrocks 2004) of starch preserved in the soil of a pit feature and from a sediment core

Box 8.8 (continued)

most of the starch introduced to the site represents food remains. In contrast, the very high density of starch in the earliest layer at FAO is mainly represented by small starch granules, but their frequency decreases through time. In contrast, the density of large granules remains relatively constant through time. To explain the initial differences in the sites followed by similarity in the most recent periods, Therin et al. (1999) compared the starch trajectory with patterns generated by studies of the stone artifacts at these sites.

Integrating Starch Results

Studies of the way stone tools were made and used have concluded that changes in the nature of activities at FAO and FRL reflect shifts in the regional pattern of land use (Fullagar 1992; Fullagar et al. 1998; Torrence 1992; Kealhofer et al. 1999; Therin et al. 1999). In the earliest period, when people were moving from place to place relatively frequently, sites such as FAO were occupied for short periods of time. This explains the higher proportion of vegetation (small granules) and smaller proportion of food remains (larger granules) in the early period. In contrast, FRL, which is located adjacent to a stone tool quarry, began its life as a place where stone tools were manufactured in advance of use and carried over large areas. Through time, as mobility decreased, the quarry at FRL was abandoned and the site was used mainly for domestic purposes. For this reason the starch assemblage at FRL gradually becomes more like the domestic assemblage at FAO and the quantity of food processed at FRL rises.

Despite these trends, there are other minor variations in the frequencies of starch which are difficult to interpret, possibly because they are related to variations in the spatial patterning of activities on the sites. A much larger area would need to be excavated to test this possibility. What is more important is that the basic structure of the starch data is remarkably similar to changes in the stone tools. By integrating various sets of information together, it is possible to use rather gross descriptions of the starch assemblage to monitor changes in the way places were used in the past. This is especially true if a regional perspective can be obtained.

Box 8.9. Starch As An Indicator Of Activity Areas

Wendy Beck and Jane Balme

The aim of this pilot study was to examine whether starch distribution could reflect past cultural activities at archaeological sites. To do this, we compared the horizontal distribution of starch within the sediments of Petzkes Cave, a large (150 m^2) sandstone rockshelter in northern New South Wales, Australia, with the spatial patterns of material introduced through post-depositional processes and key physical characteristics of the site.

Starch Analyses

The site was excavated to a depth of 3 cm within dark grey sand dating to about the last 1,500 years. Eighty-six sediment samples of 2 g each were selected from a range of 25 cm squares across the site (Vinton 1995). Starch granules were extracted using caesium chloride (CsCl) (Fullagar et al. 1998: 52; Atchison and Fullagar 1998: 115; Therin et al. 1999: 451). Two slides from each sample were prepared and all starch granules observed were counted.

The caesium chloride extraction method is not without problems, one being the destruction of starch granules during the extraction (cf. Box 8.3) and another that the CsCl tends to recrystallise during the procedure. However, these systemic errors would equally affect each sample. Since we are concerned with relative rather than absolute numbers of granules, these effects were not important.

Results

Figure 8.9 shows that there is a low background presence of starch across most of the site. The number of starch granule numbers in each sample ranged from 1,107 to 53 per 2 g sample. The mean number of granules per sample is 244 per 2 g sample, which is higher than starch frequencies recorded for sediments at Jinmium rockshelter (Atchison and Fullagar 1998: 118–19). The maximum diameters of 72 percent of the starch granules were over 10 μm. The most frequent shape was angular, followed by round, sub-round, and elliptical. The frequency of both size and shape varied, which suggests potential for identification of the plants represented in the starch assemblage.

The most likely sources of starch in the shallow shelter deposits we excavated are from the *in situ* decomposition of leaves, seeds, fruits, and underground storage organs which grew naturally at the site, were brought into the site by humans, or were introduced through human and animal faeces. The macroscopic plant remains recovered by the excavation were mostly wood and leaves (19 and 67 percent by weight respectively) (Vinton 1995) derived from a fig tree (*Ficus rubiginosa*) growing over rocks at the shelter entrance. While it is possible that some of the plant material (especially leaves and seeds) containing starch may have blown into the site, the differences between the macroscopic plant distribution and the starch granule distribution at Petzkes Cave (Figure 8.9) suggests that post-depositional processes were negligible.

Visual comparison of the distribution of starch within the cave when compared to pH, moisture, ceiling height, and sediment compaction, suggests that the physical properties were not the main determinants of the spatial distribution of the starch assemblage (cf. Balme and Beck 2002). Figure 8.9 shows there is little association between the distribution of starch and the dripline. Instead, the densest accumulation of starch is at the front of the shelter near the major opening (southern end). This region is well lit from the south and east and is on the flat shelter floor above the relatively steep talus slope. Ethnographic observations summarised by Walthall (1998) concerning the

Box 8.9 (continued)

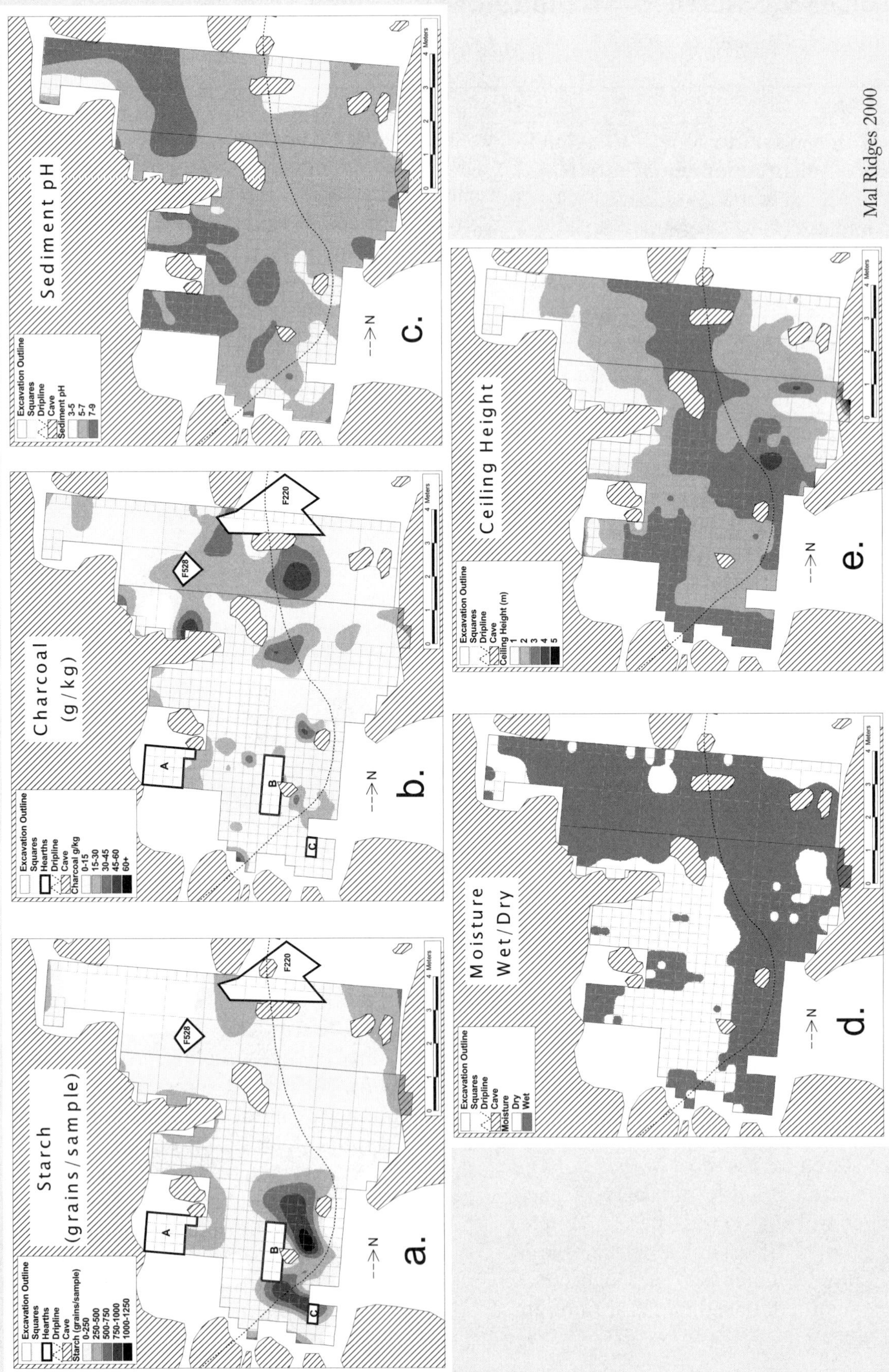

Figure 8.9. Spatial distribution of starch granules and other materials at Petzkes Cave. *Reproduced with permission from Balme and Beck 2002: figure 2*

at the site of Rangihoua Bay, New Zealand, confirms site use as a location for the cultivation of introduced crops (Box 8.11). Starch identified as sweet potato was also recovered within a large stone mound at another site in New Zealand, but identifiable starch was not present in other smaller stone features (Horrocks et al. 2004a). The authors concluded that the large mounds may have been used for the cultivation of sweet potato whereas the others were used for growing gourds, represented in one case by pollen.

Coprolites

The Horrocks et al. (2002, 2004a) studies described in Box 5.2 have also made an important innovation in ancient starch studies because it successfully extracted starch granules from human and canine coprolites. Sweet potato and bracken fern were identified from coprolites excavated from several sites in New Zealand using a version of Therin's protocol (Box 8.4). The identification of the starch granules is very important because they provide the first direct evidence for the presence of these important food sources in New Zealand during the prehistoric period. The study also demonstrates that starch granules can be well preserved within coprolites. In fact, the excellent preservation of starch granules is especially notable given the highly corroded condition of pollen and spores recovered from the same coprolites (Horrocks et al. 2004a: 257).

MOVING AHEAD

To increase the potential of this approach, analyses of starch extracted from sediments will require continued improvements in methodology, especially in terms of protocols for extracting starch with or without other macrofossils and in the development of more powerful keys for identification. Previous studies have shown that experiments using modern starch and/or modern settings provide excellent testing ground for new approaches and techniques.

The increasing number of significant archaeological results generated by the pilot studies con-

Box 8.9 (continued)

use of rockshelters by people in Papua New Guinea, the Australian desert, and South Africa's Western Cape indicate that this front central position beneath the dripline, where there is good daylight, is most frequently used for cooking and maintenance activities.

Apart from this dense patch of starch at the southern front of the shelter, the other pattern obvious from Figure 8.9 is the accumulation of starch around barriers (walls at the rear and front of the shelter) and large boulders in the centre and southern end of the shelter. While it is conceivable that some physical action such as wind and water flow might cause these accumulations, the fact that they surround the barriers rather than accumulate on either side, which might be expected if there was a physical source, suggests that the cause of the spatial patterning might be human behavior such as sweeping.

Conclusions

Overall, the processes affecting the distribution pattern of plant remains at Petzkes Cave seem to be different from the trampling that has spread the stone artifacts (Theunissen et al. 1998). In contrast, starch is likely to remain *in situ* in areas where foot traffic is high (measured here by sediment compaction and ceiling height). Although trampling and other sources such as water may cause some redistribution of the plant remains, perhaps contributing to the background scatter of starch, they have not obscured the general pattern of starch distribution at Petzkes Cave. Given these results, we therefore interpret the spatial patterning of starch granules extracted from the sediment as representative of past human use of space at the rockshelter. It may be possible to refine our interpretations of dense patches of sediment starch as plant processing areas (Balme and Beck 2002), when the starch granules have been identified to species.

Box 8.10. Investigation of Activity Areas Through Starch Analysis: A Case Study From Copán, Honduras

Michael Haslam

The identification of archaeological household activity areas can be problematic. Reliance on identifiers such as artifact types or relationships between structural remains (e.g., Webster et al. 2000: 101) can distort our reconstructions of activity areas by placing too much emphasis on movable objects and archaeological guesswork (Parnell et al. 2002: 379). The use of a relatively immovable class of material as an indicator of activity may lead to a more accurate reconstruction of the human use of an area. As shown in the following case study, microscopic starch granules are potentially an ideal source of information, as they are often abundant and are quite resistant to subsequent maintenance activities, especially when compared to artifacts.

The site of Copán is located in western Honduras on the southeast periphery of the Maya cultural area. The Copán Valley was first occupied sometime before 1,000 B.C. by indigenous non-Maya peoples and reached its population peak at A.D. 600 to 800 (Fash 1986). The local inhabitants relied heavily on *Zea mays* (domesticated corn or maize) throughout ancient times and continue to do so today. Currently, maize cobs are dried in the field, then taken to the domestic area and the kernels ground manually (Figure 8.10). The resulting flour consists of essentially millions of *Zea mays* starch granules which are readily transferred to the soils surrounding the grinding area.

In 1999, 150 stone artifacts from inside and surrounding a pit in an inferred domestic context at Copán were analysed for microscopic use-wear and residues (Haslam 1999). A variety of lithic types including obsidian, chert, and basalt, dated from 100 B.C. to 30 A.D., were examined. Sediment samples were not available because of import and quarantine restrictions, but the soils adhering to the unwashed artifact surfaces acted as a useful sediment trap.

Figure 8.10. Maize is still hand-ground daily in rural houses in the Copán Valley. Note the dried maize cobs hanging from the roof, and the earthen floor in Don Saturnino and Doña Lucia's house shown here. *Photo by Michael Haslam*

ducted so far demonstrates that the analysis of starch granules extracted from sediment samples has a robust and exciting future. As with all ancient starch research, the most conclusive studies have been those which demand a low level of identification (presence/absence or density) or for which only a small number of plant taxa are required. The analysis of starch extracted from sediments is also especially important for identifying the presence of plants that do not produce pollen or phytoliths or for interpreting the function of particular archaeological features. In addition, starch analyses of sediments are likely to expand because they provide an excellent opportunity for monitoring human behaviour at the scale of a whole landscape. Along these lines, the extraction and analysis of other types of plant microfossil in conjunction with starch is likely to provide the best way forward.

Box 8.10 (continued)

During microscopic analysis of the sediments adhering to the artifacts, an abundance of *Zea mays* starch granules (identified by shape, size, and surface characteristics) was observed. Concurrent and complementary use-wear analyses differentiated these granules from maize starch residues resulting from the use of the artifact. Only those starch granules directly associated with used edges, ridges, or surfaces were classified as having resulted from use. All other granules were considered to be 'taphonomic' starch, a term used here to avoid the negative connotations of 'contamination,' which has been used to describe residues unrelated to use (cf. Barton et al. 1998: 1233; Atchison and Fullagar 1998).

A number of conclusions can be drawn from finding a high concentration of taphonomic starch in the sediments. First, the presence of the starch lends weight to the hypothesis by Hall and Viel (1998) that the pit in which many of the artifacts were found had a domestic function. It also supports the ethnographic analogue for maize being ground in a household context. The study therefore provides new avenues for the detection and interpretation of archaeological household activity areas in the Maya region and elsewhere.

Finally, the results emphasise the utility of analysing ancient starch granules from sediments, provided the context and mode of preservation of the granules is taken into consideration. In particular, the *in situ* analysis of sediment residues on artifacts can provide the advantages of bulk soils analysis with the addition that there is a direct association between artifact and residue that relates directly to the archaeological context. In conjunction with the critical use of ethnographic analogy, it should be possible to use the study of starch granules in soils to draw meaningful conclusions about past activities.

Box 8.11. Starch Residues In A Sediment Core And Earth Oven

Mark Horrocks

Analysis of plant microfossils (starch residues, pollen, and phytoliths) in a sediment core taken from a small swamp (0.5 ha) at the Maori site of Rangihoua Bay on the North Island of New Zealand demonstrated that several introduced crops had been cultivated at this location (Horrocks et al. 2004b; Horrocks 2004). The swamp has a network of prehistoric ditches with slightly raised drier areas between them (Figure 8.11). A nearby eroded soil profile with buried stones and macroscopic charcoal was also sampled (Figure 8.12). Starch residues were extracted from the core and soil samples following the protocol in Horrocks (2005), which used a modified version of Therin's heavy liquid separation protocol (Box 8.4).

The lower part of the core comprised lake sediments which were overlain by peat with an upper thick layer of fine allocthonous gravel which was deposited after the first people had arrived at this location. Pollen from the lowest levels represented a conifer-hardwood forest that commenced c. 7,500 BP. A sharp decline in pollen of tall trees accompanied by high charcoal concentrations and increases in pollen of several indigenous taxa commonly linked to large-scale deforestation by Polynesian settlers was observed in the uppermost peat section of the core. Starch residues (starch granules and xylem cells) of sweet potato root and taro corm (Plate 56) as well as pollen and phytoliths from the paper mulberry plant coincided with the trends in the peat and continued upwards into the gravel layer. These cultigens were among the six introduced species which were under cultivation in New Zealand at the time of European contact. The presence of known volcanic tephras established a firm chronology in the upper part of the core which showed that deforestation and associated cultivation of introduced plants had occurred after c. 650 BP but prior to European arrival.

In addition to the ditch system in the swamp, many of the hill slopes at Rangihoua and adjacent bays have prehistoric linear depressions which are thought to be associated with Maori agriculture. The presence of a prehistoric earth oven, as indicated by the buried stones and macroscopic charcoal in the soil profile, was confirmed by the discovery of taro starch residues in that deposit. The starch residues from Rangihoua Bay have helped identify specific purposes for landscape features and provide direct evidence for prehistoric multicropping at the site. These data represent the first concrete evidence for making a secure link between these ancient landscape features in New Zealand and Maori agriculture.

Figure 8.11. The swamp at Rangihoua Bay, previously an early Polynesian garden, now in pasture. The swamp was originally a small lake formed between a Holocene sand dune (left) and a much older Pleistocene dune (arc, right). The swamp drains into the stream behind it which discharges into the bay (shoreline, left). The ditches running across the swamp are 20–25 m long. *Photo by Mark Horrocks*

Figure 8.12. Soil profile at Rangihoua Bay, exposed by erosion. The presence of taro starch residues confirmed that the buried stones and macroscopic charcoal represented a prehistoric earth oven. The fence post is 10 cm across. *Photo by Mark Horrocks*

CHAPTER 9

STARCH ON ARTIFACTS

Richard Fullagar

Artifacts, particularly stone tools, appear to operate as a 'starch trap' by shielding starch from decay by microorganisms and providing suitable conditions for long-term preservation (cf. Chapter 5). This situation provides opportunities for two related types of analysis. First, one can study how these artifacts were used in the past by integrating analyses of starch granules with other approaches, such as the study of use-wear or additional residue types. Second, identifications of starch granules on artifacts aids the reconstruction of past plant use and diet. Case studies using the latter approach were reviewed in Chapter 7. This chapter focuses on how starchy residues can be used to inform on artifact use.

Finding out about how tools were used has been central to many archaeological questions. A wide range of artifacts, including spear tips, sickle blades, grindstones, pots, and other implements, represent a wide range of behaviour, such as how people got their food, what they ate, and their craft activities. As discussed in Chapter 4, starch granules on their own or in association with other types of microscopic evidence may be indicative of many kinds of deliberate activity, such as plant food or fibre processing, woodwork, hafting, or of products derivative of plants such as glues or resins used as binding agents. Chapter 10 will present examples of how starch found on artifacts may also be indicative of processes to modify raw foods such as cooking or fermentation. This chapter outlines methods for studying starch that has been preserved as residues on artifacts. After a review of the range of possible applications of this approach, methods for collecting artifacts in the field and laboratory procedures for retrieving starch are described. Finally, the chapter discusses the basic principles for interpreting residues through an integrative approach involving the nature of starch residues, use-wear traces, and experimentation.

Residue analysis generally, and starch granule identification in particular, supplements a functional study of stone tools by providing potentially more precise interpretations. This approach, however, works best when integrated with additional information. As with the sediment studies discussed in the previous chapter, starch is best employed in conjunction with a range of other variables. Simply removing starch from the surface of an artifact is not enough to ensure valid interpretations of past activity because the history of how the starch came to be preserved on the tool can be quite complex. For this reason, experiments designed to help associate microscopic observations with larger scale behavioural inferences are required.

The degree of detail required in the analysis of starch will vary according to the type of research question addressed. For example, some questions may demand knowledge about the on-site processing of a particular plant species. This could be obtained by means of a very detailed study of starch granules on various artifacts. Other hypotheses might propose that a particular site functioned as an implement production centre rather than as a habitation site. To answer other archaeological questions, it may be sufficient to determine which tools were used or not used.

An integrated approach to interpretation demands consideration of the life histories of an artifact. For example, residues may not necessarily be equated with the dominant tool function. Spear points may be used to peel roots and residues may arise from incidental contact with a variety of clingy starchy materials. As discussed in Box 9.1, a stone axe can provide a serendipitous opportunity to reconstruct unrelated events like the most recently cooked meal.

SIGNIFICANT CONTRIBUTIONS

The potential of starch residues to contribute important information about plant use and artifact function is continually being expanded into many exciting areas of research, time periods, and

Box 9.1. Starch On The Axe

Tom Loy

In one of the highest passes through the Alps, the most important and in many ways the most puzzling mummy of the late twentieth century was discovered in 1992 (cf. Egg et al. 1995; Barfield 1994; Spindler 1994; Loy 1994b, 1998a, 1998b). The Ice Man of Haslabjoch (dubbed Ötzi) is famous and archaeologically significant because all his tools, clothing, and personal effects were found intact: a man frozen in both ice and time. As well as frozen artifacts and mummified tissue, even microscopic particles and organisms have been superbly preserved, free from the damaging effects of soil as is usually the case with archaeological deposits.

Rarely do we encounter a single moment in prehistory. More often the artifacts recovered represent refuse, worn-out, or forgotten tools. In some cases burial goods can provide a glimpse into the individual past, but they represent special offerings, not the totality of an individual hunter's toolkit. Ötzi is puzzling as well because he cannot be placed precisely in the cultural landscape of 5,300 years ago. He is not associated with his village since he perished in a pass high in the Alps. In order to find out who he was and what he did in life, his body, his tools, and his possessions are the only place to look.

There has been much speculation about Ötzi's life and his occupation. Many detailed studies have been undertaken using methods ranging from CAT scans to the analysis of pollen trapped in his clothing. Over 60 international groups continue to study the minutiae of his remains. Late in 1993 I was invited to analyse his tools for traces of use and organic residues. I examined just about everything he carried: his bow, quiver, copper axe blade, flint-tipped 'dagger,' two arrows, stone and bone tools, and raw materials. Surprisingly, his copper axe contained abundant starch residues.

Methods

Using an Olympus BHSM incident light microscope, each artifact was examined at magnifications ranging from 50 to 1000 diameters. I first used a simple colourimetric screening test to identify proteins, chlorophyll and divalent cations. Subsequently, immunological methods were used on aqueous samples removed from the tool surfaces to identify specific proteins. Artifacts were mounted directly on the microscope stage and supported in position by modeling compound covered in cling-wrap or plastic wrap. The film was changed often to prevent contamination. During the examination in Mainz, no gloves were used in order to avoid contamination with modern starch powder. The tools were manipulated with forceps, hemostats, and other implements. The examination of the tools was carried out in a laboratory room in which archaeological iron chariot fittings were being cleaned and restored. Samples of dust were examined and contained no materials that would be confused with authentic residues.

Overall, the organic residues consisted of blood, collagen, and other animal products consistent with functions and tasks related to hunting and butchery. The overwhelming conclusion I reached was that Ötzi was an experienced hunter, capable of making and carefully maintaining his hunting gear.

Copper Axe

One of the most spectacular of his tools, and the one which caused Spindler to predict that Ötzi was not a medieval mummy, was an axe with a yew wood handle and a copper blade (Figure 9.1). Examining the axe blade revealed, as expected, the presence of woody fibres trapped in irregularities along the sharpened bit. Nevertheless this artifact presented a complex set of residues, made further complicated by

Box 9.1 (continued)

the presence of copper oxidation compounds. The axe blade had been removed from its haft and had been cast, but no contaminants from the casting were observed. Macerated cellulose clumps were abundant over both surfaces. Woody fibres were observed near the edge, consistent with the inferred function of woodworking. Collagen fibrils and bundles were observed near the poll (or butt) end of the blade, and about one-half of the distance from the bit to the hafted area. It is likely that this collagen is from the leather lashings.

Figure 9.1. Ötzi's copper axe showing the areas of discolouration from pitch, protein, and plant residues. *Photo by Tom Loy*

Red blood cells were observed in two locations, corresponding to the discoloured zones. A cluster of red blood cells in a classic *en rolleaux* (stack of coins) formation was observed near the lateral edge of the bit. These cells were associated with a polygonal shrinkage crack pattern typical of thick proteinaceous residues. Hair fragments and muscle tissue were observed on both sides. The presence of animal blood and tissue suggests strongly that the axe was used in animal butchery, a practice still current in hunting societies where small axes and hatchets are used to disarticulate big game.

Most surprising was the observation of abundant discontinuous starchy residue patches on all parts of the axe blade, *including areas which would normally have been under the haft and along the bit itself.* The grains are large and ovate with a circular cross-section (Plate 57). The arms of the interference cross rotate upon rotation of the polariser plate but are partial or show indistinctive interference. The starch granules are aggregated and mixed with cellulose fragments. The aggregations are stained a similar copper green as was the proteinaceous deposit.

The plant origin of the starch remains unknown because we lack good comparative material appropriate for starchy foods in the Alps. It is clear, however, that the large size of the granules preclude them from being derived from grass seeds. The lack of fracture of the granules further suggests that they are not from an extensively milled flour. Gregg indicates that in the Neolithic of central Europe the only regularly available tubers were wild garlic, wild onions, and cattails (*Typha* sp., cf. Box 4.3), although it is likely there was a wider variety of wild plants associated with the alpine and subalpine environments of the Alps.

Why is there starch on an axe? The presence of starch granules and plant tissue, if limited to the exposed cutting edge and blade, might suggest the use of the axe in harvesting or food processing. The granules show the optical effects of heating (indistinct cross). Because the granules on the axe blade have been heated to a point below the temperature at which they would form a gel, and because there is cellulose tissue intimately associated with the granules, it is reasonable to infer that they represent the remnants of cooked food.

In contrast, the distribution of the starch granules indicates that they came into contact with the axe blade while it was removed from its haft. Experimental use of similarly hafted Pacific Northwest Coast stone adze blades demonstrated that rehafting the blade is a regular task. As a result of impact while in use, the blade is loosened and the lashing thongs stretch. At some point it becomes necessary to remove the lashings, reseat the blade, and tie it up with

geographical regions. A series of significant archaeological questions has been addressed by analyses of starchy residues, including tool function, settlement patterns, type and nature of plant use, and the history of plant domestication.

Tool Function

Until researchers began to branch out into other areas in recent years, the major aim of studying starchy residues was to reconstruct how tools were used in the past (Box 9.2). Beginning with the microscopic techniques introduced by Semenov (1964), functional studies based on microscopic wear traces have developed into a specialised field within archaeology (e.g., Hayden 1979; Kamminga 1979; Keeley 1980; Fullagar 1986, 2005). It is only recently that biochemical and morphological studies of associated organic residues have enlarged the specificity and types of functions that can be reconstructed (e.g., Fox et al. 1995; Fullagar et al. 1996; Brass 1998; Field and Fullagar 1998; Kealhofer et al. 1999; Hardy 2004; Pearsall et al. 2004; Piperno et al. 2000, 2004; Rots and Williamson 2004).

Studies about past tool functions can contribute to an understanding of broader behavioural questions. For example, Barton and White (1993) employed residue analyses to test the relationship between social interaction and tool function at Balof Cave in New Ireland. They noted that an increase in the proportion of artifacts made with imported stone was contemporaneous with the introduction of a new animal prey. They asked whether these two introductions represented a new regional social system of interaction. The results did not support a connection between obsidian and animal processing since starch residues from tubers were just as common on the more recent obsidian tools as they had been on older stone artifacts made primarily from local materials (Figure 9.2).

Sometimes just answering the question of whether an artifact has been used for a particular function can be quite powerful. As discussed in Chapter 7, Perry's (2001, 2004, in press) research on stone flakes thought to have been incorporated into specialised equipment for processing the

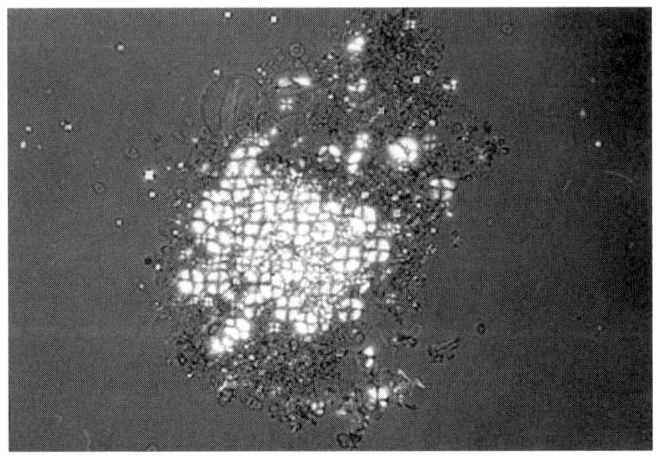

Figure 9.2. Starch granules extracted from residue on an obsidian flake from a level at the site of Balof 2, Papua New Guinea dating to c. 3,000 BP; viewed under cross-polarised light. *Photo by Huw Barton*

Box 9.1 (continued)

wet leather thongs which will shrink tight as they dry. The presence of starch granules on the bit and other exposed areas of the blade indicate that the axe was not used after the deposition of the starchy residue.

In light of the evidence, the conclusion is that axe was rehafted either during a period when Ötzi was eating, or shortly thereafter while he had moist food residue on his hands. This event may have occurred shortly before his death, and certainly before he had an opportunity to use the axe and thus remove the starchy residue from its bit. Because no other food was found with his remains except a single sloe berry, the starch residues show that Ötzi had eaten relatively recently prior to his death.

Residue analysis of the axe blade revealed that in addition to the expected use of woodworking, butchery was also performed. The starch granules found are not related to the functional use of the axe, but rather they provide important indicators of diet and tool maintenance.

Box 9.2. History Of Starch Research On Stone Tools

Richard Fullagar

Although starch granules have been noted by archaeologists for almost a century (Chapter 2), the presence of microscopic starch residues on stone artifacts has only been recognised for about 25 years and systematic study of these has only just begun. The recognition of starch granules as tool residues arose in part as a result of the intensive study of artifacts from archaeological sites with outstanding preservation (e.g., Shafer and Holloway 1979; Loy 1994a; Samuel 1996a, 1996b, 2000). Of equal importance were high-magnification studies of microscopic use-wear that suggested residues of various kinds could contribute to, or interfere with, observations of other microscopic traces such as polish (e.g., Anderson 1980; Keeley 1980; Kamminga 1982; van Gijn 1990; Hurcombe 1992) and, moreover, that starch granules in particular survive in association with utilised edges (Fullagar 1986).

In 1976, Briuer published a paper (based on his doctoral dissertation) in which he described microscopic plant and animal residues on 37 stone tools from two dry rockshelters in Arizona, USA. Although he did not report the presence of starch granules on specific artifacts, he did refer to the use of iodine-potassium-iodide as the typical stain for starch granules. John Cockerham (State of California Department of Justice Crime Lab) and Flora Scott (UCLA plant anatomist) employed this stain to identify 'plant tissue' from the artifacts (Briuer 1976: 478). Inspired by Briuer's (1976) work, Shafer and Holloway (1979) selected 11 flaked stone artifacts (5,000–2,000 years old) from Hinds Cave in Texas, USA, for residue analysis. Organic material was removed from each specimen and was then mounted in glycerol and studied at x200 and x400 magnifications. Starch granules were observed on seven specimens. Heavy accumulation of starch granules on one specimen with little edge damage suggested to these archaeologists that the starch originated from soft areas of plants (e.g., leaf or tuber). These early observations prompted further research on where starch granules might be found. On what kinds of raw materials? On what tool types? In what environments?

During the early 1980s Thomas Loy and Richard Fullagar found starch on a variety of flaked stone, ground stone, shell, and glass tools in various ethnographic and archaeological contexts in Oceania and North America. Loy (Loy and Nelson 1986) had observed starch granules on artifacts using cross-polarised light. In 1983 he discussed his results with Fullagar who subsequently reported starch granules on a variety of ground and flaked stone artifacts from Australia and Papua New Guinea (Beck and Fullagar 1985; Fullagar 1986, 1988, 1989, 1992, 1993a, 1993b). Since then, starch granules have been observed in association with use-wear in a variety of contexts by archaeologists working on materials primarily from Australia and the Pacific region as summarised in Table 1.1 (e.g., Brown 1988; Higgins 1988; Hall et al. 1989; Brass 1993, Barton 1991; Loy et al. 1992; Barton and White 1993; Boot 1993; Loy 1994a; Richter 1994, 1996; Robertson 1994, 1996; Wallis 1994; Skelton 1996; Barton et al. 1998; studies in Fullagar 1998; Haslam 1999; Conway 2000). Attempts to identify the plant taxa represented by the starch granules and to relate them unequivocally with processing particular plants have been difficult because of the paucity of reference material for nondomesticates. Loy (1994a) has argued very strongly that particular kinds of tubers could be identified on the basis of starch granule morphology and that others could be eliminated on the basis of granule shape and size together with analysis of raphides.

Stimulated by Loy's pioneering research, others have examined starch granules on ground stone artifacts, smooth cobbles, and

currently important economic plant manioc, demonstrates that residue analyses to test tool function can be very powerful at dispelling long-entrenched myths.

Starch residues have also helped archaeologists identify functions that might have been overlooked, as in the case of seemingly nondiagnostic artifacts or when unaltered stones have been used. Use-wear analysis is essential in these cases, but is often highly dependent on the presence of residues since casual or short-term uses do not always created interpretable traces. Good examples of the value of residue analyses are the combined use-wear/residues studies of obsidian artifacts from sites on Garua Island, Papua New Guinea, by Barton et al. (1998) and Kealhofer et al. (1999). These used starch in the former case and both starch and phytolith residues in the latter. The studies found that many unretouched flakes with very scanty evidence for use-wear had starchy residues whose composition was distinct from the surrounding soils. The identification of pounders at the site of Jinmium in northern Australia (Atchinson and Fullagar 1998) and at various sites in Queensland, Australia (Hall et al. 1989), also highlights the importance of starch residues for identifying the range of uses for this class of material and other cultural objects that might have been used in food or plant processing (cf. Fullagar and Field 1997). As reported in Box 9.3, the recovery of starch and pollen residues on a number of hand-moulded clay objects, from the prehistoric village of Poverty Point in Mississippi, has helped to demonstrate that these unusual objects were probably used as heat retainers for cooking.

Archaeological research on the island of New Britain, Papua New Guinea, has attempted to reconstruct the history of mobility and settlement patterns during the past 10,000 years or so using a wide range of approaches including the study of starch residues. On the basis of an integrated use-wear and residue study in which starch played a prominent part, Fullagar (1992) documented a change through time from sites represented by single-purpose activities to multipurpose base camps. From these data he hypothesised that land use shifted from a highly mobile to a more sedentary pattern. Kealhofer et al. (1999) used starch residues within a test of this proposition. They found that in the early period starch residues were

Box 9.2 (continued)

other nonflaked stones (e.g., Field and Fullagar 1998; Piperno 1998, in press; Piperno and Holst 1998; Piperno et al. 2000, 2004; Pearsall et al. 2004; Perry 2002a, 2004, in press). Fullagar with Jones (2004; Fullagar 1986) argued that bracken ferns (*Pteridium esculentum*), which were present in the deposits, were a likely source of starch granules on stone fragments from the Inner Cave at Rocky Cape in Tasmania, southeastern Australia, dated to c. 7,000 BP. Reference collections were not adequate to unequivocally demonstrate this preliminary identification. Except for the microscopic study, these stones might not otherwise be recognised as artifacts. Similarly, the macroscopically unmodified cobbles from northwestern Australia studied by Atchison and Fullagar (1998) had microscopic smoothing/pitting but no other indication apart from the presence of starch granules that these were indeed artifacts. Aboriginal women today still use similar unmodified river cobbles to process water lily seeds, fruit, and tubers.

As knowledge of starch analysis has grown, research has begun to focus on plant use in addition to elucidating details of tool function. Field methods for dealing with the collection of artifacts and laboratory procedures for extracting starch granules are now well established, and those commonly applied are reviewed within this chapter. For example, in South America Piperno's studies (1998, in press; Piperno and Holst 1998, 2004; Piperno et al. 2000) have focused on the identification of a limited range of economically important plants with great success, and these have stimulated important results in other countries (e.g., Perry 2002a, 2004, in press; Pearsall et al. 2004; Irarte et al. 2004) (cf. Chapter 7).

Box 9.3. Poverty Point Objects

Linda Scott Cummings

The prehistoric village of Poverty Point is built on a bluff overlooking the Mississippi River swamplands in the Upper Tensas Basin–Macon Ridge area of northeastern Louisiana. The artificial mounds and embankments, which were built around 3,730 BP, form a huge crescent against the bank of the river, are among the largest earthworks in eastern North America, and are definitely the largest pre-agricultural mounds in the region (Gibson 1996). The mounds are formed by six concentric artificial earth embankments separated by ditches or swales and may have stood 1 to 2 m in height and some 40 to 60 m apart, with the ends of the outermost ridges separated by 1,204 meters. The village inhabitants are thought to have built their homes along the tops of the embankments, possibly housing up to 500 to 1,000 people (Gibson 1996).

Scattered throughout the site are large numbers of hand-shaped clay objects, so numerous that they have been named Poverty Point Objects (PPOs). PPOs come in various shapes and sizes. They are primarily described by their shapes, such as biscuit, sphere, melon, biconical, biconical-grooved, cross-grooved, and cylindrical-grooved (Figure 9.3). Some PPOs were further embellished using reeds or other hollow stems to make holes or decorations on their surface.

The area around Poverty Point contains few rocks and it has been suggested that the PPOs may have been used as heat retainers to cook or boil foods in a container or earth oven. Furthermore, Gibson (1996: 19) has argued that these objects may be part of a sophisticated technological solution used to regulate cooking temperatures. He noted that if the number of PPOs used in an earth oven is held constant. Then the temperature will vary when using different shapes (Gibson 1996: 19).

To help determine the function of these items and clarify if, in fact, they were in direct contact with foodstuffs to facilitate cooking, a sample of 13 PPOs was investigated microscopically for the presence of pollen, phytoliths, and starch granules.

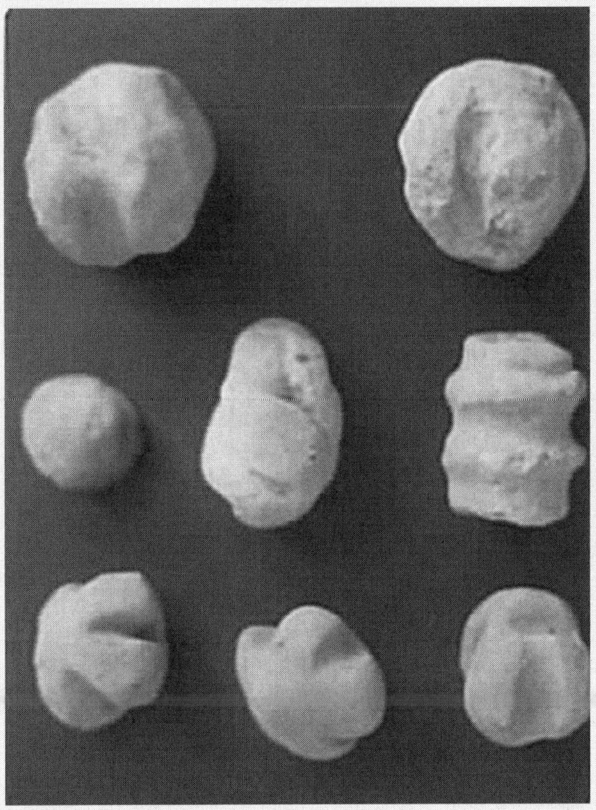

Figure 9.3. Poverty Point Objects. *Photo courtesy of Louisiana Division of Archaeology*

Method

Each Poverty Point Object was first dusted off using commercially available pressurised air, after which it was sonicated in reverse osmosis water and Triton™, a surfactant that facilitates removal of residue from objects. The solution containing minute quantities of debris liberated from the PPOs was processed by sieving through 150-micron mesh to remove rootlets and larger items of debris. Then the samples

Box 9.3 (continued)

were centrifuged at high speeds for relatively short times to allow the clay particles to stay in suspension and be decanted. The remaining sediments were dried and then pulverized. Sodium polytungstate (density 2.3) was used for the flotation process to recover pollen, phytoliths, and starches. The samples were mixed with sodium polytungstate and centrifuged at 2,000 rpm for 5 minutes to separate organic from inorganic remains. The supernatant containing pollen and organic remains was decanted. Sodium polytungstate was again added to the inorganic fraction to repeat the separation process. The supernatant was recovered and diluted with distilled water, then centrifuged at 3,000 rpm to concentrate the organic fraction in the bottom of the tube. The samples were then acetolated for 3 minutes to remove any extraneous organic matter. After rinsing the pollen, phytolith, and starch-rich organic fraction, a single microscope slide was made from each sample to evaluate the phytoliths. The remainder of each sample then received a short (10- to 15-minute) treatment in hot hydrofluoric acid to remove any remaining inorganic particles.

Results

Starch granules, pollen, and phytoliths were recovered from the surface of the PPOs. The majority of the starch recovered was identified as belonging to the lotus family (*Nelumbo* sp.) (Figure 9.4), although other nondiagnostic starches possibly belonging to several plant groups including cattail (*Typha* sp.), grasses, and roots of the family Apiaceae were also present. Pollen indicated several plant species might have been associated with the PPOs including *Artemisia* sp., possibly from wormwood, and Low-spine Asteraceae which may have included marsh elder and sumpweed, known for their edible seeds. Other species pollen recovered included Cyperaceae (sedges), Malvaceae (mallow), and Liliaceae (lily). Phytoliths recovered from the PPOs represented primarily grasses and were interpreted to have been introduced from the sediments in which the PPOs were buried.

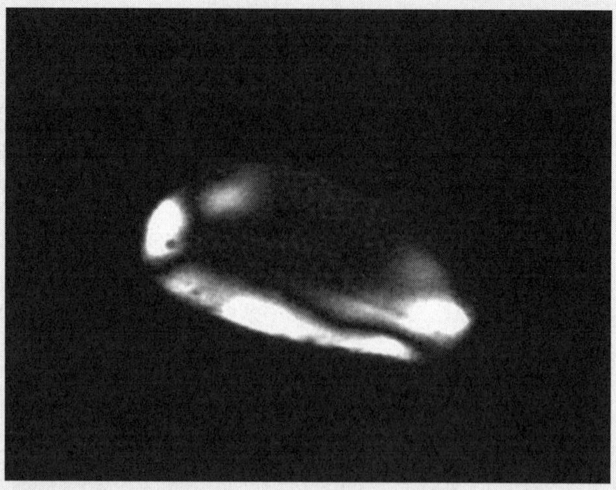

Figure 9.4. Starch granules from *Nelumbo* sp.; viewed under cross-polarised light. *Photo by Linda Scott Cummings*

Lotus root appears to have been the main plant food processed with the PPOs, its starch granules occurring on 7 of the 13 objects (54 percent) that comprised the sample. These starches were also recovered from a wide variety of PPO shapes including biconical, biconical-grooved, sphere, melon, cross-grooved, and cylindrical-grooved. PPO shape may not be all that important for many cooking methods, such as boiling or by directly adding to raw foods in a pot, even though work by Gibson (1996) has suggested they could function in that way when used within an oven. Certainly lotus root was an important food for people living at Poverty Point, and one that would not have shown up in traditional archaeobotanic studies. Other elements of their diet of cooked plant food probably included a wide variety of edible greens and seeds. The results of this microfossil study, which included identification of starch, pollen, and phytoliths, has shown that the moulded clay objects from Poverty Point did function as heat retainers. They represent an ingenious addition to the monumental architectural achievements on the banks of the Mississippi River, some 3,700 years ago.

highly correlated with retouched, probably hafted tools that had been transported around for use in multipurpose activities. In contrast, the later assemblages were characterised by unretouched flakes that had been used very lightly and then discarded. This shift was explained in terms of a trend toward more sedentary communities. These results are supported by previous studies by Fullagar (1993a, 1993b) that investigated the presence of hafting on particular stone artifacts with prominent retouched stems. His study compared the differential spatial patterning of starch and phytoliths and showed that the starch, probably from palms, was associated with the used end of the tool, whereas palm phytoliths indicated the presence of a haft. The ongoing research in New Britain illustrates the power of integrating starch analysis into a wider study of tool use.

Starch studies also allow opportunities for investigating incidental contact with other plant-based organics such as adhesives used in hafting. The identification of hafting as a component of tool function is important because it is thought to confer advantages in efficiency, power, and precision of tool use (Odell 1994; Rots 2003). Such studies may also help determine the source of the adhesive or binding agent such as resin. Sophisticated biochemical methods can be used to determine which plant species the resins are derived from, but starch granules embedded in resin may offer a much less destructive and expensive way to identify resins and other binding agents (see Box 9.4).

Plant Use

Since starch is often concentrated within residues, artifacts provide an excellent place to search for evidence of past plant use. The pioneering study by Loy (1994a; Loy et al. 1992) of starch residues from two kinds of taro on stone tools dating to c. 28,000 years ago in the Solomon Islands signaled the possible importance that transportation of plants outside their natural range may have played in the early stages of management that led to domestication (cf. Box 1.2). Further evidence for an independent centre of domestication in the Pacific region has been provided by Fullagar et al. (2005), who have identified granules of taro and yam on stone tools from the site of Kuk, Papua New Guinea. This evidence supports the results from studies of pollen, phytoliths, diatoms, insects, and archaeological features that some form of plant management and possibly cultivation was present from the early Holocene (Box 1.3; Denham et al. 2003, 2004a, 2004b).

Moving forward in time, recovery of starch from ceramic potsherds in the Pacific has been used to detect the use of the cultivated taro, *Colocasia esculenta*, and, very importantly, its direct association with Lapita pottery (c. 3,300 BP) for the first time (Box 9.5; Crowther 2005). An exploratory study recovered a combination of both native, unmodified starch granules and gelatinised material from the interior of some ceramic sherds, suggesting that some ceramic vessels may have been used to store or prepare rhizomes before consumption. Whereas the Crowther (Box 9.5; 2005) study treated the potsherds in much the same way as residue studies on stone tools, Horrocks and Bedford (2005) used the artifacts as a sediment trap. They scraped trapped sediment and pottery dust off sherds recovered from a well-preserved Lapita site in Vanuatu and then used density separation (Chapter 8) to extract pollen, starch, and phytoliths. Since they found abundant evidence for gelatinisation in terms of 'discoloration, swelling and loss of birefringence' (Horrocks and Bedford 2005: 70), it is possible that starchy foods had been prepared in some of the Lapita pots. The relatively poor state of the granules meant, however, that the potential for identification was limited to the family of Araecae (aroids) which had to have been imported to Vanuatu. The results are important because there has been some controversy about whether Lapita colonizers brought agricultural practices with them or were foragers. These two new studies demonstrate the potential of ceramic residues for studying the history of plant use (cf. Juan-Tresserras 1998; Chapter 10 for use of ceramic residues to study plant processing and food preparation).

Studies of starchy residues on stone tools from other parts of the world have also challenged ideas about the nature and timing of plant domestication. For example, Piperno and Holst (1998; Piperno et al. 2000) extracted starch granules of manioc and arrowroot from stone artifacts within a rockshelter in Panama dating to a much earlier date than domestication of these foods had been

Box 9.4. Resins, Traces Of Use, And Hafting

Jeff Parr

Resins from a variety of plant species were used by Australian Aborigines as fixatives for hafting (e.g., Cribb and Cribb 1987; Stanner 1933; Thomson 1949; Berndt 1951), a process in which a handle is attached to a stone tool (Flood 1995). This was often achieved by moderate heating of the resin which was then fashioned into place (Cribb and Cribb 1987). Moreover, resin is known to survive for long periods of time in a variety of archaeological situations. Well-stratified archaeological deposits containing stone tools bearing different and/or specific resin species could provide a history of resin use and tool manufacture over time. In addition, the trading of items, including resins, within Australia is well documented (Cribb and Cribb 1987; Stanner 1933; Thomson 1949; Berndt 1951). By identifying the species of the resins found on artifacts, mapping their distribution, and comparing these data to known exchange systems, important information on trade, prestige, and function could be established.

Previous methods employed to identify fresh and archaeological resins have concentrated on the general appearance of the raw material itself (Boot 1993) or used chemical analysis in the form of chromatography (Bowden and Reynolds 1982; Rosenfeld et al. 1981; Boot 1993; Welch 1997). Both techniques have limitations. For example, the identification of resins by their appearance alone is open to misinterpretation because many resins are similar in colour and texture. Various chromatography techniques can achieve satisfactory results with some modern resins (Bowden and Reynolds 1982), but they are time consuming and can be costly, particularly if a number of samples need to be analysed. Moreover, it is difficult to produce chromatographs that are comparable to those of archaeological samples, because the heating of some resins has been shown to cause loss of volatiles that changes their chemical signatures (Parr 2003b).

A number of researchers have reported the presence of starch granules within resins adhering to stone tools (Fullagar 1986; Boot 1993). Starch granules in fresh resin samples have also been observed in *Angophora floribunda*, *Grevillea pteridifolia*, *Triodia irritans*, *Xanthorrhoea australis* (Parr 2003b), *Myoporum montanum*, *Callitris collumellaris*, and *Brachychyton populneus* (Boot 1993).

A pilot study was conducted to determine if starch granules were contained within resin samples of a number of different species of *Xanthorrhoea* and to assess the feasibility of differentiating between the resin from these species through the morphological attributes of their starch granules. A total of 45 starch granules from 14 resin samples from vouchered plants of 10 different *Xanthorrhoea* species were assessed: *X. australis*, *X. brevistyla*, *X. drummondii*, *X. johnsonii*, *X. malacophylla*, *X. nana*, *X. preissii*, *X. reflexa*, *X. semiplana*, and *X. thorntonii*. Samples were ground, mounted on slides, and viewed using an Olympus BH.2 light microscope fitted with a polarising filter at x500 magnification. Starch granule length and width were recorded using an ocular micrometer at x1000. Other attributes recorded included morphological characteristics such as granule shape, style of the polarising cross, hilum position, and the presence or absence of lamellae (after Nägeli 1858 [cf. Table 7.1]; Meyer 1895; Reichert 1913).

All of the *Xanthorrhoea* resin samples contained starch granules which could be differentiated by one or more of the recorded attributes (Figure 9.5). While the sample size was small, the results strongly indicated that many starch granule attributes were species specific for these resins. Although overlap in both morphological attributes and size occurred

expected. The process of domestication is also illuminated by the presence of starch granules from barley and wheat on a ground stone artifact from Israel as much as 12,000 years before the plants were genetically modified (Piperno et al. 2004). Going back even further in time, starch research has contributed to the controversy about the origins of modern human behaviour. Van Peer et al. (2003) discovered starch and phytoliths on two grinding slabs dated perhaps as early as 180,000 BP. The existence of a complex method for plant processing at this date is suggested by these researchers to provide confirmation for the emergence of modern behaviour within Africa.

Box 9.4 (continued)

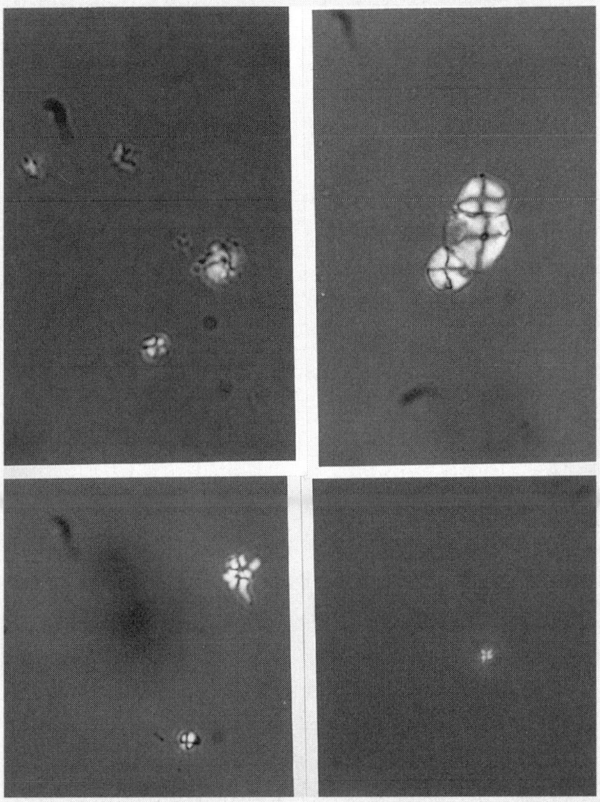

Figure 9.5. Starch granules extracted from resin samples: top left, *Xanthorrhoea nana*; top right, *Xanthorrhoea australis*; bottom left, *Xanthorrhoea malacophyla*; bottom right, *Xanthorrhoea johnsonii*.

for one starch granule in each of the species *X. reflexa* and *X. brevistyla*, it may simply reflect the natural phenomenon of growth. Sugimoto et al. (1987) suggest that starch granules can increase in size and change morphologically over time (cf. Chapter 2).

Those species represented by two resin samples (i.e., *X. johnsonii*, *X. preissii*, *X. reflexa*, and *X. thorntonii*) offered the most compelling support for the identification of resins through starch morphologies. In each case samples were collected from individual plants in different locations and with the exception of *X. thorntonii* all had starch granules with the same attributes in both samples of their corresponding species. Choosing samples from different locations made the possibility of uniform edaphic environments and coincidental attribute regularity less likely. The *X. preissii* samples were located about 100 kilometres apart. Nevertheless, they contained starch granules that were identical in morphology and size and yet totally unique from all other samples.

The successful morphological analysis of starch granules is not subject to the limitations of the chromatographic technique and does not require manipulation such as the preheating of fresh samples (Parr 2003b). Finally, since the morphological analysis of starch granules requires only basic laboratory skills such as the preparation of slides and the use of microscopy, it is relatively cheap and readily accessible to many practitioners.

This study confirms that starch granules are present in all of the *Xanthorrhoea* resins examined and that at least some of the recorded starch attributes appear to be species specific. The approach can be broadened because starch granules are not uncommon in the resins of other plant genera known to be used as fixatives for hafting. The comparative analysis of starch morphologies may be a methodology which can be applied to all ethnographically described hafting resins. Further research into the attributes of starch granules found in resins could prove to be rewarding as an efficient and cost-effective means by which species identification for archaeological samples could be achieved.

Box 9.5. Taro Processing And Early Lapita Pottery

Alison Crowther

Identifying human plant use in Pacific prehistory is hindered by poor preservation of intact macrobotanical remains, due to the region's humid tropical environment. As a result, archaeologists have relied on indirect indicators to reconstruct early subsistence in Oceania, particularly during the Lapita period (c. 3,300–2,200 BP). Putative plant-processing artifacts, land-use and erosion patterns, and linguistic reconstructions of plant terms have been taken to indicate that Lapita peoples practiced horticultural production involving taro and yam (Kirch 1997: 192–226; Spriggs 1996: 532; 1997: 88), but botanical remains of the crops have not been found in a Lapita context to lend direct support to this hypothesis.

An exploratory residue analysis was conducted on a sample of 20 undecorated Lapita potsherds from one test pit of the Kamgot site, Anir Group, New Ireland, dated to c. 3,300 BP. The study aimed to obtain direct evidence for the use of key plant foods in Pacific prehistory by investigating the relationship between plant processing and the subsistence-related use of this type of pottery. Specifically, the research goals were to determine whether use-related plant residues had survived on these potsherds and, if possible, to identify the species of origin (cf. Crowther 2005).

During the analysis, the interior and exterior of each sherd and any broken edges were examined at a variety of magnifications in brightfield incident light using an Olympus BX60 metallographic microscope. Microscope slides were then prepared for analysis in transmitted light following the protocol set out by Loy and Fullagar in Box 9.8. Sediment samples from an associated test pit were also analysed for comparison. To identify the starchy plant residues to species level, three key variables were recorded: mean starch granule size, starch granule shape, and the presence of raphides (Box 7.7). The compilation of a reference collection of 24 major plant foods from the Pacific region was also a key element in this analysis.

The results of the study indicate that the vessels from which the sherds were originally derived may have been used to process taro (*Colocasia esculenta*). All potsherds bore an abundance of starch granules that were 1 to 6 μm in size and essentially spherical in shape (Plate 58). The larger granules were spherical-angular (i.e., they displayed surface facets), which is typical of taro (Loy 1994a: 104; Plates 3, 4). Starch grains on nine of the 20 artifacts were cooked, as indicated by their diffuse extinction crosses. They resemble starch granules from experimental cooking of taro (Plate 59). The presence of cooked starch was consistent with patterns of charring and sooting observed on the same sherds. Starch grains were largely absent from the sediment samples. Whisker raphides (Plate 54), ranging in length from 10 μm to 60 μm, were observed on all surfaces of the pottery and in the sediments, indicating that extensive processing of *C. esculenta* had occurred.

These residues represent the first direct evidence for taro processing in undecorated pottery vessels by Lapita peoples. They also corroborate other archaeological and linguistic evidence. Given *C. esculenta*'s important ethnobotanical role in Melanesia and its attested use 28,000 years ago in the Solomon Islands, these new findings are reasonable. Further, while the peoples living today in the region where the pottery was found bake aroids in earth ovens, the analyses show that in prehistory pottery was used to undertake this task. Finally, this research demonstrates that microscopic examination of pottery can enhance our understanding of early plant processing and consumption in the Pacific.

Assessing the role of plants in early hunter-gatherer societies has often been a difficult problem because of the scarcity of macrofossil evidence and the implicit assumption that chipped stone artifacts are more likely to have been used to process animals than plants. Williamson's (2004; Box 7.4) study of Rose Cave, South Africa, shows the power of starch residues in bringing the role of plants to the foreground and in the potential for assessing site function in mobile societies. Similarly, Fullagar (1992, 1993a) showed that in contrast to predictions based on ethnographically documented cases of shaving and cutting human skin in Papua New Guinea, obsidian tools had been used to process plants in the distant past. As shown in Box 9.6, another new and different perspective has been offered for the site of Cuddie Springs, Australia, by combined analyses of starches and phytoliths (Fullagar et al. in press). Previously the major contribution of the site for understanding early settlement in Australia was the abundance of faunal material. In contrast, the confirmation for seed processing as much as 30,000 years ago during especially dry climatic conditions helps place the megafaunal remains into a wider context of human adaptive strategies.

FIELD TECHNIQUES

Although research questions will establish the kinds of artifacts required for residue analysis, procedures for collecting in the field should follow the same basic procedures. Given what is known about starch granule preservation, dispersal, and likely contaminants, as discussed in Chapter 5, material collected during archaeological excavation needs to be handled in particular ways in order to maximize the quantity and quality of the information that can be gained from a residue study.

In most situations it will be unnecessary to collect all stone artifacts for residue analysis. A sample based, for example, on technological and morphological types, the range of size classes, and raw material types, will suffice. For certain research questions only a particular class of stone implement will be investigated. For most research questions, it is important to sample a selection of artifacts during excavation. The numbers appropriate to a study will vary, depending upon research aims, but it can be noted that some studies found relatively high proportions (i.e., up to 20 percent) of flaked stone artifacts (greater than about 5 mm maximum dimensions) had evidence of use-wear or residues from tool use (Fullagar 1986; Odell 1996; Vaughan 1985; van Gijn 1990). There will of course be special circumstances. For example, grinding stones are relatively scarce in Australian stone assemblages, mostly less than one percent by number. Consequently, at the site of Cuddie Springs, all grinding stones and grinding stone fragments are being studied for use-wear, starch granules, phytoliths, blood, and other traces (Field and Fullagar 1998; Fullagar and Field 1997; Fullagar et al. in press; cf. Box 9.6). A further consideration when designing a sampling strategy is the time required for the microscopic component of a residue study because this normally takes a minimum of one hour per artifact just for initial observation.

Ideally, one should collect at least some artifacts *in situ* from each excavation level/layer/spit/unit/context and transfer these to sealed plastic bags with minimal handling. They will then form a control sample for evaluating the potential for more detailed studies, especially if procedures such as washing or sieving are used on other excavated artifacts.

The likelihood of modern contamination raises the need to take into account excavation/digging procedures: e.g., wet sieving causes much more damage to residues than dry sieving. The general aim should be to eliminate, as far as possible, any contaminants by limiting handling and to be aware of likely associations between artifacts and features or processes unrelated to use. Ideally, artifacts should be removed *in situ* from excavation trenches after relevant contextual details have been recorded, rather than collect from sieves. Starch-free gloves, plastic bags, or other—ideally plastic—instruments should be used to retrieve tools with their adhering soil and also for collecting associated sediment samples. To prevent contamination from modern hands, artifacts collected for analysis should be recovered using plastic tweezers and placed directly into a plastic bag. Contact with metal should be avoided, because metallic residues can be transferred to the artifact surface and complicate later interpretations of function. When sampling, it is important to allow some

Box 9.6. Starchy Seeds And Megafaunal Extinction At Cuddie Springs

Judith Field, Richard Fullagar, Beth Gott, and Lisa Kealhofer

Excavations at Cuddie Springs, located on the central northern plains of New South Wales, Australia, have provided paleoecological and archaeological records of Aboriginal occupation and resource use from at least 30,000 years ago (Furby 1995; Field and Dodson 1999) (Figure 6.10). Use-wear and residue analyses of stone tools show that animal flesh and plant foods (including starchy grass seeds and fern sporocarps) were processed at the site. The utilisation of seeds is also indicated by the presence of phytoliths and starch granules on a range of grinding stones of distinctive shapes, similar in shape to those used ethnographically to process grass seeds (Fullagar and Field 1997; Field and Fullagar 1998). The timing of the introduction of seed-grinding is much earlier than previously thought (Smith 1986, 1988), but makes sense because it coincides with climatic indicators (pollen and sediment analyses) which show that the environment was becoming drier, grasses were more abundant, and several large animals were on the road to extinction.

Microscopic study of a sandstone fragment with smoothed surfaces (Figure 9.6a) revealed distinctive use-wear and residues including developed polish, starch granules, phytoliths, and other traces of plant tissue (Figure 9.6b). Phytoliths and abundant starch granules were observed in the sample extractions removed from the used surfaces and no starch granules were found in the samples from the unused surfaces. Starch granules ranged in size from 3 to 4 microns in diameter to over 35 microns in diameter (Plates 62, 64). The starch granules had survived standard phytolith extraction protocols including nitric acid treatment and zinc bromide heavy liquid separation (cf. Kealhofer et al. 1999).

When compared to specimens from a reference collection of plants from the same area (cf. Box 6.4), the starch granules in the archaeological sample show marked similarities in morphology and size range to the Panicum (Yakirra) grass types (small starch granules, 3–4 microns diameter) (Plate 64) and also the Nardoo fern (large starch granules, ~35 microns diameter) (Plate 62). The phytolith assemblage contains a range of grass species which supports the starch findings (Fullagar

Figure 9.6. (a) Grindstone CS6023 from Cuddie Springs showing the ground surface from which starch granules were extracted (Plates 61, 63); (b) Use-wear on the surface of grindstone CS6023 is consistent with seed grinding, x50. *Photos by Judith Field*

sediment to remain attached to the artifact. Sediment samples from each context should be collected for later comparison with residues. Sample bags need to be individually labelled. Contact between artifacts and staples, ink, or paper labels should be avoided.

It is appropriate to undertake artifact cleaning and sediment removal from artifacts in the laboratory so that sediments can be retained for further analysis. Particular strategies are needed to devise methods for measuring the effects of pre- and post-depositional processes, excavation, post-excavation handling, and other ways in which residues not associated with tool use become attached to artifacts (Box 9.7). It is possible to assess likely transference of starch from surrounding soils to the surface of artifacts using the starch itself as an indicator. One such approach is illustrated in Box 9.7, which has relied on starch granule counts, but assessment of granule size and granule morphology may also be more informative (e.g., Atchison and Fullagar 1998).

LABORATORY TECHNIQUES

For the purposes of microscopic study, a tool can be considered analogous to an archaeological site. As such, it requires similar care in recording and sampling, as well as attention to particular research goals (Loy 1993). General principles of conservation should also apply. Consequently, descriptions and visual records should come before cleaning, removals, peels, and chemical tests. Samples of the residue rather than total removal from the artifacts are usually preferred. Residues can be studied in terms of their stratigraphic position as well as their spatial location on the tool. A range of equipment needed is shown in Figure 6.14.

Figure 9.7 presents an appropriate sequence of analysis of residues from artifacts if the main aim is to understand tool function. It is also important to remember that as with excavation, the study of residues involves the collection of multiple sources of information, often simultaneously. In addition to starch granules, the study of other kinds of residue, such as phytoliths, tissues, blood, etc., together with information about use-wear provides a much richer picture than focusing on one form of evidence by itself (e.g., Fullagar 1993a, 1993b; Kealhofer et al. 1999; Box 9.6). In addition, one needs to consider data from different sources in order to evaluate the life history of the tool and the possibility that some residues may have occurred

Box 9.6 (continued)

et al. in press). The absence of starch granules that could be identified as Acacia (e.g., Figure 2.8), a very important resource according to ethnographic accounts, was not expected. Although seeds from this plant could be roasted in the fire, in the recent past they were also ground.

The stone raw material for the grinding stones at Cuddie Springs comes from a quarry about 100 km to the northwest, so it is not surprising that these important implements were used fairly intensively to process several kinds of plant. The preparation of Nardoo (*Marsilea* sp.), a small fern with starch-filled sporocarps (Figure 6.12), may represent a secondary use for grindstones. The identification of grass seed and Nardoo indicates that at 30,000 years ago climatic changes may have initiated exploitation of a wider range of locally available plant foods at a time that is coincident with the extinction of a range of large animals. The Aboriginal solution to increasingly harsh and uncertain environment had two aspects. First, they expanded their dietary focus to incorporate a range of plants that require energy-intensive processing, deploying a new technology (seed grinding) so they could maintain life in a harsher and depleted environment (cf. Edwards and O'Connell 1995). Second, the social impact of these changes must have been far-reaching. Perhaps, this is the time (from 30,000 years ago) when we see the beginnings of complex social networks with fluid boundaries, highly visible in Aboriginal desert culture today.

Box 9.7. Testing For Contamination On Tool Edges

Huw Barton

The decomposition of plants releases starch granules into the soil. As a consequence of processing starchy plants, humans can also contribute starch granules to the soils of archaeological sites (cf. Chapter 4, Box 4.6). Objects that could have been in contact with plant materials or the starchy hands of tool users may bear starchy residues. With so much starch available within the environment, it is imperative to know whether the starch observed on artifacts is the result of tool use or of secondary contact with soils containing starch.

This case study outlines the results of an experiment designed to address the problem of testing for contamination on artifacts. It seems reasonable to argue that (1) there should be more starch granules on the edges of used stone artifacts than in the surrounding sediments, and (2) the amount of starch on used artifacts should be higher than that found on unused artifacts. In contrast, if starch granules result from secondary contact with soils containing starch, then the abundance on the tool surface should be equal to that in the soils and there should be no relationship between the abundance of starch granules and intensity of tool use.

The sample for this experiment was drawn from a collection of obsidian artifacts and associated soil samples recovered from the excavation of an open site (FAO) in Papua New Guinea (Torrence et al. 2000). During excavation, each artifact, including the soil adhering to it, was carefully lifted by trowel or plastic bag and placed directly into a plastic bag. Artifact and soil were then allowed to air-dry before the bag was sealed for storage. Paper labels were stapled on the outside of the plastic bag to guard against contamination by fibers and cellulose tissue. All artifacts were bagged individually to avoid cross-contamination. Consequently, during the analysis artifacts will have only come into contact with plastic bags, plastic gloves, or in some cases a metal trowel. This collection technique provided three different contexts from which the quantity of starch granules could be determined.

- *Soil samples.* Soil taken from the archaeological deposit within one metre of the collected artifacts.
- *Bag sediment.* Loose soil in the artifact collection bag that was, at the time of collection, adhering to the surface of the artifact. In cases where the majority of sediment was still adhering to the artifact, the plastic bag was rubbed gently to remove loose sediment. This deposit was thought to be the most likely source of local contamination for starch granules not associated with tool use.
- *Residue.* The final sample was taken directly from the artifact itself. At this stage all loose sediments had been removed leaving only firmly bound sediments, or none at all, on the artifact surface. Where sediments were still visible, they appeared as a light, dusty coating. Removing these sediments and any potential starchy residues deposited during tool use required an additional step. Each artifact was placed in a sonic bath and cleaned in distilled water for one minute. The solution was evaporated in a kiln at 25° C leaving a fine sediment. A low temperature was chosen to avoid gelatinisation (cf. Chapter 2).

Starch was extracted from all three sample types using a heavy metal liquid separation technique (Barton et al. 1998; Fullagar et al. 1998). Permanent slide mounts were made of each sample. These were examined under a light microscope and counts of starch granules recorded. Finally, sample counts were standardised to one gram of soil weight.

Next, artifacts were passed on for use-wear analysis. This stage of the study was

Box 9.7 (continued)

conducted as a blind test so that tool use could be determined independently of the starch extraction results. Artifact use was scored from lowest to highest confidence (1 to 3) given the quality and quantity of the use-wear. Results are presented in Table 9.1.

If post-depositional contamination of artifacts had occurred at this site, we would expect a roughly equal distribution of starch granules across used and unused artifacts, but this is definitely not the case. In contrast, there is a very strong correlation between the presence of use-wear on artifacts and high counts of starch granules. Both the *bag sediments* and the *soil samples* contained considerably lower quantities of starch granules than the used artifacts. Likewise, unused artifacts contained similar quantities of starch granules to both the *bag sediments* and *soil samples*. The results led to two major conclusions: (1) starchy residues on artifacts from FAO are the result of use and were not created in a post-depositional context; and (2) minor contamination does occur at FAO since starch granules were present on all artifacts studied.

Nevertheless, it has been shown that contamination can be easily recognised and accounted for because the quantities of starch on the majority of the unused artifacts were the same very low orders of magnitude found in the surrounding soils and bag sediments (cf. Barton et al. 1998).

Table 9.1. Results of starch extraction and use-wear analyses from tools and sediments at FAO.

Sample		Sonic Cleaning			Bag Soil			Use-wear		
Tool number	Soil layer	Soil weight (g)	Starch count	Starch/g of soil	Soil weight (g)	Starch count	Starch/g of soil	Use	Confidence level	Material worked
M104	1	0.01	30	3000	0.05	6	120	yes	3	plant
M105	1	0.03	5	166	0.1	11	110	no	3	none
M106	1	0.01	16	1600	0.1	4	40	yes	1	plant
M112	4	0.01	20	2000	0.11	8	72	yes	3	plant
M119	4	0.05	4	80	0.1	8	80	no	3	none
M121	4	0.01	15	1500	0.1	3	30	no	3	none
M125	4	0.01	2	200	0.1	8	80	no	3	none
M129	6	0.01	16	1600	0.14	7	50	yes	1	plant
M131	6	0.01	14	1400	0.1	7	70	yes	1	plant
M132	6	0.01	20	2000	0.1	10	100	yes	1	plant
M133	6	0.01	19	1900	0.23	43	187	yes	1	plant
M135	6	0.01	6	600	0.13	6	46	yes	2	plant
M432	6	0.15	4	52	0.5	16	32	no	3	none
M433	6	0.05	8	320	0.5	8	16	no	3	none

Soil samples

Soil sample	Excavation layer	Soil weight	Starch count	Starch/g of soil
2	1	0.5	26	52
13	4	0.5	9	18
22	6	0.5	131	262

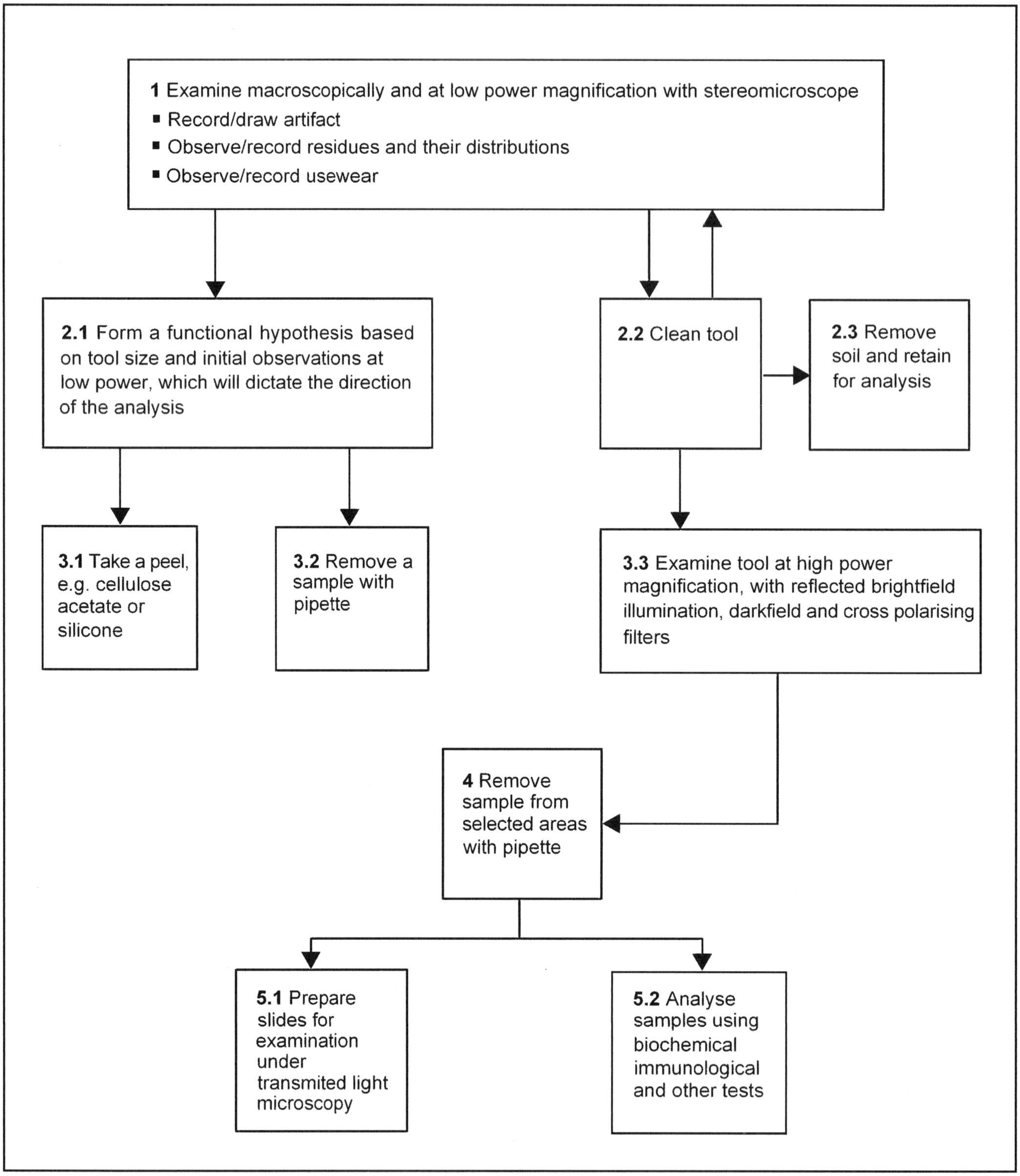

Figure 9.7. Flow chart for analysis of starch on artifacts. *Illustration by Richard Fullagar and Huw Barton*

only incidentally to the main patterns of tool use.

In cases where the goal is simply to obtain residues that might provide evidence for the presence or absence of plants at a site, rather than the more elaborate study of tool functions, then a simpler path of analysis may be appropriate. In these situations less attention can be paid to other sources of information such as use-wear traces or the spatial patterning or structure of the residues. Instead, the analyst will move more quickly to step 4 in Figure 9.7, although the early procedures involving cleaning and location of the residues will be important.

As with the study of use-wear, the best procedure for residue analysis is to begin on a macroscopic level before proceeding to low magnification observations with a stereomicroscope. Subsequently, one will move gradually to higher magnifications with a compound microscope (see Chapter 3 for a review of the relevant microscopes). It is important to take into account that certain features and structures are more easily visible under different lighting conditions and magnifications. For example, the spatial distribution of residues across a grinding surface is more readily assessed at low magnification (up to about x50) with oblique external lighting, although individual starch granules will not be visible under these conditions. A metallographic microscope (with vertical incident lighting, brightfield/darkfield and polarising filters) is necessary for good quality starch research that incorporates use-wear analyses (cf. discussion in Chapter 3).

Cleaning

Artifact cleaning, particularly with strong acids and bases, can seriously harm or destroy organic residues as well as alter the preservation and character of microwear and so must be avoided or undertaken carefully and in gradual stages of increasing harshness. Consequently, it is best for the first assessment of residue distribution and preliminary analysis to take place before cleaning and before any use-wear analysis or other technological analysis.

The first step in the study of starch granules utilizes a macroscopic and low-power investigation with a hand lens and stereomicroscope to examine the entire tool (Figure 9.7, step 1). These initial observations may lead to an initial inference about tool-use (Figure 9.7, step 2.1), which will demand a strategy for further microscope work or the need for some kind of cleaning. If sediment obscures a clear view of the surface, it may be necessary to remove some by rubbing the artifact gently inside a plastic bag (to remove thick blocky lumps) or with a soft, dry nylon brush (a brush made from animal hair may introduce unwanted contaminants) to remove loosely adhering sediment (Figure 9.7, step 2.2). The sediment should then be bagged separately. Firmly adhering, clayey sediments may require washing. This should be done by immersing the tool in pure water within a plastic container that floats in a sonic cleaner for about 10 to 20 seconds. The sediment removed by cleaning can also be kept for further analysis as required (Figure 9.7, step 2.3). Experimentation may be needed to determine whether more or less time in the sonic bath is required, with care being taken not to remove all sediment. If the artifact is large, it may be advisable to sample different parts separately (e.g., hafted end versus point). This can be done by holding the tool with a small part immersed in the floating container during sonic cleaning. Tools that cannot fit into the sonic cleaner, like the large seed-grinding dishes of Central Australia, require different methods, such as careful brushing, pipette removals, or a series of peels, as discussed below. Caution should be taken when using the sonic bath with very friable materials, such as old bone or pottery, because surface features such as decoration or paint could be removed with even a short exposure.

Selection of Extractions

Once the major macroscopic features are clear enough, the artifact should be sketched (taking care not to touch it with the pencil or pen) and then examined at high magnification (Figure 9.7, step 3.3). On the basis of these observations, an inference will be made about tool use on the basis of various forms of use-wear (scarring, rounding, polish, beveling, and striations) and the presence of residues. Initial recording of organic residues at step 3.3 should include the distribution of material, position in relationship to the worked edge, and association with use-wear. It is important to note whether there are any patterns visible in the general distribution of the residue, e.g., evidence of smearing or directionality, since these can indicate the mode of use when the residue was deposited. Recording color and structure (e.g., does it appear 'resinous,' dry and crumb-like, or sticky) is also particularly useful. The presence of structurally distinctive residues such as plant cells (these should be readily visible at x100 magnification and higher), starch granules, hair, feather barbules, phytoliths and other similar microfossils, if present, should also be recorded at this stage. These observations of use-wear and residue distribution will direct further investigation of the tool.

Specific areas of the artifact may now need to be targeted in order to test hypotheses based on the observations collected so far. Analysis under reflected light is not always very informative about structure, although starch granules are often identified at this stage. Residue removals are likely to be required at this stage (Figure 9.7, step 4) to help further determine their composition. They can also be the basis of studies focused on plant identification. For example, the identification of residues might help determine the nature of worked material where use-wear can only inform on the intensity of use, i.e., soft material or hard material. Differential use-wear patterns can indicate hafting (e.g., see Odell 1994; Rots 2003) and an inference based on use-wear might be testable by comparison with the distribution and frequency of starch granules between these areas (e.g., Fullagar 1993a; 1993b). A sample of the residue may help to identify what plant it was derived from. Each of these inquiries will require removal and mounting of starch samples (Figure 9.7, step 5.1) with sufficient sample volume to undertake other tests if feasible (Figure 9.7, step 5.2). Throughout these procedures, it remains vital to record the locations of the residues and use-wear on the artifact. The next section describes the steps necessary to remove organic residues from an artifact surface.

Pipette Extractions

For large artifacts that cannot be examined under a metallographic microscope, it will be necessary to extract residues by pipette or remove them with a peel. These techniques are also essential for detailed studies of starch granules observed on small artifacts. Although starch granules are usually visible directly on artifact surfaces under reflected light, it is difficult to record detailed or quantitative information in a way that they can be identified to plant species through comparison with starch granule reference collections. Identification of granules is best carried out from viewing under transmitted light microscopy, which usually requires small samples (e.g., 10–20 microlitres) to be removed in aqueous solution and mounted on glass slides. It is preferable to target specific, visible, and already recorded residues rather than extract all residues from the entire tool. It is best to begin with small samples and enlarge them gradually.

The procedure for removing samples of residue from an artifact surface is presented in Box 9.8. First, a nylon-tipped pipette is used to deliver 10 to 20 microlitres of pure distilled water onto the residue to be extracted. The residue is agitated with the nylon tip and then drawn back into the tip. The solution can then be transferred to a microscope glass slide and sealed with a cover slip. The corners of the cover can be fixed with nail polish, so that when the water evaporates, more water can be delivered by pipette to the edge of the cover slip (Figures 9.8–9.11).

Sometimes the frequency of starch granules is so low that it is necessary to remove larger samples, perhaps by immersing the entire artifact edge in a water-filled container floating in a sonic bath. Samples can be taken by pipette from the resulting liquid, but more commonly the sample is dried and the starch is extracted from the sediment in the same way as for soil samples (cf. Chapter 8; Barton et al. 1998; Kealhofer et al. 1999; Pearsall et al. 2004).

The methods and order of sampling described above are most appropriate for small, chipped stone artifacts and in situations where the observation of use-wear is important. Other studies focused purely on plant identification may slightly alter the sequence. Also grindstones can present different problems from chipped stone. For example, Piperno and Holst (1998) and Piperno et al. (2000, 2004) used the point of a fine needle to remove starch residues directly from small cracks and crevices in grindstones. Pearsall (2004: 427–28) collected a series of samples in four stages beginning with gentle dry brushing of loose adhering

Box 9.8. Residue Extraction

Tom Loy and Richard Fullagar

The following protocol describes the steps necessary to extract a residue from the surface of an artifact. This method will be applicable for a wide range of materials such as glass, stone, bone, and shell. As noted below, it can be difficult to remove material from very porous substances with this method because the water droplet is rapidly absorbed into the material. Experimentation will help determine the best method of residue removal for particular cases.

1. Apply between 20 and 60 µl of ultrapure water to the area of interest using a variable volume pipette (Figure 9.8).

Figure 9.8. Pipette extraction from the surface of an obsidian tool. *Photo by Stuart Humphreys, Australian Museum*

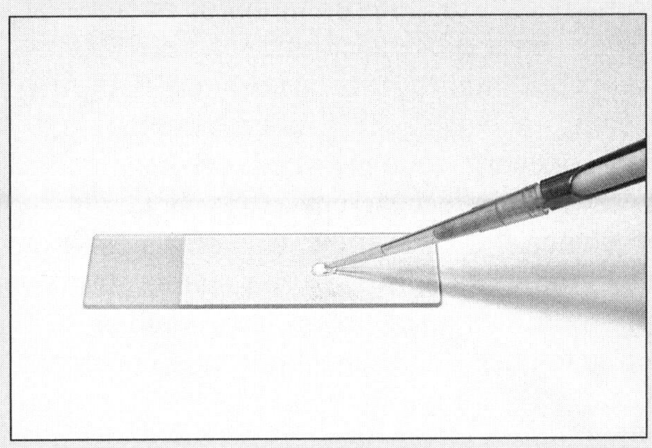

Figure 9.9. Residue extract applied to a microscope slide. *Photo by Stuart Humphreys, Australian Museum*

2. Leave the undisturbed drop on the surface for a few minutes. Next, using a disposable pipette tip, agitate the drop and scratch the surface of the artifact with the tip to loosen and break up the residue. It may be necessary to repeat this step. Remove the drop with the pipette, attempting to withdraw as much residue and water as possible. Porous sandstone, scoriaceous basalt, and other similar materials can cause the water droplet to spread and dry rapidly. If this is the case, allow the first application to dry and soften the residue, then apply another equal amount of water and immediately draw the water back into the pipette tip. Repeat this action until some particulates are observed in the withdrawn solution. (This method also works on bone that is friable or highly degraded.)

3. Transfer the residue from the pipette to the centre of a cleaned microscope slide and allow it to dry at room temperature (Figure 9.9). Spread the drop to avoid aggregation of soil particles and organic mixtures, which obscures viewing.

4. Thoroughly clean and then place a #1 thickness cover slide over the dried residue and fix it to the slide surface by applying a small drop of nail polish at each of the corners (Figure 9.10). If the coverslip is applied directly to the droplet containing the extracted residue, the starch grains may be ejected to the very edge of, if not outside, the coverslip. Application of the coverslip to the dried residue extract, although slightly more time consuming, results in the retention of starch grains in the final slide mount.

sediment (Sediment 1), then viewing the surface under low-power magnification to identify any areas with visible residues. These *in situ* samples were removed with distilled water, which was applied with a variable volume pipette, and then were placed directly onto a microscope slide (cf. Perry 2004: 1073–74). The next phase of sampling involved scrubbing the grindstone surface in cool, distilled water until no further sediment could be removed (Sediment 2). The sequence was finally terminated by sonicating (placing in an ultrasonic bath) each tool. Since material will vary, researchers may need to experiment to find the appropriate methods (cf. Barton et al. 1998) but the basic principle of slowly increasing the aggressiveness should always apply. Whichever method you follow, bear in mind that too aggressive sampling could permanently alter areas with use-wear/residue by scratching the surface or smearing residue deposits that could confuse other specialists solely concerned with the interpretation of microwear.

Potsherds present additional difficulty because, in addition to their occasionally large size and complex shape, the residue sample usually consists of charred and browned organic materials that obscure admixed starches. Compared with the more standard direct application of water to the surface, the scraping method is superior. Using a fresh, sterile scalpel blade, scrape a small area (3–5 mm) and allow the charred residue to be deposited directly onto the microscope slide, or, if not possible, onto a clean, nonstatic surface. The residue may then be prepared for slide mounting as described previously.

Peels

A range of compounds used by dentists to take tooth impressions are suitable for high-resolution impressions, or 'peels', of stone tool surfaces (Figure 9.7, step 3.1). These products include acetate or silicone, polyvinylsiloxane, and other compounds, many of which are available from dental suppliers. The peels have also been used to document use-wear and record use-polish at various stages of development (e.g., Fullagar 1991: 20). Peels usually need only be about one centimetre square and a few millimetres thick, and can also be used to remove residues from artifact locations not easily viewable under a microscope (Figure 9.11). The peels can then be easily transferred and mounted on the stage of a reflected light microscope for viewing (Figure 9.7, step 3.3). The use of peels is especially valuable when very large artifacts are involved, since these cannot be directly examined under the microscope. Peels have also proved worthwhile for investigating use-wear with the Scanning Electron Microscope. A recent study of bone tools from Broken Hill, Africa, took high-resolution silicon elastomer moulds of worked areas. These were then coated with platinum. The production of a silicon copy avoided the problem of permanently altering the tool surface and eliminated charging effects induced by the electron beam (Barham et al. 2002: 6).

Box 9.8 (continued)

5. Refer to Box 6.5 for suggestions on different mounting media and associated protocols. Table 6.4 contains some useful data on the refractive index (ri) of different mounting media that will influence the visibility of some features. Again experimentation may be required.

Figure 9.10. Nail polish applied to corners of a coverslip. *Photo by Stuart Humphreys, Australian Museum*

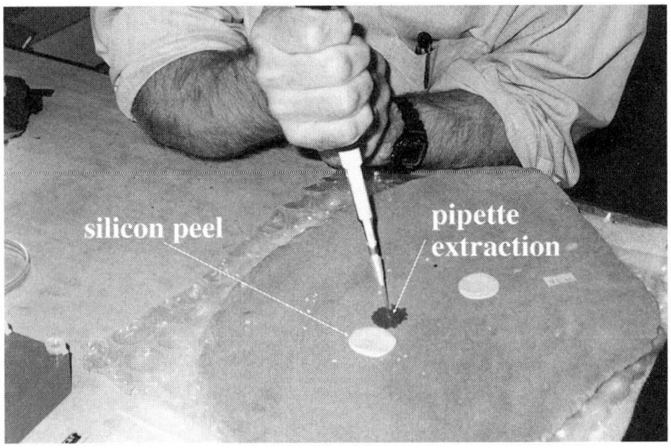

Figure 9.11. Richard Fullagar extracting residue from a grinding stone with a pipette. Two silicon peels have also been applied to the artifact. *Photo by Rebecca Conway*

INTERPRETATION THROUGH EXPERIMENTATION

The presence of starches is only one of a number of variables that are used to interpret the nature of tool use. The interpretation of artifact function requires the integration of a combination of techniques involving the shape of the artifact, the type and location of residues, and the pattern and condition of various use-wear indicators such as macro- and microscopic scarring, striations, and polishes. Each archaeological context raises unique problems and demands a specific research design to interpret the evidence adequately through understanding the range of factors, including the life history of the artifact, which might have had an effect on the use-wear and residue traces. It is important to be aware that not all organic residues result from tool use. As with many aspects of archaeological interpretation, collaboration among a number of specialists, some of whom are experts on the identification of particular residues and some on use-wear analysis, can radically improve the specificity and accuracy of the results.

Analysts need to interpret the origin, survival, and distribution of starch observed on tool surfaces in order to determine how and whether these are related to tool use. The mere presence of starch is not by itself an adequate measure of tool function. To infer tool function, archaeologists use an integrated approach which makes use of two main sources of information: (1) reference collections for experimental use-wear/residues and (2) patterns of use-wear/residues on the tools themselves. Since the interplay between experiments and functional interpretation of tools is critical for accurate assessments of artifact use, this process is discussed at length.

An important methodological issue that has not yet been fully considered is how and why residues accumulate where they do. Observations of flaked stone tools often reveal that residues accumulate in greatest abundance several millimetres away from the working edge that had the most intensive contact with the use material (Shafer and Holloway 1979: 396). It seems likely that different kinds of processing, such as cutting, shaving, pounding, and grinding, will cause variation in the degree to which starch granules will stick to tool surfaces. A variety of experiments should therefore be designed to assess the relationship between tool function and starch survival, residue distribution, stone type, and other variables. Understanding how these interact will assist in determining sampling strategies for the study and removal of residues from archaeological artifacts.

The need for experiments also arises in at least three analytical stages of archaeological use-wear/residues analysis. Experiments Stage 1 represents the best way to learn microscope techniques. In this case elementary tool-use experiments are conducted under controlled conditions and the tools are observed under low- and high-power light microscopy. Once one has a good idea of what to look for and what microscope skills are needed, it is then appropriate to undertake a preliminary examination of the tools, describe the technology, and record use-wear and residues (see below). Study of the archaeological tools will prompt new questions so that it may be necessary to design further controlled experiments (Experiments Stage 2) which reveal properties of the particular raw materials (perhaps the range of stone types or tool designs) and starchy resources that might have been exploited, as suggested by regional ethnography or local biogeographic data. Re-examination of the tools with this new information will probably require that the initial interpretations of tool function be modified. Finally, it may be useful to try and replicate the particular patterns of use-wear and residues as suggested by the modified interpretations (Experiments Stage 3).

Experiments Stage 1

In order to gain familiarity with microscopes and to understand processes of residue accretion, it is useful to begin with a fresh, unretouched stone flake and a simple task that will not produce very much edge damage or surface modification. Slicing raw potato or any other soft tuber with a fine-grained stone flake is an ideal first. The flake should be small: e.g., about 5 cm long, a few cm wide, and less than 1 cm thick. A flake of about this size is very easy to observe under any stereomicroscope with oblique lighting, as well as under a metallographic or compound microscope with vertical incident illumination (Figure 9.13).

A second simple experiment is to scrape a more dense starchy tissue such as palm pith or bamboo (Figure 9.14), using a flake similar to the one used to scrape potato. This task will produce more edge damage and use-polish than the previous experiment with raw potato (Figure 9.15). The edge scarring, striations, polish, rounding, and residues of both starchy materials (soft and hard) as well as the residues created can then be observed microscopically. The simple controlled laboratory experiments are designed to keep all but one vari-

Figure 9.12. Peeling the cortex off a taro (*Colocasia* sp.) with an obsidian flake. *Photo by Robin Torrence*

Figure 9.14. Peeling bark from a *Hibiscus* sp. stick with an obsidian flake. *Photo by Peter White*

Figure 9.13. Residue adhering to an obsidian flake used to scrape a soft tuber (*Colocasia* sp.). *Photo by Robin Torrence*

Figure 9.15. Residue and edge damage resulting from sawing bamboo with an obsidian flake. *Photo by Robin Torrence*

able (the starchy material) constant so one can examine the effects. Despite the controls, the cause of variation in wear patterns may be complex. In this case, for example, variation in use-wear will be due to the differences in tissue densities, moisture content, and especially the different density of phytoliths. Moisture content or phytolith abundance is directly proportional to the rate of polish formation (Kamminga 1979; Fullagar 1991).

Other questions and variables can be examined as well. For example, is the abundance or frequency of particular residues related to intensity of use or a by-product of the tool material? Determining the relationship between the surface relief of a tool and attachment of starchy tissue could be assessed by repeating the same experiment (e.g., slicing yams) with very smooth (glass, obsidian, quartz) and very rough (sandstone, granite, quartzite) raw materials.

Experiments Stage 2

Following initial experiments and familiarisation with microscopes, an examination of archaeological artifacts can lead to new questions. These will demand more detailed experimental study either of particular use-wear/residue patterns or of particular tool forms. At this stage experimental programs can help resolve apparent contradictions if, for example, particular forms of use-wear typical of a proposed function for archaeological tools are inconsistent with the apparent residues or their distribution. For example, abundant phytoliths and starch combined with an absence of any use-polish could imply very light contact with a siliceous material. The pattern might indicate contact with a starchy siliceous plant, perhaps a haft or container. Further experiments could perhaps demonstrate how rapidly polish can form through contact with highly siliceous plants. Replication using a range of starchy plants might confirm that the proposed function is consistent with the use-wear and residues.

Many tool-use experiments have been undertaken to investigate particular wear patterns and the associated starchy residues. They include starch residues on tools used to process some woody plants and many starchy foods like grass seeds, yams, taro, cassava, palm pith, nardoo fern, bracken fern, and nuts (e.g., Fullagar 1986; Specht and Fullagar 1988; Hall et al. 1989; Robertson 1994, 1996; Atchison and Fullagar 1998).

Experiments Stage 3

The third stage of tool experiments provides the fine tuning of interpretations. It assumes familiarity with stone raw materials, technology, design, and use-wear/residue patterns on the archaeological specimens. The primary purpose of experiments at this stage is to replicate the archaeological use-wear/residue patterns. If an interpretation is to have any validity, it should be possible to replicate the wear patterns and test whether other possible tasks develop overlapping use-wear/residue patterns. Replications of processes, like controlled laboratory experiments, provide further information about residue forms and distribution on tool surfaces, and may also suggest the likelihood of contamination from other sources (cf. Chapter 4). Replications may be carried out in a variety of circumstances, including ethnoarchaeology and active participation with modern stone-using groups (e.g., Miller 1979; Specht and Fullagar 1988; Atchison and Fullagar 1998; White et al. 1977). The extant ethnographic literature or data from museum collections can also suggest hypotheses about how residues get onto tool surfaces and, in turn, may prompt further experiments (Fullagar et al. 1992; cf. Chapter 4).

Stage 3 experiments should consider relationships between the wear patterns and the residues. Scraping plant tissue, for example, usually causes striations at right angle to a working edge, in contrast with slicing which usually causes striations that are oriented parallel to the working edge. At the same time, residues may build up at different distances from the working edge depending upon the mode of tool use and the characteristics, such as hardness, of the worked material (cf. Figures 9.13, 9.15). Pounding causes distinctive crushing and some smoothing of tool surfaces in contrast with grinding actions that tend to produce more extensive polish development. Starch granule residues can have distinctive distributions and may be modified in peculiar ways by particular forms of processing (see Box 4.5). Combining the two forms of evidence, use-wear and residues, will provide a much stronger inference about past tool functions.

As an example of this approach, a project to investigate past and present resource-use and settlement history in the Kimberley region of Australia conducted several ethnographic studies to observe how food was processed by Aboriginal people living in the region (Head and Fullagar 1991; Fullagar and Head 1998; Atchinson et al. 2005; cf. Fullagar et al. 1996). In one case study detailed records were made about the way water lilies were processed (Atchison and Fullagar 1998; see Box 4.6) to test the hypothesis that archaeological implements had similar wear patterns and starch granules. In this case, similar wear patterns were produced, but the starch granules of the water lily (*Nymphaea gigantea*) had an average diameter of about 5 microns, significantly smaller than the starch granules found on archaeological specimens. Atchinson and Fullagar (1998) concluded that the tool use in the past was probably the same as in the experiments, but that different plants had been processed.

LIFE HISTORIES OF TOOLS

Once the technological, use-wear, and residue variables have been recorded and the starch granules have been identified to a relevant taxonomic level, a range of issues still needs to be considered prior to a final interpretation of tool function. For example, are there alterations to the residues (peculiar distribution, crushed granules, or chemical alterations), which we might expect as a consequence of tool action? Or do they derive from incidental contact or post-depositional processes? To address a range of problems like these, we also need to consider the total range of contact with starchy tissues that might have occurred during the life history of each tool. Many of these will have been incidental to the primary use of the tool:

- Extraction/collection getting raw material (e.g., digging sticks);
- Production or manufacture (e.g., pressure flaking, anvil, hafting processes);
- Storage/carrying (e.g., bark, baskets, and other containers);
- During use (e.g., incidental contact with starchy food);
- Sheaths and other hafting devices (e.g., palm pith handles);
- Dirty/sticky hands;
- Rejuvenation, resharpening, re-use;
- Place of discard (e.g., in middens).

At each stage in the life history of a tool there are many potential contaminants that may contribute to residues. Digging sticks or pressure flakers may introduce wood or bone residues, which might confuse interpretation of tool function. Studying the distribution of such residues on artifacts is likely to resolve this kind of problem. The context of discard may also introduce potential contaminants. For example, artifacts buried in shell middens are likely to have shell residues (Attenbrow et al. 1998). Careful study of shell particle distribution, frequency, and associated sediment samples is likely to discriminate between primary use and secondary association. A good example of the methodology used to accomplish this is the study by Barton et al. (1998; cf. Box 9.6) in which the nature and distribution of use-wear on artifacts was used to show that an unmodified piece of stone which had abundant starch residue on it had probably not been used as a tool, but had obtained the residue through contact within a midden containing starchy material.

Artifacts get carried around and placed in such a wide variety of situations that some traces could be completely misleading. For example, Aboriginal men in northern Australia carried stone spearheads around in their hair, tied in a bun that also held various items including food. Bark wallets were also used to transport toolkits. Moreover, certain items were regularly used for minor tasks rather than the main function of the tool. In his study of residues on the Iceman's toolkit, Tom Loy (1994b, 1998a, 1998b) has shown that while the primary functions of a hatchet head were woodworking and butchery, starch residues found on a flat surface were from a starchy gruel or porridge (see Box 9.1). On the basis of starch distribution and identification, Loy infers that the Iceman had, perhaps inadvertently, smeared the starchy remnant of his last cooked meal on his axe.

WHERE NEXT?

Future research into starch residues adhering to artifacts could benefit from broadening the type of applications and data sets, but archaeologists also need to address potential limitations in the approach. Two ways forward are required. First, one should always compare and contrast the evidence provided by starch with contextual archaeological evidence from other sources of data: e.g., location and spatial context of the findspot, other traces of use, technological variables (manufacture, design, raw material), and likely contaminants and contact materials. Second, tools can be used to process a variety of materials for many purposes: e.g., food preparation, manufacture of other artifacts, decoration of items, etc. To understand the ways that residues are formed as a consequence of these and other activities and to predict the nature of incidental contacts with other materials, actualistic research is required (cf. Chapter 4).

Ethnography and ethnohistory can provide useful examples, but controlled tool-use experiments will always form the backbone of residue research. Some of the problems that still require attention in further research, however, are the effects of applying the tools to a number of different starchy plant sources or of multiple tasks. A solution is to replicate a series of simple tasks and then gradually create more complex life histories for tools. The nature of the starchy residues can then be studied in terms of the content, quantity, granule characteristics (e.g., intact or damaged), and spatial patterning of the starch on the tool. Well-documented ethnographic collections, often stored in museums under stable environmental conditions for a century or more, would provide an excellent source of information for studying preservation of tool-residues originating from a variety of conditions, tasks, and life histories.

To date most research on starch as tool residues has focused on stone artifacts and particularly those used in food preparation, but a large range of additional behaviour, including craft activities involving the preparation of adhesives, medicines, and other nonfood products, could be studied. The approach described here, especially when combined with techniques discussed in the next chapter, is also applicable to many other types of artifacts: e.g., wooden spoons, bowls or serving trays, bamboo knives, pottery, stone mortars, basketry, or metal objects. Clearly, the potential of studies using starchy residues has only begun to be tapped.

The stages of observing, recording, and identifying starch on artifacts proposed here demand a sequence which is designed to maximise the information and minimise contamination and loss of detail. They are intended to be a general guide and are not meant to be overly prescriptive since each archaeological research project will have its own special needs. The most important lesson that all projects can learn from is that the successful analysis of starch residues will benefit greatly from an integrated approach in which a wider set of evidence including use-wear, other types of residues, tool design, archaeological context, etc., are combined with and balanced by the starch data.

CHAPTER 10

MODIFIED STARCH

Delwen Samuel

Modified starch derived from ancient remains has an unusual capacity to provide detailed insights about the processing of starchy resources, but to date this potential has only been realised in a few key cases. Modified starch is produced when the morphological or physico-chemical structure of native starch is disrupted in some way, such as during food preparation (cf. Chapters 4, 5). Since it is more susceptible to organic decay than native starch granules, ancient modified starch is only likely to be preserved in particular conditions, such as in arid or otherwise unusually protected situations.

Since processing methods affect native starch morphology in specific and predictable ways (e.g., Boxes 4.2, 4.4), these morphological changes provide a link between exploited starchy resources, artifacts and installations involved in food preparation, and the ways in which both ingredients and tools were used. As a consequence, studies of ancient modified starch can contribute to an understanding of ancient food technology and of variations in cuisine among different social groups.

Analysis of modified starch may also be relevant to ancient nutrition. As discussed in Chapter 5, processed starch may be classified according to how easily it is digested by humans (i.e., rapidly digestible, slowly digestible, or resistant). Currently, a high intake of rapidly digestible starch is considered detrimental to health (Roberts 2000). In ancient times, with less efficient processing technologies and much more active lifestyles, increasing starch digestibility may have been an advantage.

Starch food staples can be bland and their monotony may depress appetite before sufficient calories have been consumed. The taste of traditional foods may be improved by various accompaniments (Mintz 1985; Christensen 2002) or by using different methods to process the raw starch staple. Analyses of ancient modified starch could indicate how the problem of palatability was resolved in the past.

Just as evidence for native starch granules on artifacts can indicate past usages (Chapter 9), the detection and analysis of modified starch can improve our understanding of the function of food-processing equipment. The application of heat is one of the most common ways to modify starch. Fire installations such as cooking pits, hearths, and ovens may have come into contact with heated starchy material and may therefore retain traces of modified starch. The containers, tools, and installations used to make prepared starchy foodstuffs, such as pottery, troughs, and baskets, may also bear starch residues from various stages of food preparation, as depicted in the general starch pathways model in Chapter 4 (Figure 4.1).

Most research on the physical and chemical changes observed when starch is modified has been done on cereals, especially wheat and barley, but other grains such as maize, sorghum, and rice have also been studied. Far less data is available on the behaviour of starch from underground storage organs (although see review in Hoover 2001), but the general principles of starch alteration derived from cereals should apply more widely to all types of starch.

PRESERVED FORMS OF MODIFIED STARCH

The processes that make native starch more digestible for humans render modified starch highly vulnerable to the type of microbial breakdown discussed in Chapter 5. It is thus unlikely that ancient modified starch will survive if it is not protected from degradation by charring or by desiccation. It is therefore important to consider the archaeological contexts in which modified starch might be recovered.

Discrete Desiccated Macroremains

Among the most recognisable ancient starchy prepared foodstuffs are discrete desiccated macroremains. These are coherent foods that are not attached to another object. When complete or nearly whole, such as ancient bread loaves, they are quite recognisable. If fragmented, they may be recognised as food residues but a more precise identity may be unclear. They may incorporate whole or fragmented plant material such as seeds, grains, and root fragments which can help to identify the residues as processed foods. Discrete macroremains may be the intended final foodstuff, such as bread loaves found in ancient Egyptian tombs and temple foundation deposits, or they may be intermediate products of the food processing sequence, such as starch-rich, chaffy lumps produced during ancient Egyptian brewing.

Figure 10.1. Thick encrustation of desiccated residue on the ceramic base of an unslipped siltware, closed form vessel, c. 1350 B.C., excavated from a rubbish dump at the Amarna Workmen's village, central Egypt. *Photo by Delwen Samuel*

Attached Desiccated Residues

Cooks and those who wash up are familiar with traces of foodstuffs that cling to the surface of utensils. For example, when discrete starch foods such as rice or potatoes are boiled, fine particulate matter including modified starch collects at the surface of the water and adheres in a ring to the inner surface of the cooking vessel. If the cooking water boils over, the particulate matter may be distributed in a thin irregular deposit over the lid and outer surface of the cooking vessel or on a supporting structure. Overcooking, especially of thick, starchy, porridge-like foods, may lead to a dense mass encrusted on the interior base of the pot. This crust can be difficult to clean off even with prolonged soaking. If the container was not cleaned or was discarded shortly after food preparation, the residues may be preserved in arid conditions. Utensils used to stir or transfer cooked starchy foods may also end up in the archaeological record with residue still adhering to them. Another source of attached desiccated residue may be sediment remaining after evaporation of a liquid in an abandoned or deliberately deposited container. Uncooked starchy mixtures can also form persistent encrustations.

Figure 10.2. Thin desiccated smear on a red slipped siltware jar, c. 1350 B.C., excavated from the lower levels of the main rubbish dump at the Amarna Workmen's village, central Egypt. *Photo by Delwen Samuel*

Identification of attached desiccated residues may be affected by the quantity of surviving residue, its appearance (colour, texture), and also awareness by excavators that such residues may be recovered. Thick encrustations with obvious plant tissues embedded in them are the most likely to be recognised (Figure 10.1). Thin smears, such as that illustrated in Figure 10.2, are often covered in a coating of sediment and may well be overlooked, mistaken for dirt, and consequently inadvertently scrubbed off as part of routine finds processing. Some residues will only be microscopic (Box 9.5). The use of Congo Red dye may be helpful for identifying small samples of processed starch still adhering to artifacts (Lamb and Loy 2005).

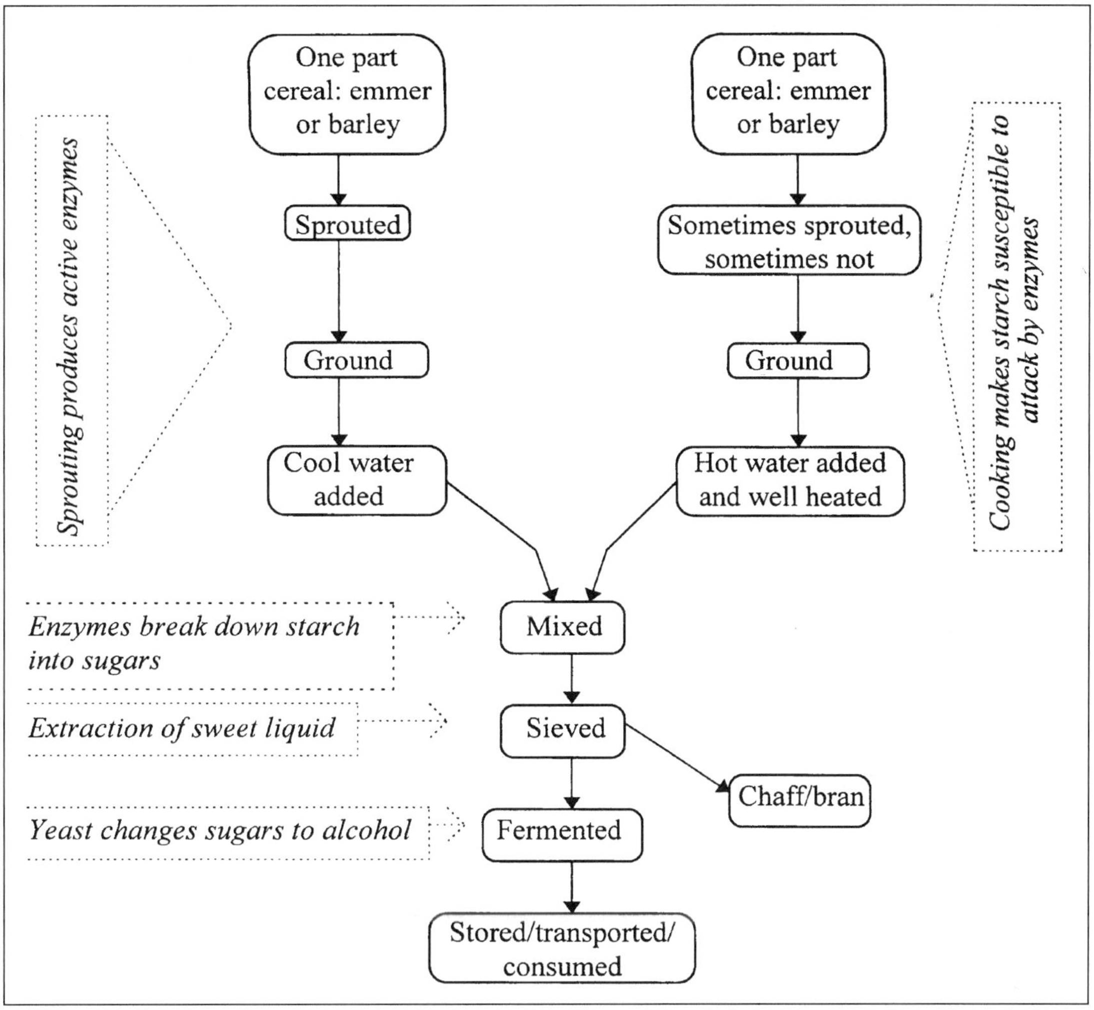

Figure 10.3. Analysis of starch granules has helped reconstruct the two-part process used for brewing ancient Egyptian beer. *Diagram by Delwen Samuel*

Charred Residues

Another route by which modified starch can be preserved is through charring. For example, Hather (1993) has shown that native starch can survive in a chemically or microscopically recognisable form within a charred matrix. Charring occurs through exposure to heat or by direct or close contact with fire. The precise temperatures required depend on the time of exposure and the nature of the material, but in general, temperatures above about 180° C can char organic material over a period of time (Boardman and Jones 1990; Hather 1993). Although charred starchy residues may be considerably more difficult to recognise and identify than desiccated materials, they could potentially survive in a wide range of taphonomic conditions, just as charred seeds do (Pearsall 2000: 240).

Accidental severe overcooking can produce large quantities of scorched or charred material. The burnt food may remain adhered to the cooking container or may have been scraped out and then preserved as scattered discrete fragments. Charred foodstuffs can also result from intentional discard of unwanted or excess food scraps into the fire. These may have been leftovers from a meal but may have been considered unpalatable for some reason and so may not be representative of the regular diet. Catastrophic charring accidents can result in the preservation of complete foods. For example, the volcanic eruption at Pompeii preserved whole bread loaves in the baker's precinct (Mayeske 1979).

Additional Sites of Preservation

Modified starch granules, such as those described by Babot (Box 4.4) may persist on ground stone tools. The identification of starch on this class of artifacts has concentrated on native starch (e.g., Atchison and Fullagar 1998; Pearsall et al. 2004; Piperno 1998: 1875; Piperno et al. 2000; Box 9.4) but modified granules have also been observed (Dolores Piperno personal communication, 2002).

A final suggestion for site of preserved modified starch is on heavily used porous preparation or cooking surfaces. For example, I have found a few isolated non-birefringent but iodide-staining starch granules (see Chapter 3, Box 7.2) on the clay lining of a bread-baking *tannour* from New Kingdom Egypt (Samuel, unpublished data). Further investigation is needed to find out whether modified starch has survived in circumstances apart from those subjected to desiccation or charring.

RETRIEVAL, INITIAL ANALYSIS, AND SAMPLING

The choice of appropriate retrieval methods for particular archaeological contexts depends on the nature of the residue. If residues have been found or are expected on potsherds or other artifacts, it is important that such objects are not washed and are handled with extra care to avoid damage to residues. When desiccated or charred discrete residues are recognised during excavation, they should be carefully placed in plastic bags labelled with their provenience. Fragmented charred starchy remains may be retrieved if water flotation has been used to recover plant remains (Pearsall 2000: 153).

Visible residues have been found in museum collections, but the quality of accompanying information is variable. For example, there are several hundred ancient Egyptian bread loaves in museum collections worldwide, but many are derived from excavations characterised by rapid clearance and very basic record-keeping. Others were purchases derived from illicit activities in the nineteenth and early twentieth centuries and have no provenience data. Such museum accessions may have been damaged by careless handling, by inappropriate packing such as with sawdust, straw, and cotton wool, and even by 'conservation' with paraffin wax and other materials.

As for any archaeological find, data that should be recorded for excavated samples include context identifier (such as unit number or lot/locus number), type of context (for example, pit, hearth, floor), and status of context (e.g., disturbed, mixed, undisturbed). Records for museum material should include the accession number and storage or display location. Specific details may be relevant to particular excavations or museum collections and liaison with other specialists may be required.

As with the case of the residues described in Chapter 9, first observations are best made with the naked eye or with the help of a hand lens. A low power stereomicroscope can be very helpful at this stage, but is not always available when working in the field or at a museum. As analysis proceeds, higher levels of magnification become desirable. Detailed description, accompanied by a sketch (scale included) should include overall appearance and shape, the texture of the residue, and any visible inclusions or surface features such as tissue shreds or seed fragments. Sometimes desiccated residues have a discernable scent, which should be recorded if detected. Surface markings can be important. For example, impressions made by the ancient baker with fingers and thumbs on some Egyptian bread crusts can be used to reconstruct how the loaves had been formed.

One or more small samples must be removed from visible residues to investigate the starch content (Box 10.1). There is a problem, however, because no matter how carefully the residue is examined and even when several samples are taken from the same accession, the material obtained may still not be entirely representative of the whole foodstuff. For example, if the original preparation was made with a range of ingredients that were not thoroughly mixed, some components of the residue may be overlooked. It is impossible to be certain whether this is likely to be the case, although it can be judged to some extent by examining the heterogeneity (colour, texture, identifiable ingredients) of the residue. A conservative approach to initial sampling is recommended, since additional samples may be required as analysis proceeds.

Box 10.1. Sampling Procedures For Visible Residues

Delwen Samuel

Samples should be treated with a view to long-term conservation and storage, just as with the original material. The sample may be altered or destroyed during analysis, but any remaining material and all accompanying data records should be conserved with care.

Sample location should be chosen carefully with the aid of a magnifying glass or low power dissecting microscope if possible. If the residue appears homogeneous, only one sample need be taken. If there are variations or inclusions in the residue, it is desirable to take extra samples of each type of texture, inclusion, or other variable if possible. Differences between each sample should be recorded.

The best place to take a sample of discrete desiccated or charred residue is from a freshly broken surface since it is less likely to be contaminated by materials such as dust unrelated to its original preparation. Sometimes large pores are present on the surfaces of larger residues, into which a dissecting needle or small scalpel blade can be inserted. In this case a sample can be scraped from the interior of the residue even if the exact location and appearance of the area cannot be recorded. This is useful, for example, when dealing with a whole, undamaged bread loaf. Detailed notes should be taken about sampling locations.

It is more difficult to avoid dust and other post-depositional contaminants when sampling thin coatings of residue on potsherds or other substrates. Prior to sampling, the residue can be dusted by gently blowing air over the surface, for example with a pipette bulb attached to a length of rubber laboratory hose. This will remove much of the loose dust and sediment that may be adhering to the residue. If the residue is not too thin, gentle scraping might be used to remove the top layer so that the freshly exposed surface can be sampled for further analytical work. In general, however, it is not always possible to sample thin coats of adhering residue without also collecting dust, sediment, and other possible contaminants.

Sampling

Since many have been charred, ancient desiccated starchy foodstuffs are often highly friable. Sampling starchy residues can be done with a scalpel or dissecting tool. These should be thoroughly cleaned and dried with 50 percent alcohol and disposable wipes between each sample to prevent cross-contamination. Samples may come away in one piece, but often they crumble. If a whole sample fragment can be obtained, it can be placed directly into a storage container. If the residue is friable and comes away as small crumbs and dust, this material can be saved for possible future analysis of starch or other constituents of interest. A fresh paper funnel should be used for each accession to prevent cross-contamination of samples. Plastic weigh boats, glass Petri dishes, and similar containers are not appropriate for catching residue crumbs, because they easily build up static and a considerable amount of residue may adhere to their surfaces. I have never observed contamination by paper fibres.

In general, the sample size should be kept to the minimum necessary to obtain useful information while leaving enough to repeat procedures. Depending on the size and density of the original ancient residue, a sample quantity of between 50 to 150 mg should be adequate. Weigh all the samples. Having done this, estimation can be made of appropriate subsample quantities for subsequent analyses.

The most convenient way to store small samples is in clear gelatin capsules. These are best stored in small glass vials with plastic push tops or some other type of container that can be stored upright in racks made from expanded polystyrene or wood, for example. Essential provenience information on a paper label inside the vial as well as written on the glass vial itself should accompany the gelatin capsule.

HIGH POWER LIGHT MICROSCOPY

Two methods using high power light microscopy can be applied to desiccated remains: free suspensions (Box 10.2) and thin sections (Box 10.3). Since each yields different data about modified starch and poses particular problems, ideally both techniques would be applied to each sample.

Identifying modified starch is less straightforward than for native granules because starch morphology changes when it is processed. The properties of birefringence and the polarisation cross may be partially or completely absent. Starch granules may be considerably altered in shape or completely dispersed. Identification of modified starch may only be possible by staining with iodine-potassium-iodide (Box 7.2). Free suspensions are the easiest way to apply staining (see Box 10.2). This technique also allows detection of starch-derived soluble dextrins (cf. Table 7.2).

The drawbacks of free suspensions are that the arrangement of the modified starch within the residue structure is lost and the arrangement of the suspension on the slide cannot be controlled. The preparation of thin sections may be preferred

Box 10.2. Preparing Free Suspensions Of Desiccated Residues

Delwen Samuel

Free suspensions are destructive of samples so comprehensive record keeping, including photography of the slide preparations, is essential.

Method

1. Rigorously clean and air-dry all equipment between each sample to prevent contamination. Use new slides and coverslips for each sample and clean them thoroughly before use with 50 percent alcohol and disposable wipes.
2. Select a representative subsample of material using a low power light microscope. If the sample is heterogeneous, take separate subsamples of each type of material. The ideal quantity for each slide is approximately 2 to 5 mg. The minimum amount of sample is about 1 mg. Although a smaller sample can be used, it is difficult to manipulate and can be easily lost from the mount.
3. Place the subsample on a cleaned glass slide. Carefully and thoroughly crush the ancient material with a cleaned small metal spatula. Scrape the powder into the centre of the slide.
4. Add one drop of distilled water to the crushed powder with a Pasteur pipette. Carefully cover the water suspension with a cleaned glass coverslip.
5. Examine each slide with direct transmitted light. Magnifications of between x200 and x400 are usually appropriate. Starch granules will be easily visible if present. Other constituents such as fragments of plant tissue and phytoliths may also be present.
6. Observe using polarizing transmitted light to locate birefringent and partially birefringent starch granules. Modified starch granules may look morphologically indistinguishable from native starch granules but may have partially or completely lost their birefringence. It is therefore useful to photograph the same granules with direct light and again with polarizing transmitted light.
7. Add a drop or two of iodine-potassium-iodide (IKI). Use protocol described in Box 7.2. Record the stained mount with photomicrography fairly promptly. Modified starch, particularly well-gelatinised and pasted starch, may be permanently affected by this procedure and will not retain valid information about its structure for long after the IKI is added.

Box 10.3. Preparing Thin Sections Of Desiccated Residues

Delwen Samuel

Although various embedding media are available for thin sections destined for light microscopy, one of the most flexible is London Resin White (L.R. White, http://www.emsdiasum.com/microscopy/technical/datasheet/14380lm.aspx) available from Electron Microscopy Services (EMS, http://www.emsdiasum.com/microscopy/default.aspx). L.R. White allows very thin (10 microns and less) sections to be made, which can be observed with light microscopy and with electron microscopy if required. Sectioning starch granules can be challenging (although ancient granules are not as difficult as modern granules), but this resin holds the granules well. This box provides a general outline of the procedure, together with comments that apply specifically to ancient residues. For further details and practical assistance, it is advisable to seek help from an experienced microscopist.

Sampling

About 20 to 50 milligrams of each sample are needed for embedding. It is a good idea to fix and embed two or three fragments from each residue sample.

General Method

Day One: Embedding in Agar and Fixation

Because ancient samples are often highly friable and the biomolecular structure often highly soluble, an agar-embedding step prior to the normal fixation and resin-embedding procedure is advisable. Experimentation with the most suitable method for specific ancient samples might be necessary. One procedure is suggested.

- Prepare 2 percent agar solution. Heat to melting point as necessary.
- Cool agar to about 50° C.
- Place items to be embedded on clean glass slide.
- Carefully place a drop of agar over each item.
- Immediately place in fixation solution. Do not allow agar to soak into items.

Choose an appropriate fixation medium (refer to EMS data sheet). During fixation, modern samples are often placed on a rotator or given similar gentle agitation, but this is not ideal for fragile material. If constant mechanical agitation is not used, gentle agitation by hand from time to time may be helpful.

Day Two: Dehydration

A graded ethanol series is recommended for L.R. White. Acetone must be rigorously excluded.

Day Three: Resin Embedding

Embed samples with resin according to method used.

Sectioning and Staining

When the acrylic resin has hardened sufficiently, the specimens are best sectioned with a glass knife. The appropriate thickness of the thin sections may vary according to the sample but 4 to 10 microns are probably best. Overly thin sections will not take up stain as well as modern material. The sections can be floated onto a distilled water drop on untreated, thoroughly cleaned glass slides and dried on a hot plate. The starch will not be affected by the heat because it is fixed within the acrylic resin. It is important to dry the sections quickly so that they adhere evenly, without wrinkles. When dry, the sections will continue to adhere to the slides even when liquid stain is applied. Wherever possible, use serial sections to allow comparison of structures within the food matrix.

because this technique shows the forms of nondispersed modified starch and allows detection of dispersed starch, while preserving the arrangement of the modified starch within the residue matrix. The relationship of starch to other components can provide valuable information about the original processing methods of the starchy ingredients. Unlike free suspensions, thin sections can be stored indefinitely after preparation. Both free suspensions and thin sections allow the detection of nonstarch microcomponents in the ancient residues, such as shreds of plant vascular tissue or seed and fruit pericarp fragments.

The disadvantages of thin-sectioning ancient residues over free suspensions are that it is more time consuming, demands much more skill, and requires specialised equipment. Collaboration with an experienced microscopist is therefore beneficial. Unlike the preparation of modern material described by Field (Box 6.6), thin sections of ancient residue require more precise preparation due to the small volume of material and the need to stabilise highly friable samples. To target structural relationships, ancient residues must be sectioned more thinly than modern reference material. Even modern native starch granules may lose their birefringence when sliced so thinly. The technique is destructive, but, like free suspensions, only tiny quantities are required. Among the repertoire of analytical methods, this method is probably best left until last.

Charred starchy residues require special treatment for high power light microscopy work. The black, opaque structure of the charred material means that, unlike desiccated material, starch cannot be observed directly. The charred sample must first be disaggregated and any surviving starch isolated. During the starch isolation procedure, any dispersed and soluble starch within the charred residue will be lost, but a proportion of native or slightly distorted granules may be recovered. Box 10.4 outlines a suggested preparation technique.

SCANNING ELECTRON MICROSCOPY

The scanning electron microscope has been usefully applied to the analysis of charred residues (Hather 1993; Paz 2001; Willcox 2001), but is most informative for desiccated residues (Samuel 1996a, 2000). For ancient starch analysis, magnification up to 2,000 times is generally sufficient. The intensity of the electron beam at high magnifications may destroy the starch, depending on the machine and settings used. Techniques to prepare modern organic material for scanning electron microscopy are often lengthy, but because they are already desiccated, ancient remains need no special preparation. Box 10.5 presents a suggested procedure for preparing visible residues for the SEM.

CASE STUDIES

Arid areas are most likely to provide evidence for ancient modified starch, although few studies have been made to date. Two case studies focused on ancient Egyptian bread and beer have demonstrated the potential of this approach. Bread and beer were the key staple foods of ancient Egyptian society

Box 10.3 (continued)

Differentiate the starch with drops of iodine-potassium-iodide (IKI) (see Box 7.2). Sections from modern specimens can be immersed in IKI for 30 seconds but ancient material requires up to 10 minutes to develop good colour. Rinse sections briefly in water, mount in water, and cover with a coverslip. Record by photomicrography. As for free suspensions, the best colour film currently available for starch work is Ektar 100 ASA. Use a blue filter for the light source if you are using daylight film.

When study of the slides is completed, float off the coverslips carefully in distilled water, rinse out the IKI, and dry the slides. Then store the slides without coverslips. Most permanent and temporary mounting media leach out the IKI. The thin sections can be restained and examined as required thereafter.

throughout all periods. The ancient recipes for both have usually been assumed to be straightforward and uncomplicated. In contrast, an analysis of modified starch has shown that these technologies were considerably more complex than previously thought (Samuel 1996a, 1996b, 2000).

Bread

The ancient Egyptians distinguished many different types of breads and cakes. Over 40 types have been recorded (Gardiner 1947: 1196), although there is no documentary evidence describing their various attributes. Most scholarly accounts of ancient Egyptian baking describe how coarse flour was pounded or ground and then formed into dough and baked. While some attempts have been made to document bread types using artistic representations, these typologies rely entirely on form and decoration, with no information about possible variations in ingredients or preparation methods (Währen 1961). A notable study of an actual ancient bread loaf was made by Borchardt (1932).

Microscopic analysis of over 35 different loaves now in museum collections throughout the world has established that there were several ways in which ancient Egyptians made bread. In addition to the identification of previously unattested occasional ingredients, e.g., coriander, dates, and figs (Samuel 1994: 191), patterns of starch modification show there were variations in grain preparation. The simplest procedure was to mill grain either coarsely or finely and to form the flour into dough for baking. The presence of pitted and channelled starch granules in several loaves has shown that the ancient Egyptians sometimes made bread from sprouted grain (malt) (Plate 60). The enzymatic breakdown of the starch may have rendered it more digestible. Malted grain would have imparted a sweet flavour to the bread.

Box 10.4. Preparation Of Charred Material For High Power Microscopy

Linda Scott Cummings

Dissolve the charred material in a fume hood with a Schulze solution of concentrated nitric acid and potassium chlorate. Despite this harsh acid treatment, some starch does survive. To make a Schulze solution, add a small quantity, approximately 7 or 8 ml, of potassium chlorate powder catalyst to about 50 ml of concentrated nitric acid and stir to dissolve the powder. Crush some of the charred sample and place it in a centrifuge tube and then add a small quantity of Schulze solution. Place the centrifuge tube in a container of hot sand to accelerate the oxidation reaction. If the sample is particularly stubborn, add hydrogen peroxide (30 percent) *one drop at a time*. Add one drop, mix, observe the reaction. Add another drop *only* if the reaction is very slight or when the reaction ceases. If numerous bubbles are formed, this is enough catalyst to continue the reaction. Adding hydrogen peroxide can cause a very violent reaction, so this step must be performed under a fume hood and the chemical must be added very slowly. This should dissolve the charred material, freeing any starches. When the reaction is finished and most or all of the charred material is dissolved, add distilled water and centrifuge. Remove the supernatant and rinse again with distilled water. Very little sample may remain in the bottom of the centrifuge tube, so use a conical tube that allows concentration of the remains. Rinse the sample until it is neutral (generally 4 or 5 complete rinses). If, after the first rinse, you observe a lot of charcoal remaining in the sample, repeat the above steps to treat with Shultze solution. After rinsing until neutral, take an aliquot of the remaining suspension in the tube, place on a microscope slide, and examine with high power transmitted and polarising light microscopy.

The presence of large fragments of grain in ancient Egyptian bread has led some scholars to believe that milling was crude, implying a finer grade of flour was not possible using the available grinding technology. Microscopy of these large grain fragments shows they are mostly composed of well-gelatinised starch. This evidence indicates that ancient Egyptians deliberately added damp boiled grain to their dough prior to baking, because starch in raw, coarsely broken grain would hardly

Box 10.5. Preparing Samples For The Scanning Electron Microscope

Delwen Samuel

The material to be examined by SEM should be scrutinised with a low power light microscope and a subsample selected on the basis of representative colour, texture, and inclusions. Only one representative fragment need be selected for an apparently homogeneous sample, but heterogeneous residues should be subsampled for a fragment of each type of material observed, unusual inclusions, or other variations.

Once the subsample is selected, mount it on an SEM stub with precut double-sided sticky tape or discs. Double-sided carbon discs are not recommended because they are insufficiently adhesive for bumpy samples. Then gently dust the surface of each fragment by blowing air over the sample to remove dust or loose particles which might cause a build-up of electron charge or which might obscure the real surface of interest. Attaching a pipette bulb to one end of a length of rubber laboratory sink hose and a plastic pipette tip to the other end can make a suitable duster. The pipette tip can be directed with precision and the bulb gently squeezed to blow air over the mounted specimens. (A computer keyboard compressed air canister is *not* recommended!)

When the specimens are mounted, draw a sketch of the stub, take a photomicrograph (optional), and make brief notes about the structure and appearance of each fragment and why it was selected for examination. These measures will help in later interpretations of the sample such as determining the location of the SEM images.

Apply a dab of silver dag paint so that it firmly links the side of each fragment to the surface of the tape. (This paint is toxic so avoid breathing the fumes and skin contact.) This procedure prevents charge build-up, a common problem with highly friable and discontinuous ancient starchy residues. Great care must be taken at this stage because the silver dag can easily be drawn into the highly desiccated specimen by capillary action, thus ruining its surface features. The best method is to trim the end of a wooden toothpick at an angle, put a thick dab of silver dag paint at the trimmed end, and cautiously push the silver dag over the tape towards the specimen. Only a tiny amount need touch the ancient material, but it must make contact. Dry the stubs in a 30° C drying oven for at least 3 hours to drive off the solvent, or leave to air-dry overnight. Sputter-coat the stubs to a depth of about 250 to 300 nm. They are now ready for scanning.

The complexity of the structure of ancient starchy residues often means each specimen needs very thorough examination and many records should be made of the surface features observed. Current technology favours digital image capture. It is important to bear in mind that the technology of digital image retrieval and reproduction may change in the future. The micrographs are the raw data of SEM. Since these are unique and unrepeatable, the digital data therefore needs to be conserved and kept accessible.

gelatinise within the matrix of the baking bread. The boiled grain could not have been dried before mixing into the dough either, because this would have produced gritty cooked grain fragments scattered through the bread. We can therefore conclude that the boiled coarse grain fragments were intentional and would probably have imparted a nutty flavour and a pleasant chewy texture to the loaves. These would have made a welcome addition to the ancient Egyptian baking repertoire.

Other variations in preparation and baking techniques undoubtedly developed over time. With a greater sample size of ancient Egyptian loaves, we could discern the most common methods of grain preparation, determine whether different groups in society had certain preparation methods, and monitor how bread making evolved.

Brewing

For many years brewing was considered by Egyptologists to have been a relatively simple process, closely resembling modern traditional Coptic and Nubian beer making (e.g., Morcos et al.1973: 1170). This method involves heavily yeasted dough that is lightly baked into beer loaves, crumbled into a sieve, and rinsed through with water. The yeast from the lightly baked bread then ferments in the starchy suspension, producing beer. Dates are often thought to have been an essential ingredient of the ancient recipe, adding sugar necessary for the fermentation process. Analysis of modified starch granules recovered from the beer residues, however, does not bear out this interpretation.

A number of thin starchy residues have been found adhering to shallow ceramic drinking bowls (Figure 10.2). These residues contain fine shreds of chaff, showing they were derived from cereals. Other, more chaffy samples have been isolated from jars or retrieved as isolated finds (Figure 10.1). The main component of all these residues is heavily modified starch. Yeast cells have also been detected, some of which were actively growing just prior to being dried out. The pattern of starch modification from all types of cereal-derived residues included native starch, gelatinised and pasted starch, and pitted granules indicative of enzyme attack. Together, this evidence indicates that ancient Egyptian brewing involved a two-part process (Figure 10.3; Samuel 2000). First, cereal grain was divided into two batches. One batch may have been sprouted (malted) and once crushed, was very well cooked in plenty of water, perhaps to the consistency of porridge. This process gelatinised the starch. The other batch was sprouted to produce malt and then crushed and mixed with water, but was not exposed to heat. The sprouting process produced enzymes that attacked the starch granules, causing pits and channels (Plate 60). The two batches were then mixed together to create a warm, opaque liquid. Within the mixture, the unheated malted component provided enzymes that attacked the gelatinised starch from the cooked portion. This process yielded the sugars required for yeast (and possibly lactic acid bacteria) fermentation.

The ancient Egyptians may not have understood the complex biochemical processes which took place during beer making, but they would have had an excellent empirical knowledge of how to manipulate different conditions, such as time, quantities of ingredients, and degree of heating to achieve consistent and desirable results. The ancient Egyptian process is very similar to that used for a number of modern traditional sub-Saharan African beers, the most sophisticated of which is probably Sudanese *merissa*.

> Far from the coarse, rather crude foods often described, the cereal staples of ancient Egyptian diet were clearly sophisticated. These results fit much better with our understanding of other aspects of this culture and they provide new data relevant to ancient economy and social interactions. For example, the use of grain sprouting for malt production as a key cereal processing technique demonstrates considerable time and resources must have been devoted to this task. (Dirar 1993: 840)

FUTURE RESEARCH

A high priority for ancient modified starch research is the location and identification of additional data. Ancient grain cooking and malting has been identified through starch analysis by Juan-Tresserras (1998) from Bronze and Iron Age sites on the Iberian Peninsula, and other modified starches have been recovered from Egyptian food remains and papyri, from the unique find of the Austrian iceman Ötzi (Box 9.1), and from Lapita pottery in the Pacific region (Crowther 2005; Horrocks and Bedford 2005). However, there are a number of places throughout the world where modified starch can be expected to survive in reasonable quantities. These include all areas where organic material might have been rapidly desiccated and protected from decay, and where foodstuffs were provided with burials.

Although charred remains are far more widespread than desiccated material, the problems of extracting, identifying, and interpreting modified starch from them are greater. More extensive analysis may show that native and near-native starch is most commonly preserved in charred material. Despite these limitations, charred residues are likely to contain valuable records of starch exploitation, especially for those taxa that do not survive well in the archaeological record.

The primary aim of much ancient human starch exploitation was the production of modified starch (cf. Chapter 4), but ironically modified starch will always remain rare in the archaeological record. With knowledge of the conditions required for its persistence over time, it should be possible for archaeologists to target specific circumstances where it may be found. Once located, modified starch can provide unique insights into ancient life.

CHAPTER 11

LOOKING AHEAD

Huw Barton and Robin Torrence

Ancient starch analysis has made considerable advances since the earliest systematic applications by Don Ugent in the 1980s (Box 1.4). Building on his important scholarly contribution, current studies are also anchored in fundamental taxonomic research in botany (e.g., Reichert 1913; Box 7.1) and pharmacology (e.g., Greenish and Collin 1904; Wallis 1957) and benefit from extensive research on modern starch within the food industry. The rapidly growing number of studies demonstrates that the study of ancient starch has a significant and wide-ranging role to play in archaeology. Starch studies have also made important contributions by complementing and enhancing existing paleobotanical techniques based on macrobotanical materials such as seeds, nuts, and charcoal as well as plant microfossils such as pollen, phytoliths, and diatoms. There is no doubt that in the future starch research will greatly extend our knowledge about past human activities involving the use and manipulation of plants for food, drugs, ritual, shelter, tools, and perhaps much more, as well as contribute to reconstructions of past environments.

The numerous case studies reviewed throughout this book and illustrated in the boxes also demonstrate that starch research is already attacking and perhaps solving some of the important 'big questions' in archaeology that were introduced in Chapter 1. One reason for the current and potential significance of ancient starch is its remarkable ubiquity in time and space. As illustrated in Table 1.1, ancient starch does not seem to be restricted to particular time periods, regions, or climates. This means that starch has an unusual potential to contribute to fundamental research questions concerning the role of plants in human societies, an area that has long been under-researched due to the absence of obvious evidence. For example, until recently, discussions about the evolution of the human species (see Box 1.1) have generally ignored plants in favour of animal prey. However, recent findings of starch on grindstones has shown that plants may have played a part in the evolution of 'modern' human behaviour as monitored through particular types of tool use (Van Peer et al. 2003).

A second key aspect of starch is that it is frequently the only surviving evidence for many important economic species, particularly with regard to the underground storage organs that have dominated subsistence patterns in tropical areas. Studies of starch in these regions have already made a major contribution to our understanding of root and tuber domestication (e.g., Piperno et al. 2000; Denham et al. 2003; Fullagar et al. 2005) and have monitored the emergence of agriculture in regions that were earlier thought to be unimportant or marginal in that debate, such as Panama (Piperno and Holst 1998; Piperno et al. 2000), Uruguay (Iriarte et al. 2004), Solomon Islands (Loy et al. 1992), and Papua New Guinea (Denham et al. 2003; Fullagar et al. 2005). Furthermore, such studies need not be confined to the tropics, since, as shown in Table 1.1, starch granules are also frequently recovered in temperate areas. While some questions may be better served with other types of paleobotanical data, starch is frequently found on grinding stones and can therefore also contribute to the identification of economically important grains and seed crops and may help in the discrimination between wild and domesticated varieties (Piperno et al. 2004). Investigating starch granules preserved on cereal-processing equipment would be an excellent method of tracking the spread of cereal crops across Europe, for example. It might be possible to compare frequencies of use of wild grass against imported domesticates. Perhaps starch survives well in cereal possessing tools in contexts that macrobotanical remains are low in frequency or absent?

It is quite possible that ancient starch research will demonstrate that, contrary to current belief, roots and tubers had a much more important role in temperate zones in the past, particularly prior to the spread of crops based on domesticated grains (cf. Hather 1994a; Zvelebil 1994), but that their place in the diet has been overlooked due to the absence of macrofossils. For example, Marek Zvelebil (1994: 56) noted in a discussion about the lack of evidence for plant use in Mesolithic Europe that 'the presence of pounding stones and of similar groundstone equipment is widely reported from Mesolithic sites in Europe . . . but rarely are they interpreted as plant processing equipment.' Perhaps with the new techniques of starch granule analysis much of this material evidence could be reappraised leading to new interpretations about the life-ways of pre- and early agricultural groups in Mesolithic Europe and Britain.

One of the major conclusions that arise from the synthesis of current work presented in this book is that the inclusion of ancient starch into the paleobotanists' tool kit has the potential to unlock many new areas of research and uncover previously hidden stories. Thanks to the collaboration of many starch researchers and their willingness to share new techniques, this book provides the necessary methodology so that others can adopt this promising new field of study.

ARCHAEOLOGICAL CONCERNS

Following recommendations made by the Ancient Starch Research Group, this book was put together to promote modern starch research in archaeology by alerting scholars to the recent proliferation of starch studies, to stimulate additional students and researchers, and to furnish them with the requisite background knowledge to undertake their own studies. Given that starch is a plant microfossil, it is not surprising that the book covers many of the same topics as other texts dealing with paleobotanical material and particularly microfossils. Basic knowledge about the biological origins of the microfossils, methods for extracting them from the archaeological or environmental record, and knowledge about how to identify the material to a particular plant taxon are also critical for studies of pollen and phytoliths. The team of authors for this book have adopted a slightly different approach, however, probably because many of the Ancient Starch Research Group who workshopped its structure and contents had turned to starch research as a way to answer specific archaeological questions about plant use and tool function. Consequently, the book has a particularly archaeological flavour that is slightly different from many paleobotanical approaches to plant remains.

Context is critical in the interpretation of archaeological information and so in this book starch has been subjected to the same consideration as any other form of archaeological data. Modern concerns with how the archaeological record was formed have shaped the overall approach to starch analysis. Detailed chapters on how starch recovered from archaeological contexts had found its way there in the first place and why and how it had been preserved have therefore been covered in detail. As a way to understand starch pathways on the large scale as well as individual tool life histories in particular, various authors have emphasised the importance of actualistic research involving ethnoarchaeology and/or pure experimentation. The continued incorporation of an experimental component to all aspects of starch research is a strong recommendation arising from the book.

The increasing pace of publication, especially within the past five years, shows that ancient starch analysis has been launched as a successful field within archaeology. Despite the progress made recently, a number of weaknesses can be identified, particularly with regard to the understanding of pathways. In the following, we briefly review what we identify as areas that will benefit from targeted research in the future. We structure our discussion within the framework of key questions that still need to be satisfactorily resolved.

WHAT IS ANCIENT STARCH?

One area of ancient starch studies that demands further attention is the composition of the starch granules retrieved from archaeological and paleoenvironmental contexts. Are ancient starches simply unmodified granules that have been preserved or have they been altered in some way over time? Does the mechanism of their preservation help understand their persistence in the face of the ubiquitous biological agents evolved to attack starch, as described in Chapter 5? For example, Kealhofer et al. (1999) were surprised to find that what appeared to be ordinary starch granules had survived the aggressive chemicals used to extract phytoliths from soils. Their suggestion was 'that the starch grains were silicified in some way' as 'what appeared to be silicificd casts of grains surrounding gelatinised starch were observed' (Kealhofer et al. 1999: 535). In other words, had the granules been fossilised in some way before they were retrieved by archaeological excavation or, perhaps, had they been altered during the phytolith extraction technique itself?

Evidence from the fossil record has demonstrated that starch granules can be fossilised. For example, Baxter (1964) identified starch granules in Paleozoic examples of *Cardiocarpus spinatus* seeds and Wilkinson (1983) described three-dimensional fossil casts of starch from Eocene mangrove species. In both cases, the morphology of the fossil granules was well enough preserved to support a confident taxonomic identification of the plant. Baxter (1964: 422) attempted an iodine stain of the fossil material, but recorded a negative reaction; however, since this is common for plant fossils, it is not clear if the carbohydrates had survived or been replaced.

The successful use of the iodine stain to positively identify starch from ancient wheat, dried tubers, and tool residues (e.g., Hallam and Osborne 1974; Barton and White 1993; cf. Chapter 5), indicates the presence of amylose and amylopectin in at least some archaeological materials. Other structural details such as granule morphology, the presence of the hilum, and sometimes lamellae and fissures all indicate that at least some ancient starch appears to retain its modern structural composition. Furthermore, some starches retain the property of birefringence, which indicates that the well-ordered crystallites within the semicrystalline zones, as described in Chapter 2, have been preserved.

As overwhelmingly supported by the research reported in this book, much ancient starch is not demonstrably different from modern starch. Experimental work combined with chemical analyses of ancient starch is required to confirm the physical and chemical nature of the similarities. At the same time, work in this area might also reveal how it is that some starch can resist chemical degradation over these long periods. Studies investigating starch persistence in modern human faeces (Chapter 5) reveals just how resistant some starch granules are to the suite of enzymes evolved to break down starch. The recovery of ancient starch in prehistoric coprolites (Box 5.2) verifies these observations. Determination of the structure and chemistry of ancient starch lies outside the skills and experience of most archaeologists. Fruitful research in this area will most likely come from new collaborations between starch scholars and archaeological researchers.

Directly associated with the need to better describe the chemical properties of ancient starch is the need to address the question of its antiquity. Is it possible to directly date ancient starch? A constant concern of the technique raised at seminars and conferences is the need for an absolute date of a starchy residue. We believe that there is enough evidence to support the observation and claim that starch granules may persist for thousands of years in the archaeological record. For some, however, doubts will remain in the absence of unequivocal evidence such as a radiocarbon date. Because of the often small quantities of starches recovered from ancient tool surfaces and sediments, there would be considerable technical difficulties with such an experiment, but this is definitely an area of 'blue skies' research that deserves attention.

WHERE CAN WE FIND IT?

The highly varied contributions in this book demonstrate that starch granules have been recovered from a wide range of contexts. For example, starch has been shown to survive within tropical and arid zone sediments (Chapter 8), on the surface of a wide variety of materials including stone flakes (Chapter 9), grindstones (Box 9.6), copper (Box 9.1), pottery (Box 9.5), and other materials (Box 9.3). Intact granules have also been recovered from desiccated tubers (Box 1.4) and wheat (Hallam and Osborne 1974) as well as within components of other materials such as plant resins (Box 9.4), coprolites (Box 5.2), and cooked food products (Chapter 10). Modified or cooked granules also appear to have been preserved under certain special conditions, such as beneath the hafted area of the copper axe carried by Ötzi (Box 9.1) in the cold, dry conditions of the Italian Alps or within dried food products such as the Egyptian beer residues on pottery (Chapter 10).

Finding starch is really not an issue. The real problem is the interpretation of human behaviour based on the presence of, or more importantly, on key attributes of the starch assemblage. Simply identifying and describing starch recovered from a particular context is not a solid basis for interpreting site or artifact function or plant use. As part of the interpretation, one has to understand the direct and indirect paths that lead starch to be within the soil or on the artifact surface.

Studies such as the analysis of cooked starch on Ötzi's axe (Box 9.1) emphasise the need for careful consideration of tool life histories and lessons learned from ethnoarchaeology or lab-based experiments. In this case Loy (1994b) was surprised to find starch residue on all parts of the axe blade including along sections that would normally have been under the haft itself. The pattern did not make immediate sense. It is unlikely that the axe was used to process cooked food, so how did the starch find its way under the tool haft? It was only through reviewing previous experimental work that Loy realised the starch could have been smeared in this way by dirty hands involved in the task of fitting the axe blade into a new haft. This scenario emphasises that archaeologists need to think about the pathways by which starch residues may find their way onto the surface of implements, food preparation areas, cooking surfaces, locations of craft activity, or larger scale activities involving plant materials like construction, plant processing (e.g., sago extraction), and ultimately, areas of household clean-up, and other locations of food or waste disposal. By first considering these contexts, one can plan where and how to sample as well as make plausible interpretations about the starch that is recovered.

Starches often appear in soils alongside artifacts that also have retained starch granules from use-related activities. Barton et al. (1998) and Atchison and Fullagar (1998) have noted that starch deposited into the soil from one set of activities may find its way onto tools used in different activities, although both studies concluded that the two possibly different sources of starch could be distinguished in terms of differing abundances. In contrast, Williamson's experiments (Box 5.4) showed that tools buried alongside decaying starch-rich vegetable matter could acquire appreciable numbers of starch granules from such a neighbouring source, but she also demonstrated that residues formed on the experimental tools during their use could be distinguished. Extraneous starchy material on an artifact (e.g., sediment starch/decaying starch-rich vegetation) is far less likely to be confused with use-related residues within an approach that considers other factors such as use-wear, tool life history, and the context of deposition. On the other hand, Haslam (Box 8.10) usefully points out that artifacts discarded within features such as pits associated with food processing may retain use-related starches from surrounding sediments that derive from activities occurring at a larger scale than that of the artifact itself. In this case the starch can be interpreted in a useful framework, but again only after the formation processes have been considered.

Experimental programs and consideration of ethnoarchaeological information should definitely be a part of all future studies of ancient starch. Observations of starches recovered from different stages of tool use or material manufacture would almost certainly allow greater sophistication in the interpretation of past activities. In cases where direct observation is not possible, the creative use of well-documented ethnographic material held in museum collections may prove beneficial. Equipment or tools used to process or produce items of food may retain starchy residues. Sampling these items may provide useful details about the type and condition of starches that are typically deposited on items at various stages of food/tool production. The information gained from the museum research could then be applied to archaeological cases.

As our knowledge about where we can find starches and what they mean increases, so too will our knowledge of where we will not find starch. For example, does the absence of starch on a ground surface mean that it had not been used to process plants or that starch resulting from plant processing had simply not survived? To date, this latter project is seriously under-researched. Continued research on starch pathways also needs to consider when the absence of starch is a meaningful variable.

In Chapter 5 the suggestion was made by Barton and Matthews, and often raised in discussion with colleagues, that the materials to which starch adheres, such as an obsidian flake or piece of pottery, must provide a microenvironment that offers some kind of protection from the degrading activity of soil microorganisms and moisture. The observation that other types of plant microfossil such as pollen (Kelso et al. 1995) may also benefit from this type of association supports the hypotheses for starch survival. Just what the precise characteristics of this microenvironment might be has not been systematically researched and remains an open area for future and much needed investigation.

Once again experimentation seems the best way forward for testing the role of microenvironments and other models for starch preservation. Starch of many commercial species is readily available in different forms (e.g., tubers, seeds, and processed flours) and thus lends itself to controlled experimentation. Since starch granules are readily identifiable and can be counted (e.g., Barton et al. 1998), the task of determining rates of movement or degradation under different experimental conditions is relatively straightforward. Initial attempts along these lines by Lu (Box 5.3) and Therin (Box 5.5) indicate that experimentation is a productive research path, although both authors also highlight the difficulties and complexity of conducting experiments on a large enough scale with sufficient time depth. At the workshop in 2000, the Ancient Starch Research Group recommended the construction of purpose-built structures, tentatively called 'starch farms' in which stratified contexts could be constructed at a somewhat realistic scale in terms of areal extent and depth. A number of activities would be conducted there and then movement and preservation of starch monitored over at least a number of years. Experiments could be set up to reflect various taphonomic conditions, including variations in soil type, soil porosity, presence/absence of fungi, rates of water flow, or simulated conditions of long- or short-term exposure to the elements. Such experiments would greatly help increase knowledge about the likely presence and absence of ancient starch. These are all good ideas that would certainly benefit the long-term development of starch research, but they will require high amounts of funding and dedication on the part of the researchers.

WHERE ELSE CAN WE LOOK?

Although starch has already been found in a wide range of contexts, it is useful to identify potential additional settings for research. Museum collections have already been mentioned as a potential place to test ideas about starch pathways. Ethnographic material also offers excellent opportunities for extending knowledge about past processes for plant processing and tool use, especially if details of artifact function have been well recorded and the implements have been stored under cool, dry conditions. In some cases, items that have not been washed may retain visible residue deposits and even those that have had light washing will more than likely retain useful residues. For example, Huw Barton has recovered well-preserved starch granules and associated cellular tissues from wooden sago pounders collected in Sarawak by the 1955 Oxford Expedition to Borneo (Arnold 1959) and now housed at the Pitt Rivers Museum in Oxford, UK. Herbarium material and collections of plant products in ethnobotanical collections, including starch flours collected in 1856, have also yielded excellent samples of well-preserved starch granules (Box 7.6). Similar collections of material around the world have great potential for starch studies: e.g., details of plant use or tool function may be clarified; some items may hold evidence of early plant introductions, perhaps associated with known cultural expansions into new areas; and some tools may yield examples of pre-industrial cultivars or plants associated with early or putative agriculture. It may also be possible to determine the composition of food mixtures, prepared within containers thought to have been in use during important festive or ritual occasions.

One of the advantages of studying museum collections is that the wide range of material culture potentially available may also encourage archaeologists to consider a wider range than usual of artifacts or activities that might have involved starch or the transference of starchy materials. Although food preparation activities are one obvious area that could benefit from museum-based research, and have very wide application, there are other opportunities. Good examples are research into the historical use of plant-based adhesives, poisons, pigments, fragrances, and drugs, many of which might contain starch granules and thus provide clues as to their ingredients. Since granules are recovered in very high numbers, historical material provides a good opportunity to observe population variation of granule morphology within the sample. By contrast, small samples of starch granules recovered from ancient sediments and tools are often very small (e.g., Barton 2005; Barton et al. 1998; Piperno and Holst 1998). Large numbers of granules clearly derived from a single plant species have been a rare find to date (e.g., Fullagar et al. 2005; Loy et al. 1992).

The recovery of starch granules from tooth calculus (Juan-Tresserras et al. 1997) and coprolites (Box 5.2) suggests new opportunities for this new field of research. As direct indicators of an individual's diet, starch granules may be able to provide a unique insight into the activities of the individual in prehistory. Coprolites, in particular, are interesting because they provide a very short-term record of foods consumed.

CAN WE IMPROVE IDENTIFICATIONS?

The identification of starch with the aid of comprehensive reference material is already a routine operation for many ancient starch studies. For some studies the major limiting factor for accurate identifications is insufficient knowledge of starch variation between some species of plants at a regional and super-regional level. This problem will only be resolved with more comprehensive research collections. Since it is often not feasible and can be very costly for each researcher to acquire an adequate collection, collaborative efforts between research groups at a global scale need to be considered. At the very least, it would be ideal if researchers could make their reference material as widely available as possible through detailed publication or perhaps through web-based publication of photographs. This practice could greatly assist in the preparation of keys that incorporate the widest possible range of variants within them.

As discussed in Chapter 7, particular keys based on starch granules are gradually becoming available for some plant groups. Constructing a key can be a difficult procedure if the range of variation among the granules is not very large. The use of imaging techniques combined with multivariate techniques (e.g., Box 7.8) could greatly assist in the construction of keys by independently assessing which variables have more or less discriminatory power and may help shape the analyst's choice of variables for a particular key.

For some research questions variables additional to those described in Chapter 7 for level 4 identification and/or more detailed observations of the attributes will be required. It seems quite likely that different ways to observe and describe may be when a very fine discrimination among plant taxa is demanded, as for instance when monitoring the history of plant domestication or studying the development of specific varieties for different climatic conditions as plants have been introduced or spread into new environments. Further advances in starch granule identification may also be achieved with greater attention paid to the statistical description of size and shape variation within and between species. Techniques that enable three-dimensional recording of granule shape will definitely be a great advantage, increasing the likelihood of reliable species-level identifications and discrimination between species.

Current descriptions of starch rely heavily on recording size and some physical features, but the statistical validity of this approach has only just begun to be explored (e.g., Box 7.8). As shown in Chapter 7, size and shape are good discriminators of starch, especially where the known or likely sources of plants contributing to the starch assemblage are restricted, but in cases where starch sources are much wider, such as might occur within a soil sample, higher levels of precision may be necessary to produce reliable identifications. Statistical approaches (such as multivariate analyses) that can accommodate both metric and nominal variables should help in this area because they provide an independent assessment of which variables have more or less discriminatory power and may therefore help shape the analyst's choice of variables. Numerical methods for describing shape (such as Fourier analysis) may also help to improve identification by removing reliance on qualitative terms such as 'bell-shaped,' 'kidney-shaped,' 'ovate,' and so on.

The ability to reliably discriminate between cultivated and domesticated plants and closc wild relatives on the basis of starch granules would be extremely beneficial to many areas of archaeology, paleovegetation studies, and debates about the nature and origin of agricultural systems. While the tropics are in great need of such analyses because other forms of plant remains are rarely preserved and there remain exciting debates about the nature and history of human-plant interactions, a detailed study of starch granule variation within better understood systems of plant manipulation, like the domestication of grains and maize, could make a significant contribution. Preserved seeds and kernels could be sampled for starch, thereby providing a direct link between known changes in the plant phenotype and observed changes in starch granule size and morphology. Such changes, if any, might also be linked to observed variation in phytolith morphology, which has been associated with increases in the size of plant cells from wild to domesticated varieties (e.g., Piperno and Stothert 2003). By comparing the known changes in morphology and size of granules in known trajectories from wild to domesticated, one could test whether changes in granule size or morphology are likely to be sensitive to change in domesticated varieties.

Another approach to improving methods for observing and describing starch granules is to explore more thoroughly the utility of other types of illumination such as reflected light, darkfield illumination, phase contrast, and differential interference contrast. It would also be useful to assess other types of instruments such as the confocal laser and fluorescent microscopes because these tools could provide different views of starch granules and potentially illuminate different discriminatory traits. As these techniques can resolve variations in surface or internal features at very different magnifications, the coordinated use of images derived from the Scanning Electron Microscope together with light microscopy might also aid the identification of certain types of starch granules. We hope that in the future experimentation and exploration of new techniques will be a constant feature of ancient starch research.

It will also remain important to integrate starch analysis with other recordings of artifact residues such as cellular tissue, including fibres (e.g., Barton and White 1993) and xylem (e.g., Horrocks et al. 2004a), as well as other types of plant microfossil (e.g., pollen and phytoliths). This type of recording will increase the strength and detail of the overall interpretation of artifact and/ or site function.

INTELLECTUAL PATHWAYS

In recent years ancient starch research has moved on quickly and been successful because of productive communication and collaboration among many of the widely scattered researchers who have been pushing the field in new directions. Most of the basic groundwork for this new field has been completed and routine studies are now progressing. As identified in each of the chapters, however, there are still a number of areas that would benefit from future research. Solutions to these and other innovations require additional 'blue sky,' exploratory research. In order to convert the ideas into practical and reliable methods, considerable time and effort will need to be devoted to experimentation, often lengthy bouts of trial and error in the early stages. Much of the knowledge and skills required, for example starch chemistry and microscopy, are not common to normal archaeological training and practice but demand interdisciplinary collaboration. The future of ancient starch studies clearly lies in the continuation and enhancement of current intellectual pathways characterised by cooperation among active researchers together with new directions involving novel cross-disciplinary initiatives. We have found that being part of the development of ancient starch research has been invigorating, but we also eagerly anticipate the contributions to archaeology and environmental science that lie ahead.

REFERENCES

ABC Science Online 2004. Bloody tools tell hominids' tales. http://www.abc.net.au/science/news/stories/s1156792.htm. Accessed October 2005.

Acker, F., and B. Russell. 1999. Diatom Cleaning by Nitric Acid Digestion with a Microwave Apparatus. *Pennsylvania Academy of Natural Sciences, Patrick Center for Environmental Research,* Procedure No. P-13-42.

Albrecht, D. 1993. *Collecting and Preserving Herbarium Specimens*. Melbourne: National Herbarium of Victoria, Royal Botanic Gardens, South Yarra.

Anderson, P. 1980. A testimony of prehistoric tasks: Diagnostic residues on stone tool working edges. *World Archaeology* 12: 181–93.

Andrews, R., D. C. Coleman, J. E. Ellis, and J. S. Singh. 1974. Energy flow relationships in a shortgrass prairie system. In *Proceedings of the First International Congress of Ecology*, 22–28. Wangeningen: Centre for Agricultural Publication and Documentation.

Anto, J. M., J. Sunyer, C. E. Reed, J. Sabria, F. Martinez, F. Morell, R. Codina, R. Rodriguez-Roisin, M. J. Rodrigo, J. Roca, and M. Saez. 1989. Preventing asthma epidemics due to soybeans by dust control measures. *New England Journal of Medicine* 329: 1807–8.

Arnold, G. 1959. *Longhouse and Jungle*. London: Chatto and Windus.

Atchison, J. 1994. Analysis of food starch residues at the Jinmium Archaeological site, Northern Territory. Unpublished B. Env. Sc. (Hons) thesis, University of Wollongong, Wollongong.

Atchison, J., and R. Fullagar. 1998. Starch residues on pounding implements from Jinmium rock-shelter. In Fullagar 1998: 109–25.

Atchison, J., L. Head, and R. Fullagar. 2005. Archaeobotany of fruit seed processing in a monsoonal savanna environment: Evidence from the Keep River region, Northern Territory. *Journal of Archaeological Science* 32: 167–81.

Atkin, N. J., S. L. Cheng, R. M. Abeysekera, and A. W. Robards. 1999. Localisation of amylose and amylopectin in starch granules using enzyme-gold labelling. *Starch/Stärke* 51: 163–72.

Attenbrow, V., R. Fullagar, and C. Szpak. 1998. Stone files and shell fish-hooks in southeastern Australia. In Fullagar 1998: 127–48.

Babot, M. del P. 1999. Un estudio de artefactos de molienda. Casos del Formativo. Unpublished graduate thesis, Facultad de Ciencias Naturales e Instituto Miguel Lillo, Universidad Nacional de Tucumán, San Miguel de Tucumán, Argentina.

———. 2003. Starch grain damage as an indicator of food processing. In D. M. Hart and L. A. Wallis, eds., *Phytolith and Starch Research in the Australian-Pacific-Asian Regions: The State of the Art*, 69–81. Canberra: Pandanus Books, Terra Australis 19.

Backhouse, J. n.d. Recollections of the Botany of NSW (unpublished manuscript). Australian Joint Copying Project M749 roll 20. Sydney: State Library of New South Wales.

Badenhuizen, N. P. 1965. Occurrence and development of starch in plants. In R. L. Whistler and E. F. Paschall, eds., *Starch: Chemistry and Technology*, 65–104. London: Academic Press.

Baker, F., and P. N. Hobson. 1952. The selective staining of intact and damaged starch granules by Safranin O and Niagra blue 4B. *Journal of the Science of Food and Agriculture* 3: 608–12.

Baker, J. E., and S. M. Woo. 1992. Digestion of starch granules by alpha amylase. *Insect Biochemistry and Molecular Biology* 22: 529–37.

Baldwin, P. M., M. C. Davies, and C. D. Melia. 1997. Starch granule surface imaging using low-voltage scanning electron microscopy and atomic force microscopy. *International Journal of Biological Macromolecules* 21: 103–7.

Balick, M. J., and P. A. Cox. 1996. *Plants, People and Culture: The Science of Ethnobotany.* New York: Scientific American Library.

Ballard, C., P. Brown, R. M. Bourke, and T. Harwood, eds. 2005. *The Sweet Potato in Oceania: A Reappraisal*. Pittsburgh: Ethnology Monograph 19 and Sydney: Oceania Monograph 56.

Balme, J. 1991. The antiquity of grinding stones in semi-arid western New South Wales. *Australian Archaeology* 32: 3–9.

Balme, J., and W. Beck. 2002. Starch and charcoal: Useful measures of activity areas in archaeological rockshelters. *Journal of Archaeological Science* 29: 157–66.

Banks, W., and C. T. Greenwood. 1975. *Starch and Its Components*. Edinburgh: Edinburgh University Press.

Barfield, L. 1994. The Iceman reviewed. *Antiquity* 68: 10–26.

Barham, L. S., A. C. Pinto Llona, and C. B. Stringer. 2002. Bone tools from Broken Hill (Kwabe) cave, Zambia, and their evolutionary significance. *Before Farming* 2: 1–12.

Barker, G., H. Barton, P. Beavitt, M. Bird, P. Daly, C. Doherty, D. Gilbertson, C. Hunt, J. Krigbaum, H. Lewis, J. Manser, S. McClaren, V. Paz, P. Piper, B. Pyatt, R. Rabett, T. Reynolds, J. Rose, G. Rushworth, and M. Stephen. 2002. Prehistoric foragers and farmers in South-east Asia: Renewed investigations at Niah Cave, Sarawak. *Proceedings of the Prehistoric Society* 68: 147–64.

Barrau, J. 1958. *Subsistence Agriculture in Melanesia*. Bernice P. Bishop Museum Bulletin 219.

Barrau, J. 1965. Witnesses of the past: Notes on some food plants of Oceania. *Ethnology* 4: 289–94.

Barton, H. 1990. Raw material and tool function: A residue and use-wear analysis of artefacts from a Melanesian rock-shelter. Unpublished BA (Hons) thesis, Archaeology, University of Sydney.

———. 2005. The case for rainforest foragers: The starch record at Niah Cave, Sarawak. *Asian Perspectives* 44: 56–72.

Barton, H., and J. P. White. 1993. Use of stone and shell artefacts at Balof 2, New Ireland, Papua New Guinea. *Asian Perspectives* 32: 169–81.

Barton, H., R. Torrence, and R. Fullagar. 1998. Clues to stone tool function reexamined: Comparing starch grain frequencies on used and unused obsidian artifacts. *Journal of Archaeological Science* 25: 1231–38.

Baur, X., and D. Jager. 1990. Airborne antigens from latex gloves. *Lancet* 335: 912.

Baxter, R. 1964. Paleozoic starch in fossil seeds from Kansas coal balls. *Transactions of the Kansas Academy of Science* 67: 418–22.

Bayliss-Smith, T. 1996. People-plant interactions in the New Guinea highlands: Agricultural hearthland or horticultural backwater? In Harris 1996: 499–523.

Bayliss-Smith, T., and J. Golson. 1999. The meaning of ditches: Deconstructing the social landscapes of New Guinea, Kuk, Phase 4. In C. Gosden and J. Hather, eds., *The Prehistory of Food: Appetites for Change*, 199–231. London: Routledge.

Beck, W. 1985. Technology, toxicity and subsistence, a study of Australian Aboriginal plant food processing. Unpublished PhD thesis, LaTrobe University, Melbourne.

———. 1989. The taphonomy of plants. In W. Beck, A. Clarke, and L. Head, eds., *Plants in Australian Archaeology*, 31–53. University of Queensland, Tempus 1.

———. 1992. Aboriginal preparation of *Cycas* seeds in Australia. *Economic Botany* 46: 133–47.

Beck, W., and R. Fullagar. 1985. Starch granule analysis: Potential applications in Australian archaeology. Poster presented to the Australian Archaeological Association annual conference, Valla, New South Wales, December 2–6.

Becker, W. M., L. J. Kleinsmith, and J. Hardin. 2003. *Guide to Microscopy*. San Francisco: Benjamin Cummings.

Bedford, D., and T. James. 1995. *Collection, Preparation and Preservation of Plant Specimens*. Sydney: National Herbarium of New South Wales, Royal Botanic Gardens.

Berlin, B. 1992. *Ethnobiological Classification: Principles of Categorisation of Plants and Animals in Traditional Societies*. Princeton: Princeton University Press.

Berndt, R. M. 1951. Ceremonial exchange in Western Arnhem Land. *Southwestern Journal of Anthropology* 7: 156–76.

Beveridge, P. 1865. A few notes on the dialects, habits, customs and mythology of the lower Murray Aborigines. *Transactions of the Royal Society of Victoria* 6: 14–24.

———. 1883. Of the Aborigines inhabiting the great lacustrine and riverine depression of the lower Murray, lower Lachlan and lower Darling. *Journal and Proceedings of the Royal Society of New South Wales* 17: 19–24.

———. 1889. *The Aborigines of Victoria and Riverina*. Melbourne: Hutchinson.

Bhat, U. R., S. V. Paramahans, and R. M. Tharanathan. 1983. Scanning electron microscopy of enzyme-digested starch granules. *Starch/Stärke* 35: 261–65.

Binford, L., and S. Binford. 1969. Stone tools and human behaviour. *Scientific American* 220: 70–84.

Blanshard, J. M. V. 1987. Starch granule structure and function: A physiochemical approach. In T. Galliard, ed., *Starch: Properties and Potential*, 16–54. New York: John Wiley and Sons.

Boardman, S., and G. E. M. Jones. 1990. Experiments on the effects of charring on cereal plant components. *Journal of Archaeological Science* 17: 1–11.

Boot, P. G. 1993. Analysis of resins and other plant residues on stone artefacts from Graman, New South Wales. In B. L. Fankhauser and J. R. Bird, eds.,

Archaeometry: Current Australasian Research, 3–12. Canberra: Research School of Pacific Studies, Australian National University, Occasional Papers in Prehistory 22.

Borchardt, L. 1932. Ein Brot. *Zeitschrift für Ägyptische Sprache und Altertumskunde* 68: 73–79.

Borgstrom, G. 1973. *World Food Resources*. Aylesbury: Intertext.

Bowden, B. F., and B. Reynolds. 1982. The Chromatographic Analysis of Ethnographic Resins. *Australian Institute of Aboriginal Studies Newsletter* 17: 41–42.

Bowdery, D. 1989. Phytolith analysis: Introduction and applications. In W. Beck, A. Clarke, and L. Head, eds., *Plants in Australian Archaeology*, 161–96. University of Queensland, Tempus 1.

Boyd, W. E., C. J. Lentfer, and R. Torrence. 1998. Phytolith analysis for a wet tropics environment: Methodological issues and implications for the archaeology of Garua Island, West New Britain. *Palynology* 22: 213–28.

Bradbury, J. H., and W. D. Holloway. 1988. *Chemistry of Tropical Root Crops: Significance for Nutrition and Agriculture in the Pacific*. Canberra: Australian Centre for International Agricultural Research.

Bradbury, S., and B. Bracegirdle. 1998. *Introduction to Light Microscopy*. Oxford: Bios Scientific Publishers Limited.

Brasier, M. D. 1980. *Microfossils*. London: Allen and Unwin.

Brass, L. 1993. A residue and use-wear analysis of two stone assemblages from the New Guinea Highlands. Unpublished BA (Hons) thesis, Archaeology, University of Sydney.

———. 1998. Modern stone tool use as a guide to prehistory in the New Guinea Highlands. In Fullagar 1998: 19–28.

Bridson, D., and L. Forman. 1992. *The Herbarium Handbook*. Revised edition. Kew: Royal Botanic Gardens.

Briuer, F. L. 1976. New clues to stone tool function: Plant and animal residues. *American Antiquity* 41: 478–84.

Brown, I. 1996. Complex carbohydrates and resistant starch. *Nutrition Reviews* 54:115–19.

Brown, P. 1988. Residue analysis of stone artefacts from Yambon, West New Britain. Unpublished BA (Hons) thesis, Archaeology, University of Sydney.

Buléon, A., P. Colonna, V. A. Planchot, and S. Ball. 1998. Starch granules: Structure and biosynthesis. *International Journal of Biological Macromolecules* 23: 85–112.

Bye, R. A., Jr. 1986. Voucher specimens in ethnobiological studies and publications. *Journal of Ethnobiology* 6: 1–8.

Canti, M. G. 1998. The micromorphological identification of faecal spherulites from archaeological and modern materials. *Journal of Archaeological Science* 25: 435–44.

———. 1999. The production and preservation of faecal spherulites: Animals, environment and taphonomy. *Journal of Archaeological Science* 26: 251–8.

Chandrashekar, A., K. Somashekar, and A. Savitri. 1987. Optical interference studies on single starch granules. *Starch/Stärke* 39: 195–97.

Chikwendu, V. E., and C. E. A. Okezie. 1989. Factors responsible for the ennoblement of African yams: Inferences from experiments in yam domestication. In Harris and Hillman 1989: 344–57.

Christensen, H. 2002. *Ethnobotany of the Iban and Kelabit*. Forest Department, Sarawak, Malaysia and NEPCon, Denmark and University of Aarhus, Denmark.

Clapp, C. E., M. H. Hayes, and R. S. Swift. 1993. Isolation, fractionation, functionalities, and concepts of structures of soil organic macromolecules. In A. J. Beck, K. C. Jones, M. H. B. Hayes, and U. Mingelgrin, eds., *Organic Substances in Soil and Water: Natural Constituents and Their Influences on Contaminant Behaviour*, 31–69. Cambridge: The Royal Society of Chemistry.

Coil, J., M. A. Korstanje, S. Archer, and C. A. Hastorf. 2003. Laboratory goals and considerations for multiple microfossil extraction in archaeology. *Journal of Archaeological Science* 30: 991–1008.

Conway, R. J. 2000. Seed-grinders from semi-arid N. S. W.: an analysis of distributional variation based on comparison of millstones from the north far-western plains and the western slopes, incorporating morphological, petrological, use-wear and residue analysis. Unpublished BA (Hons) thesis, Archaeology, University of Sydney.

Cordain, L., J. Brand Miller, S. Boyd Eaton, N. Mann, S. H. A. Holt, and J. D. Speth. 2000. Plant-animal subsistence ratios and macronutrient energy estimations in worldwide hunter-gatherer diets. *American Journal of Clinical Nutrition* 71: 682–92.

Cortella, A. R., and M. L. Pochettino. 1994. Starch granule analysis as a microscopic diagnostic feature in the identification of plant material. *Economic Botany* 48: 171–81.

Cortella, A. R., and M. L. Pochettino. 1995. Comparative morphology of starch of three Andean tubers. *Starch/ Stärke* 47, 455–61.

Cotton, C. M. 1996. *Ethnobotany: Principles and Applications*. Chichester: John Wiley and Sons.

Coursey, D. G. 1973. Hominid evolution and hypogeous plant foods. *Man* 8: 634–35.

Coutts, P. J. F., F. Henderson, and R. Fullagar. 1979. A preliminary investigation of Aboriginal Mounds in Northwestern Victoria. *Records of the Victoria Archaeological Survey* 9.

Cribb, A. B., and J. W. Cribb. 1987. *Wild Food in Australia*. Sydney: Collins Australia.

Crowe, A. 1981. *A Field Guide to the Native Edible Plants of New Zealand*. Wellington: W. Collins.

Crowther, A. 2001. Pots, plants and Pacific prehistory: Residue analysis of plain Lapita pottery from Anir, New Ireland, c. 3300 BP. Unpublished BA (Hons) thesis, School of Social Science, University of Queensland.

———. 2005. Starch residues on undecorated Lapita pottery from Anir, New Ireland. *Archaeology in Oceania* 41: 62–66.

Cummings, J. H., and H. N. Englyst. 1995. Gastrointestinal effects of food carbohydrate. *American Journal of Clinical Nutrition* 61 (suppl.): 938S–45S.

Cunningham, G. M., W. E. Mulham, P. L. Milthorpe, and J. H. Leigh. 1992. *Plants of Western New South Wales*. Melbourne: Inkata Press.

Dawson, J. 1881. *Australian Aborigines*. (Facsimile, 1981). Canberra: Australian Institute of Aboriginal Studies.

de Bignicourt, M.-C., and F. Flieder. 1996. L'analyse des papyrus. *ICOM Committee for Conservation, 11th Triennial meeting, Preprints* II, 488–93. London: James and James (Science Publishers).

Denham, T. P., S. G. Haberle, C. Lentfer, R. Fullagar, J. Field, M. Therin, N. Porch, and B. Winsborough. 2003. Origins of agriculture at Kuk Swamp, in the Highlands of New Guinea. *Science* 301: 189–93.

Denham, T., S. Haberle, and C. Lentfer. 2004a. New evidence and revised interpretations of early agriculture in Highland New Guinea. *Antiquity* 78: 839–57.

Denham, T. P., J. Golson, and P. J. Hughes. 2004b. Reading early agriculture at Kuk Swamp, Wahgi Valley, Papua New Guinea: The archaeological features (Phases 1–3). *Proceedings of the Prehistoric Society* 70: 259–97.

Dennell, R. 1976. The economic importance of plant resources represented on archaeological sites. *Journal of Archaeological Science* 3: 229–47.

Diamond, J. 1997. *Guns, Germs and Steel: The Fates of Human Societies*. New York: W. W. Norton & Co.

Dimbleby, G. W. 1985. *The Palynology of Archaeological Sites*. New York: Academic Press.

Dirar, H. A. 1993. *Indigenous Fermented Foods of the Sudan: A Study in African Food and Nutrition*. Wallingford, UK: CAB International.

Donald, A. M., T. A. Waigh, P. J. Jenkins, M. J. Gidley, M. Debet, and A. Smith. 1997. Internal structure of starch granules revealed by scattering studies. In P. J. Frazier, P. Richmond, and M. Donald, eds., *Starch: Structure and Functionality*, 172–79. London: Royal Society of Chemistry, Special Publications No. 205.

Dreher, M. L., C. J. Dreher, and J. W. Berry. 1984. Starch digestibility of foods: A nutritional perspective. *CRC Critical Reviews in Food Science and Nutrition*. 20: 47–71.

Dziezak, J. D. 1988. Microscopy and image analysis for R&D. *Food Technology* 42: 110–24.

Edwards, D., and J. F. O'Connell. 1995. Broad spectrum diets in arid Australia. *Antiquity* 69: 769–83.

Egg, M., R. Goedecker-Ciolek, W. Groenman-van Waateringe, and K. Spindler. 1995. Die Gletschermumie vom Ende der Steinzeit aus den Ötzaler Alpen. *Jahrbuch des Römisch-Germanischen Zentralmuseums Mainz* 1992 (39): 1–128.

Eglinton, G., and G. A. Logan. 1991. Molecular Preservation. *Philosophical Transactions of the Royal Society of London* B 333: 315–28.

Englyst, H. N., and J. H. Cummings. 1987. Resistant starch, a new food component: A classification of starch for nutritional purposes. In I. D. Morton, ed., *Cereals in a European Context*, 221–33. First European Conference on Food Science and Technology. Chichester, England: Ellis Horwood Ltd.

Englyst, H. N., and G. T. Macfarlane. 1986. Breakdown of resistant and readily digestible starch by human gut bacteria. *Journal of the Science of Food and Agriculture* 37: 699–706.

Englyst, H. N., S. M. Kingman, and J. H. Cummings. 1992. Classification and measurement of nutritionally important starch fractions. *European Journal of Clinical Nutrition* 46: S33–S50.

Evershed, R. P., R. Berstan, F. Grew, M. S. Copley, A. J. H. Charmant, E. Barham, H. R. Mottram, and G. Brown. 2004. Formulation of a Roman cosmetic. *Nature* 432: 35–36.

Eyre, E. J. 1845. *Journals of Expeditions of Discovery into Central Australia and Overland to King George's Sound in the Years 1840–1*. Two volumes. London: Boone.

Faisant, N., A. Buléon, P. Colonna, C. Molis, S. Lartigue, J. P. Galmiche, and M. Champ. 1995a. Digestion of raw banana starch in the small intestine of healthy humans: Structural features of resistant starch. *British Journal of Nutrition* 73: 111–23.

Faisant, N., D. J. Gallant, B. Bouchet, and M. Champ. 1995b. Banana starch breakdown in the human small intestine studied by electron microscopy. *European Journal of Classical Nutrition* 49: 98–104.

Fannon, J. E., R. J. Hauber, and J. N. BeMiller. 1992. Surface pores of starch granules. *Cereal Chemistry* 69: 284–8.

FAO (Food and Agriculture Organization of the United Nations). 1990. *Roots, Tubers, Plantains and Bananas in Human Nutrition*. Rome: FAO.

Fash, W. 1986. History and characteristics of settlement in the Copán Valley, and some comparisons with Quirigua. In P. A. Urban and E. M. Schortman, eds., *The Southeast Maya Periphery*, pp. 72–93. Austin: University of Texas Press.

Field, J., and J. Dodson. 1999. Late Pleistocene Megafauna and Archaeology from Cuddie Springs, southeastern Australia. *Proceedings of the Prehistoric Society* 65: 1–27.

Field, J., and R. Fullagar. 1998. Grinding and pounding stones from Cuddie Springs and Jinmium. In Fullagar 1998: 95–108.

Fitt, L. E., and E. M. Snyder. 1984. Photomicrographs of starches. In Whistler, BeMiller, and Paschall 1984: 675–89.

Flieder F., E. Delange, A. Duval, and M. Leroy. 2001. Papyrus: The need for analysis. *Restaurator* 22: 84–106.

Flint, O. 1994. *Food Microscopy*. Oxford: BIOS Scientific Publishers Ltd for the Royal Microscopy Society.

Flood, J. 1995. *Archaeology of the Dreamtime*. Sydney: Angus and Robertson.

Fox, A., C. Heron, and M. Sutton. 1995. Characterisation of natural products on Native American archaeological and ethnographic material from the Great Basin Region, USA: A preliminary study. *Archaeometry* 37: 363–75.

Franco, C. M. L., and C. F. Ciacco. 1992. Factors that affect the enzymatic degradation of natural starch granules—effect of the size of the granules. *Starch/Stärke* 44: 422–26.

Fredriksson, H. 1998. Characterisation of Starch from Different Sources. Unpublished PhD thesis, Swedish University of Agricultural Sciences, Uppsala, Sweden.

French, B. 1978. *Food Crops of Papua New Guinea*. Madang: Kristen Press.

French, D. 1973. Chemical and physical properties of starch. *Journal of Animal Science* 37: 1048–61.

———. 1984. Organization of starch granules. In Whistler, BeMiller, and Paschall 1984: 183–247.

Fritzsche, C. J. 1834. Ueber das Amylum. *Annalen der Physik und Chemie* 32: 129–44.

Fullagar, R. 1986. Use-wear and residues on stone tools: Functional analysis and its application to two southeastern Australian archaeological assemblages. Unpublished PhD thesis, LaTrobe University, Melbourne.

———. 1988. Recent developments in Australian use-wear and residue studies. In S. Beyries, ed., *Industries Lithiques: Traceologie et Technologie*, 133–45. British Archaeological Reports International Series 411.

———. 1989. The potential of lithic use-wear and residue studies for determining stone tool functions. In P. P. Gorecki and D. Gillieson, eds., *A Crack in the Spine: Prehistory and Ecology of the Jimi-Yuat Valley, Papua New Guinea*, 209–23. Queensland: James Cook University, Townsville.

———. 1991. The role of silica in polish formation. *Journal of Archaeological Science* 18: 1–25.

———. 1992. Lithically Lapita: Functional analysis of flaked stone assemblages from West New Britain Province, Papua New Guinea. In J.-C. Galipaud, ed., *Poterie Lapita et Peuplement*, 135–43. Noumea: ORSTOM.

———. 1993a. Flaked stone tools and plant food production: A preliminary report on obsidian tools from Talasea, West New Britain, Papua New Guinea. In P. Anderson, S. Beyries, M. Otte, and H. Plisson, eds., *Traces et Fonction: Les Gestes Retrouves*, 331–37. Colloque Internationale de Liege, Editions ERAUL, 50.

———. 1993b. Taphonomy and tool-use: a role for phytoliths in use-wear and residue analysis. In B. Fankhauser and R. Bird, eds., *Archaeometry: Current Australasian Research*, 21–27. Canberra: Department of Prehistory, Research School of Pacific Studies, Occasional Papers in Prehistory 22.

———, ed. 1998. *A Closer Look: Recent Australian Studies of Stone Tools*. Sydney: Archaeological Computing Laboratory, University of Sydney.

———. 2005. Residues and use-wear. In J. Balme and A. Paterson, eds., *Archaeology in Practice: A Student Guide to Archaeological Analysis*, Chapter 7. Blackwell Publishing, Oxford.

Fullagar, R., and B. David. 1997. Investigating changing attitudes towards an Australian Aboriginal dreaming mountain over >37,000 years of occupation via residue and use-wear analyses of stone artefacts. *Cambridge Archaeological Journal* 7: 139–44.

Fullagar, R., and J. Field. 1997. Pleistocene seed grinding implements from the Australian arid zone. *Antiquity* 71: 300–7.

Fullagar, R., and L. Head. 1998. Aboriginal landscapes of the northwest Northern Territory, Australia. In R. Layton, P. Ucko, and D. Austin, eds., *Frontiers of Landscape Archaeology*, pp. 322–35. London: Routledge.

Fullagar, R., with R. Jones. 2004. Use-wear and residue analysis of stone artefacts from the Enclosed Chamber, Rocky Cape, Tasmania. *Archaeology in Oceania* 39: 79–93.

Fullagar, R., D. Price, and L. Head. 1996. Early human occupation of northern Australia: Stratigraphy and dating of the Jinmium rockshelter, Northern Territory. *Antiquity* 70: 751–73.

Fullagar, R., T. Loy, and S. Cox. 1998. Starch grains, sediments and stone tool function: Evidence from Bitokara, Papua New Guinea. In Fullagar 1998: 49–60.

Fullagar, R., B. Meehan, and R. Jones. 1992. Residue analysis of ethnographic plant-working and other tools from northern Australia. In P. Anderson-Gerfaud, ed., *Préhistoire de l'Agriculture: Nouvelles Approches Experimentales et Ethnographique*, 39–53. Paris: Monographie du CRA no 6, ed. du CNRS.

Fullagar, R., J. Field, T. Denham, and C. Lentfer. 2005. Early and Mid-Holocene tool use and processing of taro (*Colocasia esculenta*), yam (*Dioscorea* sp.) and other plants at Kuk Swamp in the Highlands of Papua New Guinea. *Journal of Archaeological Science* 32 (in press).

Fullagar, R., J. Field, and L. Kealhofer. In press. Grinding stones and seeds of change: Starch and phytoliths as evidence of plant food processing. In Y. M. Rowan and J. R. Ebling, eds., *New Approaches to Old Stones: Recent Studies of Ground Stone Artefacts*. London: Equinox.

Fullagar, R., J. Furby, and B. Hardy. 1996. Residues on stone artefacts: State of a scientific art. *Antiquity* 70: 740–45.

Furby, J. 1995. Megafauna under the microscope: Archaeology and palaeoenvironment at Cuddie Springs. Unpublished PhD thesis, University of New South Wales.

Furby, J. H., R. Fullagar, J. R. Dodson, and I. Prosser. 1993. The Cuddie Springs bone bed revisited, 1991. In M. A. Smith, M. Spriggs, and B. Fankhauser, eds., *Sahul in Review*, 204–10. Canberra: Department of Prehistory, Australian National University, Occasional Papers in Prehistory 24.

Gallant, D. J., H. Bewa, Q. H. Buy, B. Bouchet, O. Szylit, and L. Sealy. 1982. On ultrastructural and nutritional aspects of some tropical tuber starches. *Starch/Stärke* 34: 255–62.

Gallant, D. J., B. Bouchet, A. Buléon, and S. Pérez. 1992. Physcial characteristics of starch granules and susceptibility to enzymatic degradation. *European Journal of Clinical Nutrition* 46: S3–S16.

Gallant, D. J., B. Bouchet, and P. M. Baldwin. 1997. Microscopy of starch: Evidence of a new level of granule organisation. *Carbohydrate Polymers* 32: 177–91.

Galliard, T., and P. Bowler. 1987. Morphology and composition of starch. In T. Galliard, ed., *Starch: Properties and Potential*, 55–79. Critical Reports on Applied Chemistry, volume 13. Brisbane: John Wiley and Sons.

Gaillard, M. J., H. J. B. Birks, U. Emanuelsson, and B. E. Beglund. 1992. Modern pollen/land use relationships as an aid in the reconstruction of past land uses and cultural landscapes: An example from Southern Sweden. *Vegetation History and Archaeobotany* 1: 3–17.

Gardiner, A. H. 1947. *Ancient Egyptian Onomastica*. Oxford: Oxford University Press.

Gero, J. M. 1991. Gender lithics: Women's roles in stone tool production. In J. M. Gero and M. W. Conkey, eds., *Engendering Archaeology: Women and Prehistory*, 163–93. Oxford: Basil Blackwell.

Gibson, J. L. 1996. *Poverty Point: A Terminal Archaic Culture of the Lower Mississippi Valley*. Second edition. Baton Rouge: Louisiana Archaeological Survey and Antiquities Commission.

Gijn, A. van. 1990. *The Wear and Tear of Flint*. Leiden: University of Leiden, Analecta Praehistorica Leidensia 22.

Golchin, A., J. A. Baldock, and J. M. Oades. 1998. A model linking organic matter decomposition, chemistry, and aggregate dynamics. In R. Lal, J. M. Kimble, R. F. Follett, and B. A. Stewart, eds., *Soil Processes and the Carbon Cycle*, 245–66. New York: CRC Press.

Goldstein, D. J. 1999. *Understanding the Light Microscope*. London: Academic Press.

Golson, J. 1977. No room at the top: Agricultural intensification in the New Guinea Highlands. In J. Allen, J. Golson, and R. Jones, eds., *Sunda and Sahul: Prehistoric Studies in Southeast Asia, Melanesia and Australia*, 601–38. London: Academic Press.

———. 1981. New Guinea agricultural history: A case study. In D. Denoon and C. Snowden, eds., *A Time to Plant and a Time to Uproot: A History of Agriculture in Papua New Guinea*, 484–91. Port Moresby: Institute of Papua New Guinea Studies.

———. 1982. The Ipomoean Revolution revisited: Society and the sweet potato in the upper Wahgi valley. In A. Strathern, ed., *Inequality in New Guinea Highlands Societies*, 109–36. Cambridge: Cambridge University Press.

Golson, J., and D. Gardner. 1990. Agriculture and sociopolitical organisation in New Guinea Highlands prehistory. *Annual Review of Anthropology* 19: 395–417.

Gorecki, P. P., D. R. Horton, N. Stern, and R. V. S. Wright. 1984. Coexistence of humans and megafauna in Australia: Improved stratified evidence. *Archaeology in Oceania* 19, 117–19.

Gott, B. 1983. Murnong—*Microseris scapigera*: A study of a staple food of Victorian Aborigines. *Australian Aboriginal Studies* 1983/1: 2–17.

———. 1989. The uses of ethnohistory and ecology. In W. Beck, A. Clarke, and L. Head, eds., *Plants in Australian Archaeology*, 197–210. University of Queensland, Tempus 1.

———. 1997. VICUSE database: Victorian Plants utilised by Aborigines. Unpublished manuscript.

———. 1999. Cumbungi, Typha species: A staple Aboriginal food in southern Australia. *Australian Aboriginal Studies* 1999/1: 33–50.

———. 2000. NSWUSE. Ethnobotanical database for New South Wales. Unpublished manuscript.

Gough, B. M., and J. N. Pybus. 1971. Effect on the gelatinzation temperature of wheat starch granules of prolonged treatment with water at 50°C. *Starch/Stärke* 23: 210–12.

Greenish, H. G., and E. Collin. 1904. *An Anatomical Atlas of Vegetable Powders*. London: J & A. Churchill.

Gregg, S. A. 1988. *Foragers and Farmers: Population Interaction and Agricultural Expansion in Prehistoric Europe*. Chicago: University of Chicago Press.

Gremillion, K. J., ed. 1997. *People, Plants and Landscapes, Studies in Paleoethnobotany*. Tuscaloosa: University of Alabama Press.

Grey, G. 1841. *Journals of Two Expeditions of Discovery in North-west and Western Australia.* London: T. & W. Boone.

Guggenberger, G., E. T. Elliot, S. D. Frey, J. Six, and K. Paustian. 1999. Microbial contributions to the aggregation of a cultivated grassland soil amended with starch. *Soil Biology and Biochemistry* 31: 407–19.

Guilbot, A., and C. Mercier. 1985. Starch. In G. O. Aspinall, ed., *The Polysaccharides*, 3, 209–82. London: Academic Press.

Gunther, R. T., ed. 1959. *The Greek Herbal of Dioscorides*. New York: Hafner.

Haase, N. U., and J. Plate. 1996. Properties of potato starch in relation to varieties and environmental factors. *Starch/Stärke* 48: 167–71.

Hall, D. M., and J. G. Sayre. 1973. A comparison of starch granules as seen by both scanning electron and ordinary light microscopy. *Starch/Stärke* 25: 119–23.

Hall, J., and R. Viel. 1998. The Formative of Copán, Honduras. Paper presented at the 63rd Annual Meeting of the Society for American Archaeology, Seattle.

Hall, J., S. Higgins, and R. Fullagar. 1989. Plant residues on stone tools. In W. Beck, A. Clarke, and L. Head, eds., *Plants in Australian Archaeology*, 136–60. St. Lucia: University of Queensland, Tempus 1.

Hallam, N., and D. J. Osborne. 1974. Fine structure and histochemistry of ancient viable and non-viable seeds. In J. V. Sanders and D. S. Goodchild, eds., *Eighth International Congress on Electron Microscopy, Abstracts of Papers* 2: 598–99. Canberra: Australian Academy of Science.

Hardy, B. L. 2004. Neanderthal behaviour and stone tool function at the Middle Palaeolithic site of La Quina, France. *Antiquity* 78: 547–65.

Harlan, J. R. 1989. Wild grass-seed harvesting in the Sahara and Sub-Sahara of Africa. In Harris and Hillman 1989: 79–98.

Harris, D. R. 1981. The prehistory of human subsistence: A speculative outline. In D. M. Walker and N. Kretchmer, eds., *Food, Nutrition and Evolution*, 15–35. New York: Masson.

———. 1984. Ethnohistorical evidence for the exploitation of wild grasses and forbs: Its scope and archaeological implications. In W. van Zeist and W. A. Casparie, eds., *Plants and Ancient Man, Studies in Palaeoethnobotany*, 63–70. Rotterdam: A. A. Balkema.

———. 1996. Introduction: Themes and concepts in the study of early agriculture. In D. R. Harris, ed., *The Origins and Spread of Agriculture and Pastoralism in Eurasia*, 1–11. London: UCL Press.

Harris, D. R., and G. C. Hillman, eds. 1989. *Foraging and Farming: the Evolution of Plant Exploitation*. London: Unwin Hyman.

Harrisson, B. 1967. A classification of stone age burials from Niah Great Cave, Sarawak. *Sarawak Museum Journal* 50 (30–1): 126–200.

Hart, D. M. 2003. The influence of soil fauna on phytolith distribution in an Australian soil. In D. M. Hart and L. A. Wallis, eds., *Phytolith and Starch Research in the Australian-Pacific-Asian Regions: The State of the Art*, 83–91. Canberra: Pandanus Books, Terra Australis 19.

Haslam, M. 1999. What a Dump: Use-Wear and Residue Analyses of Lithic Artefacts from Copan, Honduras. Unpublished BA (Hons) thesis, University of Queensland.

———. 2004. The decomposition of starch grains in soils: Implications for archaeological residue analysis. *Journal of Archaeological Science* 31: 1715–34.

———. 2005. Potential mididentification of in situ archaeological tool-residues: Starch and conidia. *Journal of Archaeological Science* in press.

Hastorf, C. 1999. Recent research in paleoethnobotany. *Journal of Archaeological Research* 7: 55–103.

Hastorf, C., and V. Popper. 1988. *Current Palaeoethnobotany: Analytical Methods and Cultural Interpretations of Plant Remains*. Chicago: University of Chicago Press.

Hather, J. G. 1991. The identification of charred archaeological remains of vegetative parenchymatous tissues. *Journal of Archaeological Science* 18: 661–75.

———. 1992. The archaeobotany of subsistence in the Pacific. *World Archaeology* 24: 70–81.

———. 1993. *An Archaeological Guide to Root and Tuber Identification. Volume 1, Europe and South West Asia*. Oxford: Oxbow Monograph 28.

———. 1994a. A morphological classification of roots and tubers and its bearing on the origins of agriculture in southwest Asia and Europe. *Journal of Archaeological Science* 21: 719–24.

———. 1994b. The identification of charred root and tuber crops from archaeological sites in the Pacific. In J. G. Hather, ed., *Tropical Archaeobotany*, 51–65. London: Routledge.

Hather, J. G., and P. V. Kirch. 1991. Prehistoric sweet potato (*Ipomoea batatas*) from Mangaia Island, central Polynesia. *Antiquity* 65: 887–93.

Hawkes, K., J. F. O'Connell, and N. G. Blurton Jones. 1991. Hunting income patterns among the Hadza: Big game, common goods, foraging goals, and the evolution of the human diet. *Philosophical Transactions of the Royal Society of London*, B, 334: 243–51.

Hawkes, K., J. F. O'Connell, N. G. Blurton Jones, E. L. Charnov, and H. Alvarez. 1998. Grandmothering, menopause, and the evolution of human life histories. *Proceedings of the National Academy of Sciences (USA)* 95: 1336–39.

Hay, A. 1996. A new Bornean species of *Colocasia* Schott (Araceae-Colocasieae), with a synopsis of the genus in Malesia and Australia. *Sandakania* 7: 31–48.

Hayden, B., ed. 1979. *Lithic Use-wear Analysis*. London: Academic Press.

Head, L. 1989. Pollen. In W. Beck, A. Clarke, and L. Head, eds., *Plants in Australian Archaeology*, pp. 111–34. University of Queensland, Tempus 1.

Head, L., and R. Fullagar. 1991. Aboriginal land-use and resource management around a pastoral excision—Marralam, N.T. *Australian Aboriginal Studies* 1991/1: 39–52.

Helms, R. 1895. Anthropological Notes. *Proceedings of the Linnaean Society of NSW*, ser. 2, 20: 387–407.

Higgins, S. 1988. Starch granule differentiation on archaeological residues: A feasibility study. Unpublished BA (Hons) thesis. Department of Anthropology and Sociology, University of Queensland.

Hillman, G. 1984. Interpretation of archaeological plant remains: The application of ethnographic models from Turkey. In W. van Zeist and W. A. Casparie, eds., *Plants and Ancient Man, Studies in Palaeoethnobotany*, 1–42. Rotterdam: A. A. Balkema.

Hinojosa, M. 2000. Occupational asthma in bakers: Role of enzymes. *Revista Espanol de Alergol Immunol Clinica* 15: 32–42.

Hodgkinson, C. 1845. *Australia from Port Macquarie to Moreton Bay*. London: Boone.

Hoover, R. 2001. Composition, molecular structure, and physicochemical properties of tuber and root starches: A review. *Carbohydrate Polymers* 45: 253–67.

Horrocks, M. 2004. Polynesian plant subsistence in prehistoric New Zealand: A summary of the microfossil evidence. *New Zealand Journal of Botany* 42: 321–34.

———. 2005. A combined procedure for recovering phytoliths and starch residues from soils, sedimentary deposits and similar materials. *Journal of Archaeological Science* 32: 1169–75.

Horrocks, M., and S. Bedford. 2005. Microfossil analysis of Lapita deposits in Vanuatu reveals introduced Araceae (aroids). *Archaeology in Oceania* 40: 67–74.

Horrocks, M., S. L. Nichol, M. D. Jones, P. A. Shane, and D. G. Sutton. 2002. Spatial variation in pollen and charcoal records in relation to the 665 year B. P. Kaharoa tephra at Harataonga Bay, Great Barrier Island, northern New Zealand: Preliminary results, 155–70. In M. Jones and P. Sheppard, eds., *Australasian Connections and New Directions*. Department of Anthropology, University of Auckland, Auckland.

Horrocks, M., G. Irwin, M. Jones, and D. Sutton. 2004a. Starch grains and xylem cells of sweet potato (*Ipomoea batatas*) and bracken (*Pteridium esculentum*) in archaeological deposits from northern New Zealand. *Journal of Archaeological Science* 31: 251–58.

Horrocks, M., P. A. Shane, I. G. Barger, D. M. D'Costa, and S. L. Nichol. 2004b. Microbotanical remains reveal Polynesian agriculture and mixed cropping in early New Zealand. *Review of Palaeobotany and Palynology* 131: 147–57.

Hosking, J. R., G. R. Sainty, and S. W. L. Jacobs. 1996. Certainty and uncertainty in plant identification. In R. C. H. Shepherd, ed., *Eleventh Australian Weeds Conference Proceedings*, 464–67. Melbourne: Weeds Society of Victoria.

Huber, K. C., and J. N. BeMiller. 1997. Visualization of channels and cavities of corn and sorghum starch granules. *Cereal Chemistry* 74: 537–41.

Humphries, G. S., D. M. Hart, N. A. Simons, and R. J. Field. 2003. Phytoliths as indicators of process in soils. In D. M. Hart and L. A. Wallis, eds., *Phytolith and Starch Research in the Australian-Pacific-Asian Regions: The State of the Art*, 93–104. Canberra: Pandanus Books, Terra Australis 19.

Hurcombe, L. M. 1992. *Use Wear Analysis and Obsidian: Theory, Experiments and Results*. Sheffield: University of Sheffield, Sheffield Archaeological Monographs 4.

Hunt, W. H. 1977. A simulation model for decomposition in grasslands. *Ecology* 58: 469–84.

Hunziker, A. 1952. *Los Pseudocereales de la Agricultura Indígena de América*. Buenos Aires: Acme Agency.

Imshenetsky, A. A. 1968. Decomposition of cellulose in the soil. In T. R. G. Gray and D. Parkinson, eds., *The Ecology of Soil Bacteria*, 256–69. Liverpool: Liverpool University Press.

Incoll, L. D., G. D. Bonnett, and B. Gott. 1989. Fructans in the underground storage organs of some Australian plants used for food by Aborigines. *Journal of Plant Physiology* 134: 196–202.

Iriarte, J., I. Holst, O. Marozzi, C. Listopad, E. Alonso, A. Rinderknecht, and J. Montana. 2004. Evidence for cultivar adoption and emerging complexity during the mid-Holocene in the La Plata basin. *Nature* 432: 614–17.

Irwin, G. 2004. *Kohika: The Archaeology of a Late Maori Lake Village in the Ngati Awa Rohe, Bay of Plenty, New Zealand*. Auckland: Auckland University Press.

Isaac, G. L. 1967. Towards the interpretation of occupation debris: Some experiments and observations. *Kroeber Anthropological Society Papers* 37: 31–57.

Jackes, A. 1997. Spot the difference: Analyses of starch granules of the Dioscorea genus piloting a method in the search for differentiating and identifying features. Unpublished B. Env. Sci. thesis, University of Wollongong.

Jane, J-L., T. Kasemsuwan, S. Leas, H. Zobel, and J. F. Robyt. 1994. Anthology of starch granule morphology by scanning electron microscopy. *Starch/Stärke* 46: 121–29.

Jenkins, P. J., and A. M. Donald. 1995. The influence of amylose on starch granule structure. *International Journal of Biological Macromolecules*. 17: 315–21.

———. 1998. Gelatinisation of starch: a combined SAXA/WAXS/DSC and SANS study. *Carbohydrate Research* 308: 133–47.

Jenner, C. F. 1982. Storage of starch. In F. A. Loewus and W. Tanner, eds., *Encyclopaedia of Plant Physiology, Vol 13A, Plant Carbohydrates: Intracellular Carbohydrates*, 700–47. Berlin: Springer-Verlag.

Johansen, D.A. 1940. *Plant Microtechnique*. New York: McGraw–Hill.

Jones, G. E. M. 1984. Interpretation of archaeological plant remains: The application of ethnographic models from Greece. In W. van Zeist and W. A. Casparie, eds., *Plants and Ancient Man, Studies in Palaeoethnobotany*, 43–62. Rotterdam: A. A. Balkema.

———. 1987. A statistical approach to the archaeological identification of crop processing. *Journal of Archaeological Science* 14: 311–23.

Jones, R. A. 1994. The application of microwave technology to the oxidation of kerogen for use in palynology. *Review of Palaeobotany and Palynology* 80: 333–38.

———. 1998. Focused microwave digestion and oxidation of palynological samples. *Review of Palaeobotany and Palynology* 103: 17–22.

Jones, R., and S. J. Ellin. 1998. Improved palynological sample preparation using an automated focused microwave digestion system. In V. M. Bryant Jr. and J. H. Wrenn, eds., *New Developments in Palynomorph Sampling, Extraction, and Analysis*, 23–28. American Association of Stratigraphic Palynologists Inc., Contributions series 33. Dallas, Texas.

Jorgensen, B. B. 1983. Processes at the sediment-water interface. In B. Bolin and R. B. Cook, eds., *The Major Biogeochemical Cycles and Their Interactions*, 477–515. Chichester: Scope, John Wiley and Sons.

Juan-Tresserras, J. 1998. La cerveza prehistorica: Investigaciones arqueobotanicas y experimentales. In J. L. Maya, F. Cuesta, and J. L. Cachero, eds., *Geno: Un poblado del Bronce Final en el Bajo Segre (Lleida)*, 239–52. Barcelona: Universitat de Barcelona.

Juan-Tresserras, J., C. Lalueza, R. Albert, and M. Calvo. 1997. Identification of phytoliths from prehistoric human dental remains from the Iberian Peninsula and the Balearic Islands. In A. Pinilla, J. Juan-Tresserras and M. Machado, eds., *First European Meeting on Phytolith Research,* 197–203. Madrid: Centro de Ciencias Medioambientales del Consejo Superior de Investigaciones Cientificas.

Kamminga, J. 1979. The nature of use-polish and abrasive smoothing on stone tools. In Hayden 1979: 143–57.

———. 1982. *Over the Edge: Functional Analysis of Australian Stone Tools*. St. Lucia: University of Queensland, Anthropology Museum, Occasional Papers in Anthropology 12.

Kang, M. Y., Y. Sugimoto, I. Kato, S. Sakamoto, and F. Hidetsugu. 1985. Some properties of large and small starch granules of barley (*Hordeum vulgare* L.) endosperm. *Agricultural and Biological Chemistry* 49: 1291–97.

Kaplan, H., K. Hill, J. Lancaster, and A. Hurtado. 2000. A theory of human life history evolution: Diet, intelligence and longevity. *Evolutionary Anthropology* 9: 156–85.

Kealhofer, L., and D. Penny. 1998. A combined phytolith and pollen record for 14,000 years of vegetation change in northeast Thailand. *Review of Paleobotany and Palynology* 103: 83–93.

Kealhofer, L., R. Torrence, and R. Fullagar. 1999. Integrating phytoliths within use-wear/residue studies of stone tools. *Journal of Archaeological Science* 26: 527–46.

Keeley, L. 1980. *Experimental Determination of Stone Tool Use: A Microwear Analysis.* Chicago: University of Chicago Press.

Kelso, G. K. 1994. Pollen percolation rates in Euroamerican-era cultural deposits in the northeastern United States. *Journal of Archaeological Science* 21: 481–88.

Kelso, G. K., S. A. Mrozowski, D. Currie, A. C. Edwards, M. R. Brown III, A. J. Horning, G. J. Brown, and J. R. Dandoy. 1995. Differential pollen preservation in a seventeenth-century refuse pit, James Town Island, Virginia. *Historical Archaeology* 29: 43–54.

Kerr, P. 1959. *Optical Mineralogy*. New York: McGraw-Hill.

Kirch, P. V. 1997. *The Lapita Peoples: Ancestors of the Oceanic World.* Cambridge, Mass.: Blackwell.

Klein, R. G. 1999. *The Human Career: Human Biological and Cultural Origins.* 2nd ed. Chicago: University of Chicago Press.

Korstanje, M. A. 1996. Sobre la produccion agricolo-gandera en El Formativo: Reflexiones para el camino. *Actas del I Congreso de Investigacion Social: 'Region y sociedad en el NOA,'* 402–11. Tucuman: Faculdad de Filosofia y Letras (UNT).

———. 2003. Taphonomy in the laboratory: Starch damage and multiple microfossil recovery from sediments. In D. M. Hart and L. A. Wallis, eds., *Phytolith and Starch Research, in the Australian-Pacific-Asian Regions: The State of the Art*, 105–18. Canberra: Pandanus Books, Terra Australis 19.

Koyama, S. 1981. A quantitative study of wild food resources: An example from Hida. In S. Koyama and D. Thomas, eds., *Affluent Foragers, Pacific Coasts East and West*, 91–96. Osaka: Senri Ethnological Series No. 9.

Kraemer, H. 1907. *A Text-Book of Botany and Pharmacognosy*. Philadelphia: Lippincott.

Lamb, J., and T. Loy. 2005. Seeing red: The use of Congo Red dye to identify cooked and damaged starch grains in archaeological residues. *Journal of Archaeological Science* 32: 1433–40.

Leach, B., and J. Tait. 2000. Papyrus. In P. T. Nicholson and I. Shaw, eds., *Ancient Egyptian Materials and Technology*, 227–53. Cambridge: Cambridge University Press.

Leach, H. W. 1965. Gelatinization of starch. In R. L. Whistler and E. F. Paschall, eds., *Starch: Chemistry and Technology*, 289–307. New York: Academic Press.

Leach, H. W., and T. J. Schoch. 1961. Structure of the starch granule. *Cereal Chemistry* 38: 34–46.

Lee, R. B. 1979. *The !Kung San; Men, Women and Work in a Foraging Society*. Cambridge: Cambridge University Press.

Leeuwenhoek, A. 1719. *Epistolae Physiologicae super Compluribus Naturae Arcanus*. Epistolae 26.

Leffman, H., and W. Beam. 1905. *Select Methods in Food Analysis*. 2nd edition. London: Rebman, Ltd.

Lentfer, C. J. 2003. Plants, people and landscapes in prehistoric Papua New Guinea: A compendium of phytolith (and starch) analyses. Unpublished PhD thesis, Southern Cross University, Lismore.

Lentfer, C. J., and W. E. Boyd. 1998. A comparison of three methods for the extraction of phytoliths from sediments. *Journal of Archaeological Science* 25: 1159–83.

Lentfer, C. J., M. Therin, and R. Torrence. 2002. Starch grains and environmental reconstruction: A modern test case from West New Britain, Papua New Guinea. *Journal of Archaeological Science* 29: 687–98.

Lentfer, C. J., M. M. Cotter, and W. E. Boyd. 2003. Particle settling times for gravity sedimentation and centrifugation: A practical guide for palynologists. *Journal of Archaeological Science* 30: 149–68.

Lillford, P. J., and A. Morrison. 1997. Structure/function relationship of starches in food. In P. J. Frazier, P. Richmond, and M. Donald, eds., *Starch: Structure and Functionality*, 1–8. London: Royal Society of Chemistry, Special Publications No. 205.

Lineback, D. R. 1984. The starch granule: Organization and properties. *Bakers Digest* 58: 16–21.

Loy, T. H. 1990. Prehistoric organic residues: Recent advances in identification, dating, and their antiquity. In E. Pernicka, ed., *Archaeometry '90*, 645–56. Basel and Boston: Springer Verlag.

———. 1993. The artefact as site: An example of the biomolecular analysis of organic residues on stone tools. *World Archaeology* 25: 44–63.

———. 1994a. Methods in the analysis of starch residues on prehistoric stone tools. In J. G. Hather, ed., *Tropical Archaeobotany*, 86–114. London: Routledge.

———. 1994b. Analysis of the artefacts of the Ice Man of Hauslabjoch: Microscopy and residue analysis, Mainz. Innsbruck: Forschunginstitut für Alpine Vorzeit, Leopold-Franzens Universität. Manuscript report on file.

———. 1998a. Blood on the axe. *New Scientist* 2151: 40–43.

———. 1998b. The Ice Man. *Nature Australia* 26: 60–63.

Loy, T. H., and B. G. Hardy. 1992. Blood residue analysis of 90,000 year old stone tools from Tabun cave, Israel. *Antiquity* 66: 24–35.

Loy, T .H., and D. E. Nelson. 1986. Potential applications of organic residues on ancient tools. In J. Olin and M. Blackman, eds., *Proceedings of the 24th International Archaeometry Symposium,* 179–86. Washington: Smithsonian Institution Press.

Loy, T. H., M. Spriggs, and S. Wickler. 1992. Direct evidence for human use of plants 28,000 years ago: Starch residues on stone artefacts from the northern Solomon Islands. *Antiquity* 66: 898–912.

Lu, T. 2003. The survival of starch residue in a subtropical environment. In D. M. Hart and L. A. Wallis, eds., *Phytolith and Starch Research in the Australian-Pacific-Asian Regions: The State of the Art*, 119–26. Canberra: Pandanus Books, Terra Australis 19.

Lyman, R. L. 1994. *Vertebrate Taphonomy*. Cambridge: Cambridge University Press.

Martin, G. J. 1995. *Ethnobotany: A Conservation Manual*. London: Chapman and Hall.

Martínez Ungría, O. 1989. *Notas Acerca de la Quinoa y el Amaranto. Cultivos Indígenas en Expansión*. La Plata: Ediciones RAE.

Matheson, N. K. 1996. The chemical structure of amylose and amylopectin fractions of starch from tobacco leaves during development and diurnally-nocturnally. *Carbohydrate Research* 282: 247–62.

Mathew, J. 1899. *Eaglehawk and Crow*. Melbourne: Melville.

Matthews, P. J. 1990. The origins, dispersal and domestication of taro. Unpublished PhD thesis, Australian National University, Canberra.

———. 1991. A possible tropical wildtype taro: *Colocasia esculenta* var. *aquatilis*. *Indo-Pacific Prehistory Association Bulletin* 11: 69–81.

———. 1997. Field guide for wild-type taro, *Colocasia esculenta* (L.) Schott. *Plant Genetic Resources Newsletter* 110: 41–48.

———. 2003. Taro planthoppers (*Tarophagus* spp.) in Australia and the origins of taro (*Colocasia esculenta*) in Oceania. *Archaeology in Oceania* 38: 192–202.

Matthews, P. J., and R. Terauchi. 1994. The genetics of agriculture: DNA variation in taro and yam. In J. G. Hather, ed., *Tropical Archaeobotany*, 251–70. London: Routledge.

Matthews, P. J., Y. Matsushita, T. Sato, and M. Hirai. 1992. Ribosomal and mitochondrial DNA variation in Japanese taro (*Colocasia esculenta* (L.) Schott). *Japanese Journal of Breeding* 42: 825–33.

Mayeske, B. J. 1979. Bakers, bakeshops, and bread: A social and economic study. In The Archaeological Institute of America Washington Institution Society, The Smithsonian and sponsors, eds., *Pompeii and the Vesuvian Landscape*, 149–65. Washington, DC: The Archaeological Institute of America and The Smithsonian Institution.

McGrew, W. C. 1992. *Chimpanzee Material Culture: Implications for Human Evolution*. Cambridge: Cambridge University Press.

McNair, J. B. 1930. *The Differential Analysis of Starches*. Chicago: Field Museum of Natural History, Botanical Series 9.

———. 1965. *Studies in Plant Chemistry*. Published by the author, Los Angeles, California.

Meehan, B. 1982. *Shell Bed to Shell Midden*. Canberra: Australian Institute of Aboriginal Studies.

Meyer, A. 1895. *Untersuchungen uber die Die Stärkekörner. Wesen und Lebensgeschichte der Stärkekörner der hoheren Pflanzen*. Jena: E. Fischer. S. 318.

Microsoft Encarta 1993–7. *98 Encyclopedia*. CD ROM.

Miller, T. O. 1979. Stonework of the Xete Indians of Brazil. In Hayden 1979: 401–7.

Milton, K. 1984. Comment on Hominid diet before fire. *Current Anthropology* 25: 160–61.

Mintz, S. W. 1985. *Sweetness and Power: The Place of Sugar in Modern History*. Harmondsworth: Penguin Books.

Mitchell, T. L. 1848. *Journal of an Expedition into the Interior of Tropical Australia, in Search of a Route from Sydney to the Gulf of Carpentaria*. London: Longman, Brown, Green and Longmans.

Moore, G. F. 1884. *Diary of Ten Years Eventful Life of an Early Settler in Western Australia and a Descriptive Vocabulary. . .* London: W. S. Orr.

Moorthy, S. N., and T. Ramanujam. 1986. Variation in properties of starch in cassava varieties in relation to age of the crop. *Starch/Stärke* 38: 58–61.

Morcos, S. R., S. M. Hegazi, and S. T. El-Damhougy. 1973. Fermented foods of common use in Egypt. II. The chemical composition of bouza and its ingredients. *Journal of Science and Food Agriculture* 24: 1157–61.

Morwood, M. J., R. P. Soejono, R. G. Roberts, R. Sutikna, C. S. M. Turney, K. E. Westaway, W. J. Rink, J.-x. Zhao, G. D. van den Bergh, Rokus Awe Due, D. R. Hobbs, M. W. Moore, M. I. Bird, and L. K. Fifield. 2004. Archaeology and age of a new hominin from Flores in eastern Indonesia. *Nature* 431: 1087–91.

Moss, G. E. 1976. The microscopy of starch. In J. A. Radley, ed., *Examination and Analysis of Starch and Starch Products*, 1–31. London: Applied Science Publishers.

Nägeli, C. 1858. *Die Stärkekörner, Morphologische, physiologische, chemisch-physicalisch und systematisch-botanische*. Zúrich: Verlag F. Schulthess. S.441.

Newsom, S. W. B., and M. Shaw. 1997. A survey of starch particle counts in the hospital environment in relation to the use of powdered latex gloves. *Occupational Medicine* 47: 155–58.

Nikuni, Z. 1978. Studies on starch granules. *Starch/Stärke* 30: 105–11.

Nicholson, R. A. 1998. Bone degradation in a compost heap. *Journal of Archaeological Science* 25: 393–403.

Noda, T., Y. Takahata, and T. Nagata. 1992a. Properties of sweet potato starches from different tissue zones. *Starch/Stärke* 44: 365–68.

———. 1992b. Developmental changes in properties of sweet potato starches. *Starch/Stärke* 44: 405–9.

Oates, C. G. 1997. Towards an understanding of starch granule structure and hydrolysis. *Trends in Food Science and Technology* 8: 375–82.

Ochoa de Masramón, D. 1977. Prácticas tradicionales del cultivo del maíz en San Luis-República Argentina. *Revista del Museo de Historia Natural de San Rafael* 4: 283–306.

O'Connell, J. F., K. Hawkes, and N. G. Blurton-Jones. 1999. Grandmothering and the evolution of *Homo erectus*. *Journal of Human Evolution* 36: 461–85.

Odell, G. H. 1994. Prehistoric hafting and mobility in the North American Midcontinent: Examples from Illinois. *Journal of Anthropological Archaeology* 13: 51–73.

———. 1996. *Stone Tools and Mobility in the Illinois Valley: From Hunter-Gatherer Camps to Agricultural Villages*. Ann Arbor: University of Michigan, International Monographs in Prehistory, Archaeological Series 10.

———. 2004. *Lithic Analysis*. New York: Kluwer Academic.

Oliveira, A. B., D. C. Rasnusson, and R. G. Fulcher. 1994. Genetic aspects of starch granule traits in barley. *Crop Science* 34: 1176–80.

Ouzman, S. 1997. Hidden in the common gaze: Collective and idiosyncratic rock paintings at Rose Cottage Cave, South Africa. *Navorsings van die Nasionale Museum, Bloemfontein* 13: 225–56.

Ouzman, S., and L. Wadley. 1997. A history in paint and stone from Rose Cottage Cave, South Africa. *Antiquity* 71: 386–404.

Palmer, G. H. 1989. Cereals in malting and brewing. In G. H. Palmer, ed., *Cereal Science and Technology*, 61–242. Aberdeen: Aberdeen University Press.

Parnell, J. J., R. Terry, and Z. Nelson. 2002. Soil chemical analysis applied as an interpretive tool for ancient human activities in Piedras Negras, Guatemala. *Journal of Archaeological Science* 29: 379–404.

Parodi, L. 1991. *Agricultura prehispánica*. San Salvador de Jujuy, Argentina: Universidad Nacional de Jujuy.

Parr, J. F. 2002. A comparison of heavy liquid floatation and microwave digestion techniques for the extraction of fossil phytoliths from sediments. *Review of Palaeobotany and Palynology* 120: 315–36.

———. 2003a. A study of landscapes in Numundo region of West New Britain, Papua New Guinea, as indicated by fossil phytolith analysis. Unpublished PhD thesis, Southern Cross University, Lismore.

———. 2003b. The identification of Xanthorrhoea resins by starch morphology: Prospects for archaeological and taxonomic applications. *Economic Botany* 56: 260–70.

Parr, J. F., and M. Carter. 2003. Phytolith and starch analysis of sediment samples from two archaeological sites on Dauar Island, Torres Strait, northeastern Australia. *Vegetation History and Archaeobotany* 12: 131–41.

Parr, J., C. J. Lentfer, and W. E. Boyd. 2001a. Spatial analysis of fossil phytolith assemblages at an archaeological site in West New Britain, Papua New Guinea. In G. R. Clark, A. J. Anderson, and T. Vunidilo, eds., *The archaeology of Lapita dispersal in Oceania*, 125–34. Canberra: Pandanus Books, Terra Australis 17.

Parr, J. F., V. Dolic, G. Lancaster, and W. E. Boyd. 2001b. A microwave digestion method for the extraction of phytoliths from herbarium specimens. *Review of Palaeobotany and Palynology* 116: 203–12.

Parr, J. F., and K. Farrugia. 2003. Waste reduction and value adding during fossil phytolith extraction and palaeoenvironmental analysis of volcanic sediments and tephra using Microwave Digestion and ICP/MS. In D. M. Hart and L. A. Wallis, eds., *Phytolith and Starch Research, in the Australian-Pacific-Asian regions: The state of the art*, 19–30. Canberra: Pandanus Books, Terra Australis 19.

Parr, J. F., K. H. Taffs, and C. M. Lane. 2004. A microwave digestion technique for the extraction of fossil diatoms from coastal lake and swamp sediments. *Paleolimnology* 31: 383–90.

Pate, J. S., and K. W. Dixson. 1982. *Tuberous, Cormous and Bulbous Plants: Biology of an Adaptive Strategy in Western Australia*. Nedlands: University of Western Australia Press.

Paz, V. 2001. Archaeobotany and cultural transformation: Patterns of early plant utilisation in northern Wallacea. Unpublished PhD thesis, University of Cambridge.

Pearsall, D. 1989. *Paleoethnobotany: A Handbook of Procedures*. San Diego: Academic Press.

———. 2000. *Paleoethnobotany: A Handbook of Procedures*. Second Edition. Orlando: Academic Press.

Pearsall, D. M., K. Chandler-Ezell, and J. A. Zeidler. 2004. Maize in ancient Ecuador: Results from residue analysis of stone tools from the Real Alto site. *Journal of Archaeological Science* 31: 423–42.

Peekel, P. G. 1984. *Flora of the Bismarck Archipelago for Naturalists*. Lae: Office of Forests, Division of Botany.

Pérez, S., and A. Imberty. 1996. Structural features of starch. *Carbohydrates in Europe* 15: 17–21.

Perry, L. 2001. Prehispanic subsistence in the middle Orinoco basin: Starch analyses yield new evidence. Unpublished PhD thesis, Southern Illinois University, Carbondale.

———. 2002a. Starch analyses indicate multiple functions of quartz 'manioc' grater flakes from the Orinoco basin, Venezuela. *Interciencia* 27: 635–39.

———. 2002b. Starch granule size and the domestication of manioc (*Manihot esculenta*) and sweet potato (*Ipomoea batatas*). *Economic Botany* 56: 335–49.

———. 2004. Starch analyses reveal the relationship between tool type and function: An example from the Orinoco valley of Venezuela. *Journal of Archaeological Science* 31: 1069–81.

———. in press. Reassessing the traditional interpretation of 'manioc' artefacts in the Orinoco valley of Venezuela. *Latin American Antiquity*.

Piperno, D. R. 1988. *Phytolith Analysis: An Archaeological and Geological Perspective.* New York: Academic Press.

———. 1998. Paleoethnobotany in the Neotropics from microfossils: New insights into ancient plant use and agricultural origins in the tropical forest. *Journal of World Prehistory* 12: 393–450.

———. in press. Starch grains and root crops. In M. Zeder, D. Decker-Walters, and B. Smith, eds., *Documenting Domestication: New Molecular and Archaeological Paradigms.* Washington: Smithsonian Institution Press.

Piperno, D. R., and I. Holst. 1998. The presence of starch granules on prehistoric tools from the humid Neotropics: Indications of early tuber use and agriculture in Panama. *Journal of Archaeological Science* 25: 765–76.

———. 2004. Crop domestication in the American Tropics: Starch grain analyses. In R. M. Goodman, ed., *Encyclopedia of Plant and Crop Science*, 330–32. Monticello: Marcel Dekker.

Piperno, D., and D. Pearsall. 1998. *The Origins of Agriculture in the Lowland Neotropics.* San Diego: Academic Press.

Piperno, D. R., and K. E. Stothert. 2003. Phytolith evidence for early Holocene *Curcubita* domestication in Southwest Ecuador. *Science* 299: 1054–56.

Piperno, D. R., A. J. Ranere, I. Holst, and P. Hansell. 2000. Starch granules reveal early root crop horticulture in the Panamanian tropical forest. *Nature* 407: 894–97.

Piperno, D. R., E. Weiss, I. Holst, and D. Nadel. 2004. Processing of wild cereal grasses in the Upper Palaeolithic revealed by starch grain analysis. *Nature* 430: 670–73.

Plug, I., and R. Engela. 1992. The macrofaunal remains from recent excavations at Rose Cottage Cave, Orange Free State, South Africa. *South African Archaeological Bulletin* 47: 16–25.

Powdermaker, H. 1971. *Life in Lesu: The Study of Melanesian Society in New Ireland.* New York: W. W. Norton & Co.

Powers, A. H., J. Padmore, and D. Gilbertson. 1989. Studies of late prehistoric and modern opal phytoliths from coastal sand dunes and machair in northwest Britain. *Journal of Archaeological Science* 27: 27–45.

Prendergast, M. D. V., M. J. Kennedy, R. F. Webby, and K. R. Markham. 2000. Pollen cakes of *Typha* spp. [Typhaceae]- lost and living food. *Economic Botany* 54: 254–55.

Radley, J. A. 1968. *Starch and Its Derivatives.* London: Chapman and Hall.

Raven, P. H., R. F. Evert, and S. E. Eichhorn. 1999. *Biology of Plants.* New York: W. H. Freeman and Co.

Reichert, E. T. 1913. *The Differentiation and Specificity of Starches in Relation to Genera, Species, Etc.* Volume One. Washington: Carnegie Institute.

———. 1919. *The Differentiation and Specificity of Starches in Relation to Genera, Species, Etc.* Volume Two. Washington: Carnegie Institute.

Richter, J. 1994. A pound of bungwall and other measures. Unpublished BA (Hons) thesis, Department of Anthropology and Sociology, University of Queensland.

———. 1996. Weights and measures: Estimating food consumption from use-wear analysis. In S. Ulm, I. Lilley, and A. Ross, eds., *Australian Archaeology '95: Proceedings of the 1995 Australian Archaeological Association Annual Conference*, 161–68. Brisbane: University of Queensland, Tempus 6.

Roberts, S. 2000. High-glycemic index foods, hunger, and obesity: Is there a connection? *Nutrition Reviews* 58: 163–9.

Robertson, G. 1994. An Application of Scanning Electron Microscopy and Image Analysis to the Differentiation of Starch Grains in Archaeological Plant Residues. Unpublished B.A. (Hons) thesis, Department of Anthropology and Sociology, University of Queensland.

———. 1996. An application of environmental scanning electron microscopy and image analysis to starch grain differentiation. In S. Ulm, I. Lilley, and A. Ross, eds., *Australian Archaeology '95: Proceedings of the 1995 Australian Archaeological Association Annual Conference*, 169–82. Brisbane: University of Queensland, Tempus 6.

Rodale, J. I. 1974. *The Complete Book of Composting.* Emmaus, PA: Rodale Books.

Rosenfeld, A., D. Horton, and J. Winter. 1981. *Early Man in North Queensland.* Canberra: Australian National University, Department of Prehistory, Terra Australis 6.

Rossen, J., T. D. Dillehay, and D. Ugent. 1996. Ancient cultigens or modern intrusions? Evaluating plant remains in an Andean case study. *Journal of Archaeological Science* 23: 391–407.

Rots, V. 2003. Towards an understanding of hafting: the macro- and microscopic evidence. *Antiquity* 77: 805–15.

Rots, V., and B. S. Williamson. 2004. Microwear and residue analyses in perspective: The contribution of ethnoarchaeological evidence. *Journal of Archaeological Science* 31: 1287–1300.

Russell, E. W. 1968. The agricultural environment of soil bacteria. In T. R. G. Gray and D. Parkinson, eds., *The Ecology of Soil Bacteria*, 77–94. Liverpool: Liverpool University Press.

Ruzin, S. E. 1999. *Plant Microtechnique and Microscopy*. New York: Oxford University Press.

Sakai, W. S. 1979. Aroid root crops, acridity and raphides. *Tropical Foods* 1: 265–78.

Samuel, D. 1994. An archaeological study of baking and bread in New Kingdom Egypt. Unpublished PhD thesis, University of Cambridge.

———. 1996a. Investigation of ancient Egyptian baking and brewing methods by correlative microscopy. *Science* 273: 488–90.

———. 1996b. Archaeology of ancient Egyptian beer. *Journal of the American Society of Brewing Chemists* 54: 3–12.

———. 2000. Brewing and baking. In P. T. Nicholson and I. Shaw, eds., *Ancient Egyptian Materials and Technology*, 537–76. Cambridge: Cambridge University Press.

Sauer, C. O. 1952. *Agricultural Origins and Dispersals*. New York: American Geographical Society.

Scaglion, R., and K. A. Soto. 1994. A prehistoric introduction of the sweet potato in New Guinea? In A. Strathern and G. Sturzenhofecker, eds., *Migration and Transformations: Regional Perspectives on New Guinea*, 257–94. Pittsburgh: American Society of Anthropologists in Oceania, Monograph 15.

Schiffer, M .B. 1976. *Behavioral Archeology*. New York: Academic Press.

———. 1987. *Formation Processes of the Archaeological Record*. Albuquerque: University of New Mexico Press.

Schleiden, M. J. 1849. *Principles of Scientific Botany*. London: Longman, Brown, Green and Longmans.

Schultes, R. E., and S. von Reis, eds. 1995. *Ethnobotany: Evolution of a Discipline*. Portland: Dioscorides Press.

Scott Cummings, L., and A. Magennis. 1997. A phytolith and starch record of food and grit in Mayan human tooth tartar. In A. Pinilla, J. Juan-Tresserras, and M. J. Machado, eds., *The state of the art of phytoliths in soils and plants*, 211–8. Madrid: Mongrafias del Centro de Ciencias Medioambientales.

Seidemann, von J. 1964. Mikroskopische untersuchung verschiedener Dioscorea–stärken. *Starch/Stärke* 8: 246–53.

———. 1966. *Stärke–Atlas*. Berlin: Paul Parey.

Semenov, S. E. 1964. *Prehistoric Technology*. London: Cory, Adams and Mackay.

Shafer, H. J., and R. G. Holloway. 1979. Organic residue analysis in determining stone tool function. In Hayden 1979, 385–99.

Shannon, J. C., and D. L. Garwood. 1984. Genetics and physiology of starch development. In R. L. Whistler, J. N. BeMiller, and E. F. Paschall, eds., *Starch: Chemistry and Technology*, 2nd ed., 26–85. Orlando: Academic Press.

Sivak, M. N. S., and J. Preiss. 1998a. Occurrence of starch. In M. N. S. Sivak and J. Preiss, eds., *Starch: Basic Science to Biotechnology*, 1–12. Advances in Food and Nutrition Research, 41. San Diego: Academic Press.

———. 1998b. Physiochemical structure of the starch granule. In M. N. S. Sivak and J. Preiss, eds., *Starch: Basic Science to Biotechnology*, 13–32. Advances in Food and Nutrition Research, 41. San Diego: Academic Press.

———. 1998c. Starch accumulation in photosynthetic cells. In M. N. S. Sivak and J. Preiss, eds., *Starch: Basic Science to Biotechnology*, 139–52. Advances in Food and Nutrition Research, 41. San Diego: Academic Press.

Skelton, S. 1996. Starch Trek. The search for residue: An examination of the relationship between technological change and resource exploitation at Platypus rockshelter (KB:A70), southeast Queensland. Unpublished Diploma (Arts) thesis, Department of Anthropology and Sociology, University of Queensland.

Smith, B. D. 1998. *The Emergence of Agriculture*. New York: Scientific American.

Smith, M. A. 1985. A morphological comparison of Central Australian seedgrinding implements and Australian Pleistocene-age grindstones. *The Beagle, Occasional Papers of the Northern Territory Museum of Arts and Sciences* 2: 23–38.

———. 1986. The antiquity of seed-grinding in arid Australia. *Archaeology in Oceania* 21: 9–39.

———. 1988. Central Australian seed grinding implements and Pleistocene grindstones. In B. Meehan and R. Jones, eds., *Archaeology with Ethnography: An Australian Perspective*, 94–108. Canberra: Australian National University.

Smyth, R .B. 1878. *Aborigines of Victoria*, 2 vols. Melbourne: Government Printer.

Specht, J., and R. Fullagar. 1988. Preliminary Report on archaeological research in West New Britain Province, July–August, 1988. Unpublished report for the West New Britain Cultural Centre and National Museum, Papua New Guinea.

Spicer, R. A. 1991. Plant taphonomic processes. In P.A. Allison and D. E. G. Briggs, eds., *Taphonomy*, 71–113. New York: Plenum.

Spindler, K. 1994. *The Man in the Ice*. London: Orion Books Ltd.

Spriggs, M. 1996. Early agriculture and what went on before in Island Melanesia: continuity or intrusion? In D. R. Harris, ed., *The Origins and Spread of Agriculture and Pastoralism in Eurasia*, 524–37. London: UCL Press.

———. 1997. *The Island Melanesians*. Oxford: Blackwell.

Stahl, A. B. 1984. Hominid dietary selection before fire. *Current Anthropology* 25: 151–68.

———. 1989. Plant food processing: Implications for dietary quality. In Harris and Hillman 1989: 171–96.

Stanner, W. E. H. 1933. Ceremonial economics of the Mulluk Mulluk and Madngella tribes of the Daly River, North Australia. *Oceania* 4: 156–76.

Stark, J. R., and A. Lynn. 1992. Starch granules large and small. *Biochemical Society Transactions* 20: 7–12.

Sterling, C. 1984. The light microscope in food analysis. In W. D. Gruenwedel and J. R. Whitaker, eds., *Food Analysis: Principles and Techniques*, 175–201. New York: Marcel Dekker.

Stoddard, F. 1999. Survey of starch particle-size distribution in wheat and related species. *Cereal Chemistry* 76: 145–49.

Stone, A. C. 1911. The Aborigines of Lake Boga, Victoria. *Proceedings of the Royal Society of Victoria* 23(2): 433–68.

Sugimoto, Y., M. Yamamoto, K. Abe, and H. Fuwa. 1987. Developmental changes in starch properties of the Chinese yam (*Dioscorea batatas* Decne). *Journal of the Japanese Society of Starch Science* 34: 11–20.

Summerhayes, G. R. 2000. Recent archaeological investigations in the Bismarck Archipelago, Anir-New Ireland Province, Papua New Guinea. *Bulletin of the Indo-Pacific Prehistory Association* 19: 167–74.

Sunnell, L. A., and P. L. Healy. 1979. Distribution of calcium oxalate crystal idioblasts in corms of taro (*Colocasia esculenta*) *American Journal of Botany* 66: 1029–32.

Sussman, G. L., and D. H. Breezhold. 1995. Allergy to latex rubber. *Annals of Internal Medicine* 122: 43–46.

Tang, H. J., H. Ando, K. Watanabe, Y. Takeda, and T. Mitsunaga. 2001. Physicochemical properties and structure of large, medium and small granule starches in fractions of normal barley endosperm. *Carbohydrate Research* 330: 241–48.

Tateoka, T. 1962. Starch grains of endosperm in grass systematics. *Botanical Magazine Tokyo* 75: 377–83.

Taylor, J. P., B. Wilson, M. S. Mills, and R. G. Burns. 2002. Comparison of microbial numbers and enzymatic activities in surface soils and subsoils using various techniques. *Soil Biology and Biochemistry* 34: 387–401.

Tester, R. F. 1997. Starch: The polysaccharide fractions. In P. J. Frazier, P. Richmond, and A. M. Donald, eds., *Starch: Structure and Functionality*, 163–71. London: Royal Society of Chemistry, Special Publications No. 205.

Tester, R. F., J. Karkalas, and X. Qi. 2004. Starch—composition, fine structure and architecture. *Journal of Cereal Science* 39: 151–65.

Therin, M. 1994. Subsistence through starch: The examination of subsistence changes on Garua Island, West New Britain, Papua New Guinea, through the extraction and identification of starch from sediments. Unpublished BA (Hons) thesis, Archaeology, University of Sydney.

———. 1998a. The movement of starch grains in sediments. In Fullagar 1998, 61–72.

———. 1998b. Classification of starch derived from West New Britain soils. Unpublished report to the Australian Museum, Sydney.

Therin, M., R. Fullagar, and R. Torrence. 1999. Starch in sediments: A new approach to the study of subsistence and land use in Papua New Guinea. In C. Gosden and J. Hather, eds., *Prehistory of Food*, 438–62. London: Routledge.

Therin, M. J., R. Torrence, and R. Fullagar. 1997. Australian Museum starch reference collection. *Australian Archaeology* 44: 52.

Theunissen, R., J. Balme, and W. Beck. 1998. Headroom and human trampling: Cave ceiling height determines the spatial patterning of stone artefacts at Petzkes Cave, northern New South Wales. *Antiquity* 72: 80–89.

Thomas, S., and M. Corden. 1977. *Metric Tables of Composition of Australian Foods*. Canberra: Australian Government Publishing Service.

Thompson, L. U., and J. H. Yoon. 1984. Starch digestibility as affected by polyphenols and phytic acid. *Journal of Food Science* 49: 1228–29.

Thoms, A. 1989. The Northern Roots of Hunter-Gatherer Intensification: Camas and the Pacific Northwest. PhD thesis, Washington State University, Pullman.

Thomson, D. F. 1939. The seasonal factor in human culture. *Proceedings of the Prehistoric Society* 10: 209–21.

———. 1949. *Economic Structure and the Ceremonial Exchange Cycle in Arnhem Land*. Melbourne: Macmillan.

Tomlinson, K. L., J. R. Lloyd, and A. M. Smith. 1997. Importance of isoforms of starch-branching enzyme in determining the structure of starch in pea leaves. *Plant Journal* 11: 31–43.

Topping, D. L., and P. M. Clifton. 2001. Short-chain fatty acids and human colonic function: Roles of resistant starch and nonstarch polysaccharides. *Physiological Reviews* 81: 1031–64.

Torrence, R. 1992. What is Lapita about obsidian? A view from the Talasea source. In J.-C. Galipaud, ed., *Poterie Lapita et Peuplement*, 111–26. Noumea: ORSTOM.

Torrence, R., P. Jackson, C. Pavlides, and J. Webb. 2000. Volcanic disasters and cultural discontinuities in the Holocene of West New Britain, Papua New Guinea. In B. McGuire, D. Griffiths, and I. Stewart, eds., *The Archaeology of Geological Catastrophies*, 225–44. London: Geological Society of London, Special Publication 171.

Torrence, R., R. Wright, and R. Conway. 2004. Identification of starch granules using image analysis and multivariate techniques. *Journal of Archaeological Science* 31: 519–32.

Tuross, N., and T. D. Dillehay. 1995. The mechanism of organic preservation at Monte Verde, Chile, and one use of biomolecules in archaeological interpretation. *Journal of Field Archaeology* 22: 97–110.

Ugent, D. 1994. Chemosystematics in archaeology: A preliminary study of the use of chromatography and spectrophotometry in the identification of four prehistoric root crop species from the desert coast of Peru. In J. G. Hather, ed., *Tropical Archaeobotany*, 215–26. London: Routledge.

———. 1997. The tuberous plant remains of Monte Verde. In T. D. Dillehay, ed., *Monte Verde: A Late Pleistocene Settlement in Chile: The Archaeological Context and Interpretation*, 2, 903–10. Washington: Smithsonian Institution Press.

Ugent, D., and L. W. Peterson. 1988. Archaeological remains of potato and sweet potato in Peru. *CIP Circular* 16: 1–9.

Ugent, D., S. Pozorski, and T. Pozorski. 1981. Prehistoric remains of the sweet potato from the Casma Valley of Peru. *Phytologia* 49: 401–15.

———. 1982. Archaeological potato tuber remains from the Casma Valley of Peru. *Economic Botany* 36: 182–92.

———. 1983. Restos arqueológicos de tubérculos de papas y camotes del Valle de Casma en el Perú. *Boletin de Lima* 25: 1–17.

———. 1984. New evidence for ancient cultivation of *Canna edulis* in Peru. *Economic Botany* 38: 417–32.

———. 1986. Archaeological manioc (Manihot) from coastal Peru. *Economic Botany* 40: 78–102.

Ugent, D., T. Dillehay, and C. Ramírez. 1987. Potato remains from a late Pleistocene settlement in southcentral Chile. *Economic Botany* 41: 17–27.

Van Peer, P., R. Fullagar, S. Stokes, R. M. Bailey, J. Moeyersons, F. Steenhoudt, A. Geerts, T. Vanderbeken, M. De Dapper, and F. Geus. 2003. The Early to Middle Stone Age transition and the emergence of modern human behaviour at site 8–B–11, Sai Island, Sudan. *Journal of Human Evolution* 45: 187–93.

Van Veen, J. A., and P. J. Kuikman. 1990. Soil structural aspects of decomposition of organic matter by microorganisms. *Biogeochemistry* 11: 213–33.

Vaughan, J. G. 1970. *The Structure and Utilisation of Oil Seeds*. London: Chapman and Hall.

Vaughan, P. 1985. *Use-Wear Analysis of Flaked Stone Tools*. Tucson: University of Arizona Press.

Vinton, N. J. 1995. An analysis of the spatial patterning of starch grains extracted from the sediments at Petzkes Cave, Northern NSW. Unpublished BA (Hons) thesis. University of New England, Armidale.

Vorren, K-D. 1986. The impact of early agriculture on the vegetation of northern Norway: A discussion of anthropogenic indicators in biostratigraphical data. In K. E. Behre, ed., *Anthropogenic Indicators in Pollen Diagrams*, 1–18. Rotterdam: A A Balkema.

Wadley, L. 1991. Rose Cottage Cave: background and a preliminary report on the recent excavations. *South African Archaeological Bulletin* 46:125–30.

Wadley, L. 1996. The Robberg Industry of Rose Cottage Cave, eastern Free State; the technology, spatial patterns and environment. *South African Archaeological Bulletin* 51:64–74.

———. 1997. Rose Cottage Cave: Archaeological work 1987–1997. *South African Journal of Science* 93: 439–44.

Währen, M. 1961. Typologie der altägyptischen Brote und Gebäck. *Brot und Gebäck* 6: 1–17.

Wallis, L. 1994. A residue analysis of artefacts from Widgingarri Shelter 2, Western Kimberley, Western Australia. Unpublished BSc (Hons) thesis, University of Western Australia.

Wallis, L., and S. O'Connor. 1998. Residues on a sample of stone points from the West Kimberley. In Fullagar 1998: 149–78.

Wallis, T. E. 1957. *Analytical Microscopy*. London: J. & A. Churchill.

Walthall, J. A. 1998. Rockshelters and hunter-gatherer adaptation to the Pleistocene/Holocene transition. *American Antiquity* 63: 223–38.

Wandsnider, L. 1997. The roasted and the boiled: Food composition and heat treatment, with special emphasis on pit-hearth cooking. *Journal of Anthropological Archaeology* 16: 1–48.

Wang, T. L., T. Y. Bogracheva, and C. L. Hedley. 1998. Starch: As simple as A, B, C? *Journal of Experimental Botany* 49: 481–502.

Waters, A. G., and J. M. Oades. 1991. Organic matter in water-stable aggregates. In W. S. Wilson, ed., *Advances in Soil Organic Matter Research: The Impact on Agriculture and the Environment*, 163–74. Cambridge: Royal Society of Chemistry.

Webster, J., W. Beck, and B. Ternai. 1984. Toxicity and bitterness in Australian *Dioscorea bulbifera* L. and *Dioscorea hispida* Dennst. from Thailand. *Journal of Agricultural and Food Chemistry* 32: 1087–90.

Webster, D., A. Freter, and O. Gonlin. 2000. *Copan: The Rise and Fall of an Ancient Maya Kingdom*. Fort Worth: Harcourt College Publishers.

Welch, M. 1997. Characterisation of Xanthorrhoea Resins by HPLC for Botanical, Archaeological and Forensic Purposes. Unpublished BSc. (Hons) thesis, University of Technology, Sydney.

Wheeler, J. R., B. L. Rye, B. L. Koch, and A. J. G. Wilson, eds. 1992. *Flora of the Kimberley Region*. Perth: Western Australian Herbarium, Department of Conservation and Land Management.

Whistler R. L., and C. L. Smart. 1953. *Polysaccharide Chemistry*. New York: Academic Press.

Whistler, R. L., J. N. BeMiller, and E. F. Paschall, eds. 1984. *Starch: Chemistry and Technology*. Second edition. Orlando: Academic Press.

White, C. A., and J. F. Kennedy. 1988. The carbohydrate-directed enzymes. In J. F. Kennedy, ed., *Carbohydrate Chemistry*, 342–77. Oxford: Clarendon Press.

White, J. P., C. N. Modjeska, and I. Hipuya. 1977. Group definitions and mental templates: An ethnographic experiment. In R. V. S. Wright, ed., *Stone Tools as Cultural Markers*, 380–90. Canberra: Australian Institute of Aboriginal Studies.

White, N. G. 1978. A human ecology research project in the Arnhem Land region: An outline. *Australian Institute of Aboriginal Studies Newsletter* 9: 39–52.

Wiedemann, H. G., and G. Bayer. 1983. Papyrus: The paper of ancient Egypt. *Analytical Chemistry* 55: 1220A–30A.

Wilkinson, H. P. 1983. Starch grain casts of moulds in Eocene (Tertiary) fossil mangrove hypocotyls. *Annals of Botany* 51: 39–45.

Willcox, G. 2001. Charred plant remains from a 10th millennium B. P. kitchen at Jerf el Ahmar (Syria). *Vegetation History and Archaeobotany* 11: 55–60.

Williamson, B. S. 1997. Down the microscope and beyond: Microscopy and molecular studies of stone tool residues and bone sample from Rose Cottage Cave. *South African Journal of Science* 93: 458–64.

———. 2000. Prehistoric stone tool residue analysis from Rose Cottage Cave and other Southern African sites. Unpublished PhD thesis, University of the Witwatersrand, Johannesburg.

———. 2004. Middle stone age tool function from residue analysis at Sibudu Cave. *South African Journal of Science* 100: 174–78.

Winton, A. L. 1906. *The Microscopy of Vegetable Foods*. New York: Wiley.

Wittmack, L. 1905. Our present knowledge of ancient plants. *Transactions of the Academy of Science of St. Louis* 15: 1–15.

Wolski, N., and Loy, T. H. 1999. On the invisibility of contact: Residue analysis on Aboriginal glass artefacts from Western Victoria. *The Artefact* 22: 65–73.

Womersley, J. S. 1976. *Plant Collecting for Anthropologists, Geographers and Ecologists in Papua New Guinea*. Lae: Division of Forests, Papua New Guinea, Botany Bulletin, 2.

———. 1981. *Plant Collecting and Herbarium Development*. Rome: FAO Plant Production Protection Paper 33.

Wood, B., and M. Collard. 1999. The human genus. *Science* 284: 65–71.

Woods, J. D. 1879. *The Native Tribes of South Australia*. Adelaide: E. S. Wigg.

Wrangham, R. W., J. H. Jones, G. Laden, D. Pilbeam, and N. Conklin-Brittain. 1999. The raw and the stolen: Cooking and the ecology of human origins. *Current Anthropology* 40: 567–94.

Würsch, P. 1989. Starch in human nutrition. *World Review of Nutrition and Dietetics* 60: 199–256.

Würsch, P., S. Del Vedovo, and B. Koellreutter. 1986. Cell structure and starch nature as key determinants of the digestion rate of starch in legume. *American Journal of Clinical Nutrition* 43: 25–29.

Yamada, T., S. Tamaki, M. Hisamatsu, and K. Teranishi. 1997. Molecular change of starch granule with physical treatment: Potato starch by ball-mill treatment. In P. J. Fazier, A. M. McDonald, and P. Richmond, eds., *Starch, Structure and Functionality*, 59–67. London: Royal Society of Chemistry.

Yen, D. E. 1974. *The Sweet Potato and Oceania: An Essay in Ethnobotany*. Honolulu: Bernice P. Museum Bulletin 236.

Yiu, S. H. 1993. Food microscopy and the nutritional quality of cereal foods. *Food Structure* 12: 123–33.

Zayas, I. Y., D. B. Bechtel, J. D. Wilson, and R. E. Dempster. 1994. Distinguishing selected hard and solid red winter wheats by image analysis of starch granules. *Cereal Chemistry* 71: 82–86.

Zobel, H. F. 1988. Molecules to granules: A comprehensive starch review. *Starch/Stärke* 40: 44–50.

Zvelebil, M. 1994. Plant use in the Mesolithic and its role in the transition to farming. *Proceedings of the Prehistoric Society* 60: 35–74.

INDEX

General

Common Plant Names

Scientific plant names

LIST OF CONTRIBUTORS

Jennifer Atchison is Conservation Officer for the World Wide Fund for Nature Australia, Western Rivers Project. Formerly of the University of Wollongong, she is the author of "Archaeobotany of fruit seed processing in a monsoon savana environment: Evidence from the Keep River region, Northern Territory, Australia" (*Journal of Archaeological Science* 32: 167–81) and specialises in Aboriginal ethnobotany.

María del Pilar Babot is a member of the Instituto de Arqueología y Museo at the Facultad de Ciencias Naturales e Instituto Miguel Lillo, Universidad Nacional de Tucumán, and a Research Fellow at the Consejo Nacional de Investigaciones Científicas y Técnicas (CONICET), Argentina. She specialises in mineral and food-processing technology in Northwestern Argentina and the application of residue and microfossil analysis on artifacts.

Jane Balme is Senior Lecturer in archaeology at the University of Western Australia. She has published on various aspects of the archaeology of Indigenous Australia. Her current research is on the effect of European settlement on Aboriginal communities in northern Australia and the division of labour in hunter-gatherer societies in the past.

Huw Barton is Wellcome Trust University Fellow at the School of Archaeology and Ancient History, University of Leicester. He is the author of "The case for rainforest foragers: The starch record at Niah Cave, Sarawak" (*Asian Perspectives* 44: 56–72) and specialises in the archaeology of hunter-gatherers in Southeast Asia and the application of starch granule analysis in determining prehistoric diet.

Wendy Beck is Associate Professor in Archaeology at the School of Human and Environmental Studies, University of New England. She is the author of a number of publications on indigenous archaeology and has a special interest in the ethnography of hunter-gatherer subsistence and food-processing technology, and the taphonomy of plant remains in archaeological sites.

Rebecca Conway is Project Officer in the Anthropology Branch, Australian Museum. She manages ethnographic collections from Asia, Africa, and the Americas, and previously managed the Australian Aboriginal archaeological collections. Rebecca researched Aboriginal seed-grinding artifacts for her BA (Hons) thesis. She is the author, with Robin Torrence and Richard Wright, of "Identification of starch granules using image analysis and multivariate techniques" (*Journal of Archaeological Science* 31: 519–32).

Alison Crowther is a PhD candidate at the School of Social Science, University of Queensland, researching the use of pottery from Melanesia through residue analysis. She is the author of "Starch residues on undecorated Lapita pottery from Anir, New Ireland" (*Archaeology in Oceania* 40: 62–66) and specialises in the analysis of microscopic plant residues.

Judith Field is Research Fellow (Archaeology and Microscopy) at the Australian Key Centre for Microscopy and Microanalysis, University of Sydney. She is the author of 34 refereed articles and specialises in the archaeology of prehistoric Australia, with a particular interest in human megafaunal interactions, rainforest archaeology, and human occupation of marginal environments. Her specialised interests lie in the functional analysis of stone tools.

Richard Fullagar is Honorary Research Associate, Department of Archaeology, University of Sydney, and has published many papers on lithic use-wear and residue analysis, including, as editor, *A Closer Look: Recent Australian Studies of Stone Tools*, University of Sydney, Archaeological Computing Laboratory (1998). He has a special interest in the history of Aboriginal plant food processing in the Australian region.

Beth Gott is an Honorary Research Associate of the School of Biological Sciences, Monash University, Victoria, Australia. She is the author of "Aboriginal Fire Management in South-eastern Australia—aims and frequency" (*Journal of Biogeography* 32: 1203–8) and numerous papers and books on indigenous ethnobotany. Her field of study is the ethnobotany and land management practices of Indigenous Australians.

Michael Haslam is a PhD Candidate and Manager of the Archaeological Science Laboratory in the School of Social Science, University of Queensland. He is the author of "The decomposition of starch grains in soils: Implications for archaeological residue analyses" (*Journal of Archaeological Science* 31: 1715–34) and specialises in the microscopic analysis of residues and use-wear on stone artefacts.

Mark Horrocks is a consulting and research scientist with a background in plant ecology and forensic science, and Director of Microfossil Research Ltd, New Zealand. He is author of "Starch grains and xylem cells of sweet potato (*Ipomoea batatas*) and bracken (*Pteridium esculentum*) in archaeological deposits from northern New Zealand" (*Journal of Archaeological Science* 31: 251–58) and specialises in the identification and interpretation of microscopic plant material. His current work involves reconstructing Quaternary vegetation via pollen cores, identifying archaeological environments and early agriculture, and forensic botany.

M. Alejandra Korstanje is Adjunct Professor at the Archaeology Department and a titular member of the Instituto de Arqueología y Museo, Universidad Nacional de Tucumán, Argentina. She is the author of articles on ancient agricultural systems in the Andes and has specialized in the multiple microfossil approach to soils (particularly on phytoliths, starch, and spherulites).

Lisa Kealhofer is Associate Professor at the Anthropology/Sociology Department and Environmental Studies Institute, Santa Clara University. She is the author of numerous articles, including "Changing perceptions of risk: The development of agro-ecosystems in Southeast Asia" (*American Anthropologist* 104: 178–94), and specialises in the analysis of phytoliths to reconstruct paleoenvironments and in the archaeology of Southeast Asia.

Carol Lentfer is Research Fellow at the School of Social Science, University of Queensland. She is the author, with Roger Green, of "Phytoliths and the evidence for banana cultivation at the Reber-Rakival site on Watom Island, Papua New Guinea" (*Records of the Australian Museum*, Supplement 29: 75–88) and specialises in the analysis of plant microfossils for reconstructing paleoenvironments in Indonesia, Melanesia, and Australia and determining the shifting patterns and development of plant exploitation and domestication.

Tom Loy (1942–2005) was Senior Lecturer at the School of Social Science, University of Queensland. He specialised in the analysis of ancient DNA and residue analysis of artifacts. His most recent work included investigations of hominid tool use at Pliocene/Early Pleistocene cave sites in South Africa and the tools of the Italo/Austrian Ice Man.

Tracey Lu is Associate Professor at the Department of Anthropology, Chinese University of Hong Kong. She is the author of "Starch residue analysis of the Zengpiyan tools" (in *Zengpiyan—A Prehistoric Cave in South China*, ed. Fu Xiangou, pp. 646–51: Institute of Archaeology, Chinese Academy of Social Sciences, Guangxi Archaeological Team and the Zengpiyan Archaeological Museum [Beijing: Cultural Relics Publishing House]) and specialises in prehistoric archaeology and the origin of agriculture in China.

Peter J. Matthews is Associate Professor at the National Museum of Ethnology, Osaka. He is the author of "A field guide for wild-type taro, *Colocasia esculenta*" (*Plant Genetic Resources Newsletter* 110: 41–48) and specialises in ethnobotany and archaeobotany.

James F. O'Connell is Professor at the Department of Anthropology, University of Utah. He has written widely about the behavioural ecology of hunter-gatherers and has researched the ethnography and archaeology of hunter-gatherers in western North America, Australia, and East Africa.

Jeff Parr is Research Fellow at the School of Environmental Science and Management, Southern Cross University. He is the author of "Phytolith and starch analysis of sediment samples from two archaeological sites on Dauar Island, Torres Strait, northeastern Australia" (*Vegetation History and Archaeobotany* 12: 131–41) and a number of methodological and interpretive articles, and specialises in the analysis of phytoliths, starch grains, and other microfossils used for the understanding of paleoenvironments and plant domestication.

Gail Robertson has recently submitted her PhD thesis entitled "The use of backed artefacts in Eastern Australia: a residue and use-wear analysis" in the School of Social Science, University of Queensland. She is the author of articles on the use of SEM in starch granule analysis and the identification of archaeological feather residues. She is currently a research assistant to Dr. Carol Lentfer, examining stone artifacts from Liang Bua, Flores, Indonesia, for residues and use-wear.

Delwen Samuel is Visiting Research Fellow at King's College London, School of Health and Life Sciences. She is the author of "Brewing and baking" in *Ancient Egyptian Materials and Technology*, ed. P. T. Nicholson and I. Shaw (Cambridge: Cambridge University Press, 2000). She specialises in the technology of ancient cereal food processing and its social, economic, and nutritional effects.

Linda Scott Cummings is Director of the Paleoresearch Institute, Colorado. She is the author of numerous articles concerning both paleodiet and paleonutrition and specialises in analysis of pollen, starch, and phytoliths. In addition, she examines archaeoclimatic models and stratigraphic pollen and phytolith records to interpret the past environment in which people lived. She strives to build a picture of past life, including both the landscape on which people lived, and the plants that people used.

Michael Therin is Director of Therin Archaeological Consulting Pty Ltd, Australia. He is the author of "Starch in sediments: A new approach to the study of subsistence and land use in Papua New Guinea" in *The Prehistory of Food*, ed. C. Gosden and J. Hather (London: Routledge, 1999) and specialises in the analysis of use-wear and residues on artifacts.

Robin Torrence is Principal Research Scientist at the Australian Museum, Sydney. She has written widely about stone tools studies and prehistoric exchange and is currently researching the impact of volcanic disasters on settlement history in Papua New Guinea. Within this project she has collaborated in studies using phytoliths and starch to reconstruct prehistoric diet, land use, and environmental change.

Donald Ugent is Emeritus Professor at the Department of Plant Biology, Southern Illinois University. He is the former Editor-in-chief of *Economic Botany* and specialises in plant systematics and ethnobotany.

Bonnie Williamson is Associate Researcher at the School of Geography, Archaeology and Environmental Studies, University of the Witwatersrand, Johannesburg. She has written and co-authored several papers on organic and inorganic prehistoric stone tool use-residues. She has specialised in the retrieval of ancient DNA fragments from blood residues on the stone tools but has also focussed on the visual identification of many different types of residues.